PHYSICAL METALLURGY
Techniques and Applications

VOLUME 1

PHYSICAL METALLURGY
Techniques and Applications

VOLUME I

by K. W. Andrews

A HALSTED PRESS BOOK

JOHN WILEY & SONS
New York

First published in 1973

Published in the U.S.A.
by Halsted Press, a Division
of John Wiley & Sons, Inc.
New York

© George Allen & Unwin Ltd 1973

Volume One : ISBN 0 470–03150–6
Volume Two: ISBN 0 470–03151–4

Library of Congress Catalog Card No.:
Volume One : LC 72–11308
Volume Two: LC 72–11309

Printed in Great Britain
in 10 point Times New Roman
by William Clowes & Sons, Limited
London, Beccles and Colchester

Preface

This textbook is intended to cover the principal and well-established techniques which may be used in metallurgical laboratory practice and research. Many of these techniques are also applicable to other materials as well as metals. There are already a number of excellent texts dealing with modern physical and structural metallurgy. Naturally these aim to describe the concepts which have been developed and to show how they contribute to the understanding of the materials and to their applications. The achievements of the past three decades or more have undoubtedly greatly enlarged the subject which stands, in its own right, as an important field of science. The techniques which have made much of this achievement possible have themselves a strong physical basis which is quite often taken for granted, in face of the pressing need to obtain results of immediate metallurgical interest or relevance. The textbooks which concentrate on these aspects therefore tend to give only outlines of techniques which are not really adequate for the needs of student or qualified, professional metal scientist who is required to learn about and work on some aspect of the subject. There appears therefore to have been a gap here which the present book aims to close. It is sometimes considered that, partly because of their apparent complexity, some of the techniques are the proper responsibility of specialists and are provided for in specialist textbooks. This is not really a tenable viewpoint since progress depends on the extent to which techniques, skills and understanding can be more widely disseminated and used. There will still be scope for specialised activity and the educated metallurgist or materials scientist should always know when to consult others who are more deeply involved in a subject. At the same time subjects which seemed to be novel and often difficult at first became more familiar and so more widely applied. This process of 'de-specialisation' must clearly be one objective in a treatment of physical techniques.

The present work was originally planned as a volume in the series *Modern Metallurgical Texts*. In spite of the intended condensation and simplification required for such a text it has still reached a considerable length. This is because of the particular need to give what is intended to be an adequate account of the physical principles underlying the techniques and how these relate to the interpretation of results. It is likewise necessary to give an appreciation of the experimental aspects and to provide some examples of applications. By developing the explanations in sufficient detail but from an elementary level the author hopes that readers will develop an understanding on which they can base experience and when required build on these foundations by referring to more advanced work.

The book is expected to be used by students particularly in metallurgy or materials science at undergraduate and post-graduate levels. It should also be useful to some chemists and physicists. Others at later stages in their careers may find it useful to have an overall point of view available and to recover details of particular aspects which they may then be able to apply to their own problems. It can therefore be used as a reference book as well as a textbook. In the latter connection, a student user may find it desirable to omit parts of the material at a first reading or, in any case, to be selective. It is pointed out in Chapter 1 that a particular sequence is followed which is suitably divided between the two volumes. The somewhat different treatment and coverage of different subjects has arisen naturally from what appeared to be the particular requirements of the subject and having regard to the general conditions noted above. It need hardly be stated that it is impossible to give an entirely comprehensive bibliography and list of references. The aim has therefore been to give references to other books which are most likely to be useful, to the principal or primary papers and to cover a selection of some others.

The author is very grateful to Professor C. R. Tottle of Bath University of Technology, general editor of the series referred to, for his advice and help on various matters and for giving much encouragement. A number of other colleagues and friends have kindly examined sections of the book or chapters and have made useful suggestions most of which have been adopted. Particular thanks are due to Professor E. O. Hall of The University of Newcastle, Australia, and to colleagues at the author's laboratory, especially Mr M. Atkins, Dr T. Gladman and Dr A. G. Clegg (now at Sunderland Polytechnic) and Mr S. R. Keown (now at Sheffield University). Other comments and observations have been contributed by Mr D. Thacker, Dr E. R. Petty and Dr D. Allen-Booth of Sheffield Polytechnic. In connection with Chapter 4 of Vol 2, useful information was provided by Drs R. Rawlings and I. G. Ritchie of the University of Wales and Dr B. Mills of Salford University. Dr M. McCaig of the Permanent Magnet Association, kindly advised on the use of SI units in connection with magnetic properties. Other acknowledgements have been made for the use of certain diagrams where appropriate in the text. The author apologises for any omissions or oversights in this respect which are naturally unintended. In any case there is an overall debt of gratitude to many authors whose works have necessarily been consulted.

The whole project of preparing the manuscript occupied many evenings over several winters and I am most grateful to my wife for her unfailing patience, encouragement and practical help in checking manuscript and proofs. The use of typing and drawing facilities at the Swinden Laboratories of the British Steel Corporation has greatly advanced the preparation of the work for publication and I am much indebted to Dr F. H. Saniter for permission to use these facilities. The typescript was most ably prepared initially by Mrs J. Williams and later by Miss A. Roebuck and the drawings by Miss A. Sherwood. I much appreciate their patience and efficiency. Dr K. J. Irvine, now Head of Research at Swinden Laboratories, has kindly examined the whole manuscript and given official approval for publication. Finally, I should like to thank the late Sir David Lynch-Blosse of Allen and Unwin Ltd, who handled the various initial aspects of publication from that side and Mr David Grimshaw who took over later. Their care, patience and consideration throughout is gratefully acknowledged.

K. W. ANDREWS

Note

The sequence of chapters following the general pattern outlined in the Introductory Chapter 1 is broken at a convenient point at the end of Chapter 8. This first volume covers the thermal methods and the special techniques for melting and solidification. The remainder is then occupied by an account of basic crystallographic principles and related diffraction techniques. The last chapter then deals with other radiation techniques.

The emphasis in this volume is thus partly but not entirely upon matters relating to thermal effects of phase changes and certain properties which do not relate directly to structure, and secondly to structure at the level of atomic dimensions. Some techniques in the eighth chapter do, however, provide information relating to the grain structure in the microscopical range of dimension. The second volume then takes up the theme, at this level, and deals first with optical and electron microscopy. The electrical and magnetic properties are covered and then a chapter is devoted to internal friction techniques. The last chapter in Volume 2 is concerned with the measurement and calculation of internal stresses.

Quantites used in the text have, in appropriate cases, been converted to S 1 units. In certain situations, however, the older units have been retained, either because the argument may in fact lose some significance or the information, in the original form, is merely illustrative and would not be used for quantitative references.

Foreword

The text was commissioned originally within the Series produced by the Institution of Metallurgists and entitled 'Modern Metallurgical Texts'.

The objective of this particular book was to condense into one unit the very wide and ever-increasing field of techniques used in Physical Metallurgy. Many research monographs exist in which the author describes the expertise built up over the years in a relatively narrow field. Dr Andrews has attempted not only to cover the whole field, but to show clearly the relevance of each part to the other, a feature we believe has never before been achieved. In this connection, we believe the two volumes fill an obvious gap, which will be welcomed in all laboratories, both industrial and academic.

Dr Andrews has long ago established himself as an author with the gift of clear expression and a logical approach to problems and the presentation of their solutions. His industrial experience has already established his authority as an experimentalist, so that the combination of the two should prove acceptable to students of metallurgy and materials science, and give the necessary insight to physicists and engineers who may find themselves in need of extra reading in any particular topic covered. As Dr Andrews explains in his preface, the immense volume of material has not proved easy to condense, but it is doubtful if such condensation could have been carried out in any other way and still prove adequate.

C. R. TOTTLE

Contents

Chapter 1

SCOPE OF THE TECHNIQUES OF PHYSICAL METALLURGY AND THEIR SYSTEMATISATION

1.1 INTRODUCTORY

1.1.1

Modern knowledge about metals, alloys and other metallurgical materials, and its extension or employment in research or industry, presupposes or requires:

(a) accurate chemical analysis whether by chemical methods which have a longer tradition but are continually developing, or by the newer physical methods;
(b) determination of structure, constitution and properties by physical methods.

(N.B. Certain inorganic or mineralogical compounds must also be included as metallurgical materials.)

We are therefore interested in (b) but may refer to certain physical techniques which can be used to give chemical analysis. The information obtained by the techniques is mainly physical in character but some of it is sometimes described as 'structural metallurgy'.

1.1.2

In considering the application of physical methods four stages may be recognised:

1. The discovery of the basic physical phenomenon and its theoretical explanation, e.g. the diffraction of X-rays and the crystal structure.
2. Application as a research tool to a fundamental metallurgical investigation such as the determination of a phase diagram or the study of a martensitic transformation.
3. Application in research which presupposes the first two stages. This is the type of application one might find in an industrial research laboratory, e.g. the identification of inclusions or investigation of phases causing embrittlement.

4. Application of the method to the routine checking of a product either intermittently in a laboratory or by continuous monitoring.

It is necessary to recognise that there are no rigid dividing lines and that not all techniques reach the fourth stage. Some of the more elaborate (e.g. microprobe analysis) may only be found in some of the large industrial or consulting laboratories. Also, techniques which can be used intensively in basic research, e.g. for the determination of phase diagrams from pure components, may have applications to *ad hoc* works problems, or investigation of the constitution of some of the more complex industrial alloys. It is therefore essential for the practising metallurgist to appreciate the broad connection over the four stages, and obviously to be prepared to extend his knowledge beyond the present elementary level or to refer to a specialist in the field.

1.2 CLASSIFICATION AND INTER-RELATION OF METHODS

1.2.1

It is desirable to have some logical but simple basis for considering the techniques in relation to their material and the changes of state or processes it may undergo. Such a basis is suggested in Table 1.1. The point of this table is that the physical methods in the centre columns can be used either to provide information about the states and properties of the material as indicated on the left, or the processes by which it changes from one state to another as indicated on the right. Some of the methods—particularly perhaps those for determining vapour pressures—fall outside the scope of this book and many of the techniques referred to are mainly applicable to the solid state.

1.2.2

A more detailed schematisation of several of the physical–metallurgical techniques, particularly those yielding structural information, can be provided as in Table 1.2. The reader may find it useful to refer to these outlines from time to time in order to see where a particular technique fits in or relates to another. This is particularly necessary since, although all the techniques may not be available, the proper combination should always be sought or used when available.

1.2.3

The treatment of the various laboratory methods of physical metallurgy will therefore follow a sequence which is approximately related to the schemes in Tables 1.1 and 1.2. Chapter 2 concerns thermal methods which are generally applied over the whole range of states and phenomena and, apart from one or two possible exceptions, do not require structural knowledge. Chapter 3 refers to special melting techniques. The remaining chapters are concerned firstly

TABLE 1.1   *A Basis for the Relation of Physical Techniques to a Metal and its Phase Changes*

State or phases	Available or relevant techniques			Changes of state, etc.
	Structural	Mechanical and electrical properties	Thermal properties	
The vapour state	Condensation from vapour or vacuum deposition		Methods for determining vapour pressure. Liquid ⇌ vapour equilibria	Condensation ⇌ Evaporation Solid ⇌ Vapour Transition
The liquid state	X-ray diffraction of liquid	Viscosity, electrical conductivity	Determination of freezing and melting by cooling and heating curves. Melting and casting techniques	Freezing ⇌ Melting
The solid state Single crystals	X-ray, electron and neutron diffraction. Optical and electron microscope	Electrical and magnetic properties	Methods for single crystals	
Allotropy or polymorphism		Mechanical testing methods	Ingots, etc., produced by normal casting methods. Metals and alloys by zone refining. Cooling and heating curves. Dilatometry	Phase transformations in solid
Polycrystalline solid Precipitated phases	X-ray microscopy Microradiography Autoradiography X-ray fluorescence Micro-probe analyser	Internal friction methods		Precipitation processes
Grain boundaries and sub-grain boundaries				Diffusion and other grain boundary effects Grain growth and recrystallisation

Notes: 1. The states or phases themselves are more generally investigated by the techniques in the 'structural' category but also by methods in the middle column.
2. The phase changes on the right are more generally investigated by the techniques in the adjacent column (thermal) and centre. In some cases, however, a structural technique can be used to follow changes of state.

TABLE 1.2 *Principal Physical Methods for Determining Structure and Constitution of Metallurgical Materials*

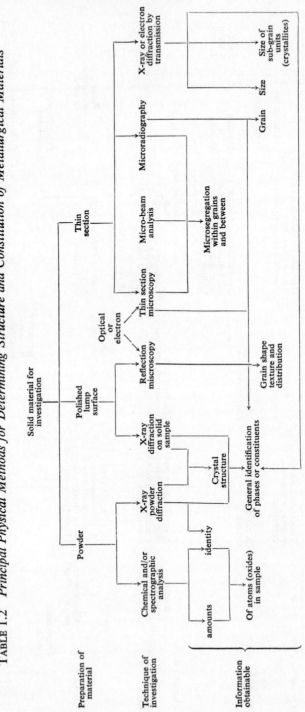

with those techniques which are used to provide structural knowledge and then properties in the solid state itself.

1.3   A NOTE ON THE MATHEMATICAL STANDARD

1.3.1

Although this book is about techniques it is clearly neither desirable nor generally possible to include detailed instructions for the use of the techniques except in an elementary way which will illustrate the principles. In most cases a deeper knowledge must and will sooner or later be achieved by an acquaintance with the technique or apparatus. What is important, however, at all times is that a student or other reader will find it an advantage to understand the physical principles behind a technique and its essential practical features. One aspect of this understanding is the meaning of the basic mathematical relationships.

1.3.2

The standard of the mathematics used here is believed to be the minimum required. It is known that difficulties are sometimes anticipated or encountered, but this need not be the case if it is considered that the potential users of the techniques of physical metallurgy will have already achieved a level in mathematics and physics which is at least equivalent to or beyond that of first year university standard: it should be higher. A particular problem is posed by crystallography, the subject matter of Chapter 4 and key to some of the techniques of later chapters. The aspects of this subject useful to metallurgists must be adequately abstracted from the whole subject and yet some small appreciation of the whole is desirable. In this connection, three-dimensional geometry is required but the concepts used are easily seen as extensions of their two-dimensional analogues and mostly involve only an acquaintance with the co-ordinate geometry of the line and plane or the circle and sphere. Diffraction phenomena (and techniques) require very little more except for an elementary idea of a sine or cosine wave. In connection with other topics the knowledge of the calculus required is quite elementary.

It is sometimes necessary in a treatment of this kind to accept a mathematical formula without proof because the proof would involve a higher standard of mathematics or concern phenomena outside the present scope, or it might be too lengthy. Quite often, however, an elementary proof can and should be given, especially when it can build on an existing foundation. It then provides a firmer foundation for understanding and using the techniques derived from the phenomenon. It is even permissible to regard some uses of symbols or equations as a very useful form of 'shorthand' or as memory aids.

A further observation concerns particularly the subject of the last chapter of Vol 2 but also applies to some topics in earlier chapters. Although it is intended to deal primarily with experimental techniques, methods are described which depend almost entirely on mathematical formulation of the problem and subsequent calculation. It is clearly an advantage if experimental support can be provided, but if this is not so, the methods are not discredited. It is therefore accepted that the use of the word 'technique' in the present context can involve widely differing combinations of experimental activity and theoretical enterpretation. The latter is involved in the physics of the method or the analysis of the results.

Chapter 2

THERMAL METHODS AND PROPERTIES

2.1 BASIC PRINCIPLES IN TEMPERATURE MEASUREMENT

2.1.1 General

The techniques of physical metallurgy are, as is implied in Chapter 1, primarily but by no means exclusively laboratory-based techniques. One of the simplest of these is generally fundamental to the remainder, viz. the measurement of the temperature at which a physical property is determined. (It is surprising to find how often the temperature is not reported.) A full account of the dependent technology of pyrometry as applied to industrial metallurgy would be outside the present scope. Interest primarily attaches here to those methods which are particularly useful in physical metallurgy:

1. In connection with melting operations
2. In thermal analysis
3. Incidental to the determination of properties.

General references [1]–[8].

2.1.2 Temperature Scales and Fixed Points

2.1.2.1 Metallurgical temperatures are generally based on the Centigrade or Celsius scale although there is some usage of the Fahrenheit scale, especially in American papers and in some metallurgical engineering (e.g. steam plant temperatures).

An ideal scale is one which would be independent of the properties of any material used to measure it, but the actual 'length' of the unit of measurement is still a matter of choice. Hence the so-called ideal gas scale, identical with the thermodynamic or Kelvin scale of temperature denoted by K or ° absolute yet still based on the centigrade interval. This conception can be justified in a very elementary way as an extension of the centigrade scale on which the (volume) coefficient of expansion of an ideal gas, between 0 and 100 °C, is given by $\alpha = 1/273\cdot16$ per degree. This is also the pressure coefficient at constant volume. At constant temperature $PV =$ constant. The combined law is:

$$PV = A(1 + \alpha t)$$

and at a temperature of $-1/\alpha$ on the centigrade scale PV would equal zero giving the choice of an 'absolute zero' at this temperature, and a temperature scale

$$t + \frac{1}{\alpha} = 273 \cdot 16 + t = T$$

$$t(°C) = T - 273 \cdot 16$$

Then

$$PV = A\{1 + \alpha(T - 273 \cdot 16)\} = (A\alpha)T = RT$$

2.1.2.2 So far we have noted two *fixed points* only on this absolute scale using the centigrade degree interval, viz. the freezing and boiling points of pure water at one atmosphere pressure. The absolute zero would be another fixed point if it could be used. A number of calibration points have been obtained, however, and these are fixed by international agreement. Table 2.1 records these and is divided to bring out the following:

(a) Points are required between 0 ° abs. (0 K) and 0 °C.

(b) Four principal points determine the scale up to 1063 °C. These points should be used to calibrate temperature-measuring devices, but in practice there are a number of difficulties and so—

(c) A number of secondary fixed points are required. Hume-Rothery *et al.* [8] have recommended those given here, but mention that the freezing points of copper and silver are lowered several degrees by small amounts of oxygen.

(d) Eutectics are much less affected by contamination. An advantage of the Cu–Al eutectic is that the Al acts as a deoxidiser, so avoiding temperature lowering due to dissolved oxygen.

(e) Temperatures above 1063 °C are determined by Planck's radiation formula and a further group of four fixed points is available. These provide calibration for temperatures up to the freezing point of platinum (1769 °C). Beyond this temperature the radiation law is used (see section 2.1.5). It is important to observe that the International Temperature Scale was revised in 1948 and temperatures reported before that date should be corrected where necessary. This is only important in the most accurate work above say 1200 °C where the difference is still only 0·5 °C. (For further details see Ref. [6], pp. 76–78, and an article by McLaren [9].)

Familiarity with the mercury-in-glass thermometer is assumed and some account will now be given of the principles of other methods for temperature measurement. The physical properties used for measuring temperature are also of intrinsic metallurgical interest.

2.1.3 The Resistance Thermometer

2.1.3.1 Because the electrical resistance of metal generally increases with temperature, this property can be used to measure the temperature. Platinum is considered to be the most suitable metal for this purpose, but copper and

TABLE 2.1 *Fixed Points for Temperature Calibration*

Group or range	Phase change and substance used		Temperature K	°C
(a) Fixed points below 0 °C	(Absolute Zero		0	−273·16)
	Boiling points:	Hydrogen	20·4	−252·8
		Oxygen	90·2	−182·97
	Evaporation:	Solid CO_2	194·7	−78·5
	Freezing point:	Mercury	234·3	−38·87
(b) Primary fixed points to 1063 °C	Freezing point:	Pure water	273·16	0
	Boiling points:	Pure water	373·2	100
		Sulphur	713·8	440·6
	Freezing points:	Silver	1234·0	960·8
		Gold	1336·2	1063
(c) Secondary fixed points	Freezing points:	Tin	505·1	231·9
		Cadmium	594·1	320·9
		Lead	600·5	327·3
		Zinc	692·7	419·5
		Antimony	903·7	630·5
		Aluminium	933·3	660·1
		Copper	1356·2	1083
(d) Some eutectics [6]	Copper–silver (71·9 wt %Ag)		1052·2	779·0
	Copper–Aluminium (8·5 wt % Al)		1310·2	1037
	Aluminium–copper (33 wt % Cu)		821·2	548
(e) Accurate fixed points from 1063 to 1769 °C	Freezing points:	Gold	1336·2	1063
		Nickel	1726·2	1453
		Palladium	1825·2	1552
		Platinum	2042·2	1769
(f) Above 1769 °C	No fixed points—radiation law used			

nickel can be used. The variation of resistance with temperature can be represented by

$$R = R_0(1 + \alpha T + \beta T^2) \tag{2.1}$$

where R_0 is the resistance at absolute zero and α and β are constants. Three standard temperatures and the corresponding resistances enable the constants to be found. Equation (2.1) is empirical and if desired the standard temperature for R_0 can be taken as 0 °C—this only alters the constants.

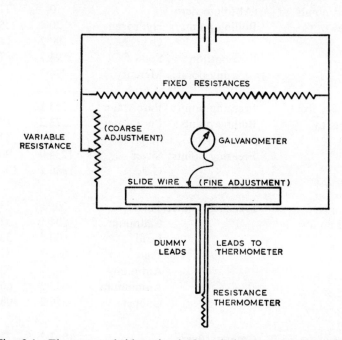

Fig. 2.1 Elementary bridge circuit for platinum resistance thermometer.

The resistance thermometer wire is non-inductively wound and coiled round an insulating support. The thermometer is contained in a silica or heat-resisting glass tube. Straining can alter the resistance and must be avoided in mounting and for this reason thermometers are annealed at a temperature higher than any which it is intended to measure.

2.1.3.2 The measurement of resistance requires a Wheatstone bridge circuit. One of the most useful of these has dummy leads in one arm of the bridge circuit and identical leads to the thermometer in another arm. In this way, changes in the resistance of the leads is compensated for—see Fig. 2.1. Methods and calibration techniques are reviewed respectively by Dauphinee [10] and by Robertson and Walch [11]. A number of automatic temperature controllers have been described [12, 13, 14]. The basic principles are outlined in Fig. 2.2.

A bridge circuit measures the resistance and if the temperature rises or falls this throws the bridge out of balance. The out-of-balance current is amplified and controls the operation of an on–off switch which, through a relay, opens and closes a series resistance thus increasing or decreasing the current through the furnace. When the balance is exact the instrument automatically switches on and off at equal time intervals (e.g. every half minute). An error signal adjusts the relative time accordingly.

The advantages of platinum resistance thermometry are:

1. It is very accurate especially where the thermal capacity of the thermometer is not important.
2. Its highest accuracy is in the range up to 600 °C but temperatures up to 1000 °C or higher can be measured.

Disadvantages are:

1. Susceptibility to strain and contamination which can, however, be generally avoided by reasonable care.
2. The large thermal capacity, which requires that the thermal capacity of the furnace and contents to be correspondingly larger.
3. Thin wire is better because it gives higher resistance which is easier to measure, and also a lower thermal capacity, but it presents a disadvantage in that there can be heating due to current from the measuring circuit. This effect can, however, be taken into account and is quite small (e.g. a fraction of a degree).

2.1.4 Thermo-electric Properties and Laws—Thermocouples

2.1.4.1 The thermo-electric properties of metals are of direct interest because they are physical properties which vary with composition and temperature, and can therefore be used in the study of alloys from the physical and constitutional point of view. Secondly the properties provide a means of temperature measurement—the *thermocouple*.

In its simplest terms a thermocouple consists of a closed loop constructed from wires of two different metals joined at their two ends—the junctions. When one junction is at a different temperature from the other a current flows round the circuit. This is known as the Seebeck effect. In practice it is necessary to open the circuit at some point to measure the current or rather the potential (voltage) that produces the current and this preferably by a potentiometric method which reduces the current itself to zero by an equal and opposite potential. The measurement thus depends on the temperature difference between the two junctions, and is used to determine the temperature of say the 'hot junction' assuming that of the 'cold junction' is known. Further details are given in section 2.1.4.5.

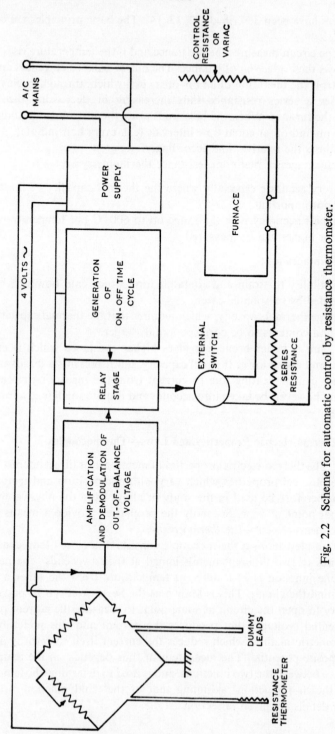

Fig. 2.2 Scheme for automatic control by resistance thermometer.

2.1.4.2 To understand the physical effects and properties involved account should be taken of the following:

1. The phenomenon of the thermo-electric e.m.f.—*the Seebeck effect*—can be considered to arise primarily from a potential difference between the two metals which are in contact. This contact potential difference varies with temperature and so if the temperatures of the two junctions are different there will be a resultant potential which causes the current to flow.

2. If, however, two metals are joined together and an electric current is passed through the junction, heat is given out or absorbed at the junction. This is the *Peltier effect*, and is reversible, i.e. if heat is absorbed when current flows in one direction, it is liberated when the current is reversed. If the direction is the same as that of the spontaneous current of the Seebeck effect then heat is absorbed at the hot junction and given out at the cold. These facts, together with other principles described later, are illustrated in Figs. 2.3(a) and (b).

3. In Fig. 2.3(a) the current is maintained automatically by the contact potentials, but in Fig. 2.3(b) there is a battery in the circuit which sends a considerably larger current round the loop. It is, however, easy to recognise that the heating or cooling at the junctions for a given current is equivalent to having an *additional* source of potential at these points for, in fact,

Heat energy (calories) × constant = heat or electrical energy (joules)
= potential (volts) × quantity (coulombs)

Hence the 'Peltier coefficient' π is defined as the heat transferred at a junction for unit quantity of electricity and so (apart from a constant) is itself an e.m.f.

If we suppose in the first instance that, without the battery, the Seebeck current is entirely due to the two Peltier effects (π_1 at T_1 K the cold junction and π_2 at T_2 K the hot junction), then the process can be treated thermodynamically as a reversible heat engine and it would follow that

$$\frac{\pi_1}{T_1} - \frac{\pi_2}{T_2} = 0 \qquad \left(\text{i.e. } \frac{Q_1}{T_1} = \frac{Q_2}{T_2}\right) \qquad (2.2)$$

Therefore
$$\frac{T_2 - T_1}{T_1} = \frac{\pi_2 - \pi_1}{\pi_1}$$

Since the resultant e.m.f. $= \pi_2 - \pi_1 = E$ say, we have $E = \pi_1(T_2 - T_1)/T_1$ so that if the cold junction is kept at constant temperature, E should increase linearly with the hot junction temperature. A straight line is, however, *not* usually obtained in practice and for many thermocouples the *empirical* equation can be supposed to carry at least one additional term thus

$$E = a + bT + cT^2 \qquad (2.3)$$

(see Fig. 2.4(a)). (Further terms in T^3.... may be required.)

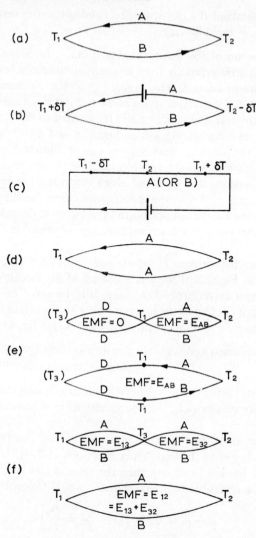

Fig. 2.3 Thermo-electric effects and laws.

(a) Seebeck effect; resultant e.m.f. $= E_{AB}$
(b) Peltier effect (heat transfer at junction equivalent to e.m.f.)
(c) Thompson effect (heat transfer equivalent to e.m.f.)
(d) homogeneous circuit (no resultant e.m.f.)
(e) intermediate metal(s); applied to cold junction connection (special case also using (d))
(f) intermediate temperatures

4. The reason for the curvature is that there must be another means by which heat is taken in or given out of the system, i.e. equation (2.2) is incomplete. This source is the *Thompson (or Kelvin) effect*, and is indicated in Fig. 2.3(c)

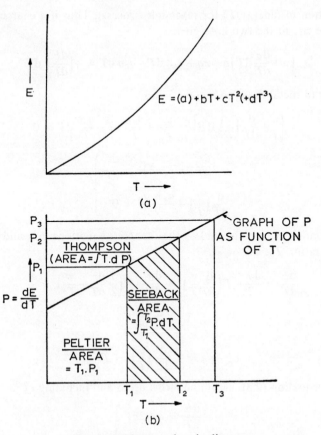

Fig. 2.4 Thermo-electric diagrams.

(a) e.m.f.–temperature relationship
(b) thermo-electric power diagram. Seebeck area = resultant Peltier $(P_2T_2-P_1T_1)$ – Thompson area

(with battery). Heat is absorbed from or liberated to the surroundings when a current flows in a wire along which there is a temperature gradient. The 'Thompson e.m.f.' is proportional to the temperature difference between any two points in a wire. It is very small but may vary with temperature so its actual contribution to the total e.m.f. in a circuit must be given by $\int_{T_1}^{T_2} \sigma \, dT$ where σ is the Thompson coefficient. This effect must be present in a thermo-couple when there is no battery and the only source of e.m.f. is the Peltier effect.

5. We may now see how equation (2.3) can be otherwise expressed in terms of the Peltier and Thompson effects separately. The simplest way is to consider only a small temperature difference (δT) and apply the first law of thermodynamics (i.e. conservation of energy) and then the second law (i.e.

29

conservation of (energy/T) for reversible process). Thus if a charge q passes round the circuit the two laws give:

$$\left(\pi + \frac{d\pi}{dT}\, dT\right)q - \pi q + \sigma_A q\ dT - \sigma_B q\ dT = q\left(\frac{dE}{dT}\right) dT$$

(σ_A refers to metal A, σ_B to metal B) and

$$\left[\frac{\pi}{T} + \frac{d}{dT}\left(\frac{\pi}{T}\right) dT\right]q - \frac{\pi}{T}q + \frac{\sigma_A q}{T}\, dT - \frac{\sigma_B q}{T}\, dT = 0$$

Therefore

and

$$\left. \begin{aligned} \frac{d\pi}{dT} + (\sigma_A - \sigma_B) &= \frac{dE}{dT} \\[2mm] \frac{d}{dT}\left(\frac{\pi}{T}\right) + \frac{\sigma_A - \sigma_B}{T} &= 0 \end{aligned} \right\} \tag{2.4}$$

These equations can be treated as simultaneous equations in π and $(\sigma_A - \sigma_B)$ (at temperature T), so

$$\frac{dE}{dT} = \frac{d\pi}{dT} - T\left(\frac{d}{dT}\frac{\pi}{T}\right) = \frac{d\pi}{dT} - T\left(\frac{1}{T}\frac{d\pi}{dT} - \frac{\pi}{T^2}\right) = \frac{\pi}{T}$$

$$\pi = T\frac{dE}{dT} \tag{2.5}$$

and

$$\sigma_A - \sigma_B = T\frac{d}{dT}\left(\frac{\pi}{T}\right) = T\frac{d^2E}{dT^2} \tag{2.6}$$

and from equation (2.3), assuming terms up to and including cT^2,

$$\left. \begin{aligned} \frac{dE}{dT} &= b + 2cT \\[2mm] \frac{d^2E}{dT^2} &= 2c \end{aligned} \right\} \tag{2.7}$$

$P = dE/dT$ is (assuming equation (2.3) is valid) a linear function of T and it is called (somewhat misleadingly) the *thermo-electric power*. It can be used as in Fig. 2.4(b) to represent the behaviour of the thermocouple with junctions at any two temperatures. The rectangular areas are T_1P_1 and T_2P_2 and their difference by (2.5) must be $\pi_2 - \pi_1$. The shaded part of this difference is

$$\int_{T_1}^{T_2}\left(\frac{dE}{dT}\right) dT = E_2 - E_1 = \text{the resultant Seebeck e.m.f.}$$

Therefore the remaining area must be the contribution from the Thompson effect, and it is in fact

$$= \int_{T_1}^{T_2} T\, dP = \int_{T_1}^{T_2} T\frac{d}{dT}\left(\frac{dE}{dT}\right) dT$$

$$= \int_{T_1}^{T_2} T\left(\frac{d^2E}{dT^2}\right) dT = \int_{T_1}^{T_2} (\sigma_A - \sigma_B)\, dT$$

(*Note:* Further details of these properties are given in many standard text-books of physics (e.g. [15, 16]) and simple apparatus can be constructed (see for example a short article in *The Metallurgist* [17]).

2.1.4.3 There are three thermo-electric 'laws' which may now be stated, and which complete Fig. 2.3.

1. The Law of the Homogeneous Circuit. An electric current cannot be sustained in a circuit of a single homogeneous metal by the application of heat alone. The Thompson e.m.f. in such a case will exactly cancel out, i.e. $\oint \sigma_A \, dT = 0$ (or $\sigma_A = \sigma_B$, Fig. 2.3(d)).

2. The Law of Intermediate Metals. If a junction is broken and a third metal is inserted there is no change in the resultant e.m.f. providing the temperature of the two new junctions is the same as that of the one replaced. Alternatively any number of metals can be placed in a circuit and there will be no resultant e.m.f. if they are all at the same temperature. Fig. 2.3(e) illustrates a special case.

3. The Law of Intermediate Temperatures. The e.m.f., in a circuit with junctions at temperatures T_1 and T_2, is equal to the sum of the e.m.f.'s in two couples with junctions at T_1 and T_3 and at T_3 and T_2 respectively.

These laws are interesting or important to metallurgists for the following reasons:

1. If a current did appear this would be evidence of inhomogeneity. It also follows that if thermocouple wires are accidently heated or cooled in between the junctions there is no net effect on the measured e.m.f.

2. Leads of inexpensive wire can be joined to the cold ends of the thermocouple wires which may be of expensive metal (e.g. platinum and its alloys). As long as the temperature of the two new junctions are constant and the circuit is completed through the one metal, the e.m.f. measured is the same as would be obtained for the single cold junction at the same temperature. This is of great importance in practical temperature measurement (Fig. 2.3(e) and section 2.1.4.5).

Secondly an important physical property is involved. So far the Thompson coefficients have been regarded as properties of the separate metals, but the Peltier coefficients as contact e.m.f.'s or properties of the junctions of pairs of metals or alloys. Now the law of intermediate metals may be stated quite generally as:

$$E_{AC}(T_1, T_2) = E_{AB}(T_1, T_2) + E_{BC}(T_1, T_2) \qquad (2.8)$$

This implies that $E_{AB} = E_A - E_B$ and so

$$P_{AB} = P_A - P_B \quad \text{and} \quad \pi_{AB} = \pi_A - \pi_B \qquad (2.9$$

31

(Note that in Fig. 2.3(e) the resultant e.m.f. is

$$[E_A(T_1, T_2) + E_D(T_1, T_3)] - [E_B(T_1, T_2) + E_D(T_1, T_3)] = E_{AB}(T_1, T_2)$$

The result (2.9) is already implicit in equation (2.6) since

$$\sigma_A - \sigma_B = T \frac{d}{dT}\left(\frac{\pi}{T}\right) = T \frac{d}{dT}\left(\frac{\pi_A}{T} - \frac{\pi_B}{T}\right)$$

Hence if the thermo-electric power of any one metal is established as a standard it can be used to determine the thermo-electric power and e.m.f./temperature relation for any other by measuring this property for thermo-couples with the standard as one element. Lead has been used for this purpose because its Thompson effect is negligible.

3. The third law is simply a consequence of the conservation of energy represented by the additivity of areas as in Fig. 2.4(b). (P has the dimension of $E = $ e.m.f. but in deducing the equations a quantity of electricity, q, was eliminated from all the terms.)

Three further observations must be made: The electric current which flows in a thermocouple will necessarily give the Joule heating effect proportional to I^2R (I=current, R=resistance). This heating is relatively small for the low currents generated in the thermo-circuit alone without additional source of e.m.f. Secondly the heat transferred produces no measurable drop in temperature—providing the thermal capacity of the medium is high relative to the thermo-electric heat at the junction which it should be. For this reason too, heat losses due to thermal conduction should be negligible, but this point may require attention in special cases.

2.1.4.4 Thermo-electric properties have been studied for their fundamental interest and as a means for investigating phase equilibria. Haughton [18] and Galibourg [19], for example, have described work on some binary alloy systems. Cobalt base alloys have been investigated in considerable detail [20]. Some systematic data is available on metals and alloys with potential uses in thermo-elements [20, 21, 22], whilst Crussard [23] has suggested a number of general principles. Thermal properties are affected by cold work, recovery and recrystallisation [24] and have been used for the study of precipitation, notably of carbon and nitrogen in α-iron [25]. There is clearly scope for further development and investigation in the use of this technique.

2.1.4.5 Thermo-electric temperature measurement is simply based on a direct use of the e.m.f./temperature relationship for the thermocouple element in use. It is quite usual to employ the arrangement referred to in connection with Fig. 2.3(e) especially when platinum and platinum alloy wires are used. In this case two cold junctions at T_1 enable the circuit to be completed by (much cheaper) copper wires through a measuring instrument. Two basic

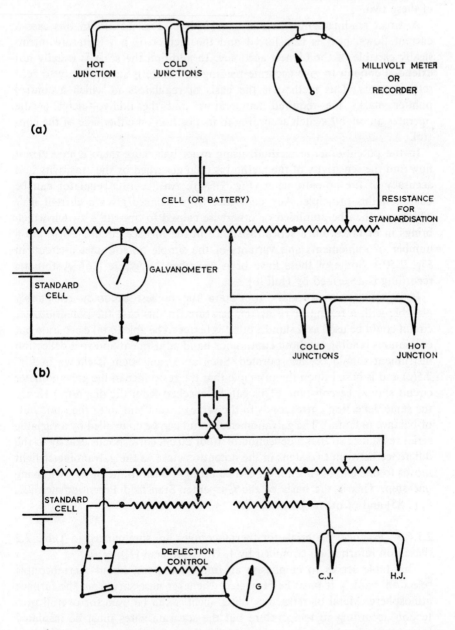

Fig. 2.5 Temperature measurement—elementary thermocouple circuits.

systems of measurement are available and a third that is really a combination of these two.

A direct reading millivoltmeter can be used (Fig. 2.5(a)). In this case a current flows, heat is transferred and the method is not therefore intrinsically capable of the highest accuracy. In addition the scale is usually not extended enough to give accurate reading and there is susceptibility to zero setting errors. This method is the basis of regulators in which a control pointer marks the required temperature and the millivoltmeter needle operates an on–off switch according to its position on either side of the control.

In the potentiometric method, using exact balancing there is no current flow and the sensitivity of the method is then governed by the sensitivity and accuracy of the potentiometer (Fig. 2.5(b)). An accurate regulator can be based on this principle. Any departure from balance gives a current flow which can then be amplified or otherwise caused to operate a switch which brings in or takes out a series resistance in the furnace circuit. There are a number of refinements and variants of the simple potentiometer circuit in Fig. 2.5(b). Some of these have been described by White [26]. Automatic recording is described by Hall [6].

It is sometimes necessary to obtain the highest accuracy reasonably possible with a falling or rising temperature. In this case the potentiometer circuit could be used as it stands, but it is better to be able to take accurate but continuous readings without continuous hand adjustment and so a deflection instrument should be incorporated. Such an arrangement is shown in Fig. 2.5(c) and is based upon the principle that the resistance in the galvanometer circuit should be constant. (This can be checked from the diagram.) Hence the same deflection corresponds to the same current and so to the same out-of-balance potential. The galvanometer circuit can be controlled by a variable series resistance so that a deflection of 10 or 20 cm corresponds exactly to the difference between two steps in the potentiometer. As the galvanometer light moves to one end of this limit it can be brought back to the other by moving one step. This is the basis of the Carpenter Stansfield Potentiometer [27, 8, p. 85] and of others [28].

2.1.4.6 Practical materials for thermocouples are summarised in Table 2.2 (based on information compiled by J. H. Woodhead [7]).

This table serves to emphasise an important point about thermocouple selection, namely it must be protected whenever necessary from the furnace atmosphere. Metal or refractory outer sheaths can be used for overall protection according to temperature but the separate wires must be insulated from each other by a suitable non-conducting refractory material, e.g. silica, which is unsuitable above 1000 °C, or alumina.

A number of other precautions must be observed. Thermocouple leads should not be near to furnace leads in order to avoid leakage or induced

34

TABLE 2.2 *Thermocouples in Common Use*

Metals used	Upper temperature (°C) Continuous	Inter-mittent	Approx. thermo-electric power ($\mu V/°C$)	Atmospheres in which couple is not stable
Platinum–platinum 10% or 13% rhodium	1500	1600	10	Metallic vapours (and Si) under reducing conditions
Chromel–alumel	1200	1300	40	Reducing or sulphurous (Depends on a *protective* oxide layer)
Iron–constantan	750	800	55	Sulphurous or damp atmospheres
Copper–constantan	350	450	45	Sulphurous atmospheres

e.m.f.'s (a.c. power is less likely to give trouble than d.c.). In fact, the whole of the thermocouple and measuring circuits should be separated and completely insulated from the rest. Atmospheric conditions, e.g. dampness, may also affect measurement, and human contact with any part of the circuit should be avoided during reading. For further details of these hazards and their avoidance, and of calibration procedures, reference can be made to the literature (e.g. Ref. [8], pp. 85–7).

2.1.5 Principles of Radiation Pyrometry

2.1.5.1 For metallurgical work involving temperatures above the range for platinum/platinum rhodium thermocouples, temperature determination depends on radiation measurement by optical pyrometry. For some purposes this technique is suitable down to 1200 °C and obviously the methods should check each other over the range 1200–1500 °C which is the practical limit for he thermocouple.

2.1.5.2 The basic principles of radiation pyrometry depend on the way in which a hot body emits radiation which increasingly spreads over the visible part of the spectrum and increases with intensity as the temperature is raised.

This phenomenon is illustrated in Fig. 2.6. For a fuller account of the explanation of the form of these curves reference may be made to the literature [29, 30]. A practical treatment is given by Bailey [31]. It is possible to gain an understanding of the essential facts by considering the equation finally proposed by M. Planck and to see how (earlier) laws or relationships follow from it.

The curves in Fig. 2.6 obey the equation:

$$E_\lambda = \frac{8\pi hc}{\lambda^5} \left[\frac{1}{\exp(h\nu/kT) - 1} \right] \tag{2.10}$$

or

$$E_\lambda = \frac{C_1}{\lambda^5} \left[\exp\left(\frac{C_2}{\lambda T}\right) - 1 \right]^{-1} \tag{2.11}$$

where C_1 and C_2 are constants.

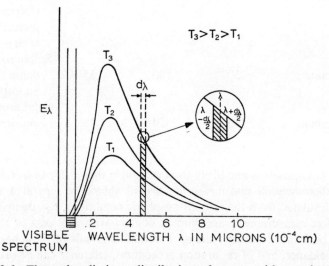

Fig. 2.6 Thermal radiation—distribution of energy with wavelength.
Note: E_λ is expressed in terms of energy per square centimetre (of surface) per cm (wavelength difference) per second (i.e. ergs cm^{-2} cm^{-1} s^{-1})

In (2.10), which is Planck's equation, h is Planck's constant, k is Boltzmann's constant, c is the velocity of light, λ is the wavelength and ν the frequency (so $\lambda\nu = c$ and $\nu = c/\lambda$). It is important to recognise that E_λ is the energy emitted per unit area, per unit wavelength, per unit time. It is therefore a power distribution function, so that $E_\lambda\, d\lambda$ is the energy between $\lambda - \frac{1}{2}\, d\lambda$ and $\lambda + \frac{1}{2}\, d\lambda$ (see Fig. 2.6). Equation (2.10) can be re-written:

$$E_\lambda\, d\lambda = (8\pi\lambda^{-4}\, d\lambda) \left[\frac{h\nu \exp[-h\nu/kT]}{1 - \exp[-h\nu/kT]} \right] \tag{2.12}$$

The expression in parentheses is the density of waves in the radiation and that in square brackets their mean energy. The product gives the mean energy of the radiation per unit volume which is E_λ. The second part of the formula arises from a consideration of the equilibrium between the radiation and the hot body—providing the hot body radiates under black body conditions (see below). A frequency ν $(=c/\lambda)$ arises from thermal oscillations in the body. These oscillations have the same frequency ν or 2ν, 3ν, etc. $(n\nu)$. The energy of the thermal oscillators cannot vary continuously but only in units of $h\nu$ which is a *quantum* of energy corresponding to the frequency ν. The probability of a thermal oscillator having energy $nh\nu$ (n is any positive integer) depends on a Boltzmann factor $\exp[-nh\nu/kT]$. Hence the sum $\sum_{n=0}^{\infty} nh\nu \exp[-nh\nu/kT]$ gives the total energy, and division by $\sum_{n=0}^{\infty} \exp[-nh\nu/kT]$ leads to the required mean energy. (The first expression is a series of the form $1 + 2x + 3x^2 + \cdots$ and the second a simple geometric progression.)

An approximation to (2.10) is Wien's law, viz.:

$$E_\lambda = C_1 \lambda^{-5} \exp[-C_2/kT] \tag{2.13}$$

and this equation is sufficient for values of λT which are not too large and so is accurate up to 4000 °C over the red part of the spectrum.

If equation (2.10) (or (2.13)) is integrated thus:

$$E = \int_0^\infty E_\lambda \, d\lambda \tag{2.14}$$

the result obviously represents the total area under the curve for a given temperature and is the total radiated energy. The integration gives the result:

$$E_0 = \sigma T^4 \tag{2.15}$$

which is the Stefan–Boltzmann law. On the other hand differentiation (and equating to zero) gives an expression for finding the wavelength for maximum radiation intensity, i.e. the peak of a curve in Fig. 2.6. The expression is

$$\lambda_m T = \text{constant} \tag{2.16}$$

If λ is in cm and T in K the constant is 0·2891 or 2891 for λ in microns (approx. 2900). For equation (2.15) σ becomes $5·71 \times 10^{-5}$ erg cm^{-2} s^{-1} (deg^{-4}). Equations (2.15) and (2.16) can be derived by thermodynamic reasoning independently of (2.10). (Refs. [25, 26] and other textbooks of physics.)

It follows that practical temperature measurement could depend on:

(a) Measurement of the total radiation according to equation (2.15).
(b) Measurement of the radiation at a fixed wavelength or a limited band of wavelengths using equation (2.11) (or (2.13)).
(c) Measurement of the wavelength λ_m at which the maximum intensity if found—equation (2.16). This will be less accurate than the other methods.

So far, however, it has been assumed that there is the maximum emission of energy from the hot body. In general, however, the full emission may not occur and there is a property of the hot surface called the emissivity ϵ and this is a fraction of the maximum possible, i.e. equation (2.15) would become

$$E = \epsilon E_0 = \epsilon \sigma T^4 \tag{2.17}$$

A corresponding *spectral emissivity* ϵ_λ is similarly required in equations (2.11) (or (2.13)). Corresponding to these emissivity coefficients we have *absorptivity* α and *spectral absorptivity* α_λ which are the absorbed fractions of the total radiation falling on a surface. (What is not absorbed is reflected unless there is some transparency when a fraction is transmitted through the material.) α is sometimes denoted by A and called (rather misleadingly) the absorptive power. *Kirchhoff's Law* states that for any two radiating surfaces

$$\frac{E_2}{E_1} = \frac{\alpha_2}{\alpha_1} \quad \left(\text{or} \quad \frac{A_2}{A_1}\right) \tag{2.18}$$

(for an elementary proof see Ref. (31)).

When $\alpha_1 = 1$, $E_1 = E_0$ since these are the maximum values for each. Hence generally

$$\frac{E}{E_0} = \frac{\alpha}{1}$$

$$E = \alpha E_0 = \epsilon E_0, \quad \text{i.e. } \alpha = \epsilon \tag{2.19}$$

It can be shown that $\alpha_\lambda = \epsilon_\lambda$ for individual wavelengths. It is a familiar notion that a material which absorbed *all* the radiation which impinged on it would be ideally black. Hence the concept of *black body radiation* to which equations (2.10) and (2.13)–(2.16) apply with $\epsilon = \alpha = 1$. Note that if ϵ_λ is the same for all wavelengths the hot body is called a grey body. If E_λ varies with λ different parts of the spectrum will be differently emphasised and a colour will emerge.

2.1.5.3 Although materials which approach the true black body behaviour are few it is possible to obtain conditions in which the total radiation measured is the maximum E_0. The requirement is a thermally insulated closed space at constant temperature. The *radiation density* in such a space is in equilibrium with the walls when energy emitted (per unit time) must equal energy absorbed $= E$. But $E = \epsilon E_0 = \alpha E_0$ so that E_0 must be the energy falling on the surface. Of the total E_0, αE_0 is absorbed and $(1-\alpha)E_0$ is reflected (assuming none escapes through the enclosure walls). The equilibrium is:

Radiation energy density $\equiv E_0 \rightleftharpoons$
absorbed (or emitted) energy αE_0 + reflected energy $(1-\alpha)E_0$

It is not possible to measure the radiation inside such an enclosure without an aperture but if its area is small compared with the total area of the enclosure walls the effect is negligible. Such black body furnaces can be used for calibrating instruments and for actual measurements of temperature such as the determination of freezing points [28].

We note that temperature measurement by radiation pyrometry assumes that the radiation is entirely *thermal radiation*. This is generally true for liquids and solids but gases may contain spectral absorption or emission lines or bands in the region.

2.1.5.4 An instrument which uses equation (2.12) is called a total radiation pyrometer. In such an instrument, radiation of all wavelengths is focused by a concave mirror or lens so that it falls on a thermocouple or thermopile, the temperature of which is a measure of the temperature of the hot body. The thermocouple or thermopile (i.e. thermocouples in series) is provided with a black receiving surface. Since lenses absorb radiation mirror instruments are preferred. The principles of such an instrument are apparent in Fig. 2.7(a). The eyepiece is required to focus on image of the hot object or surface on to

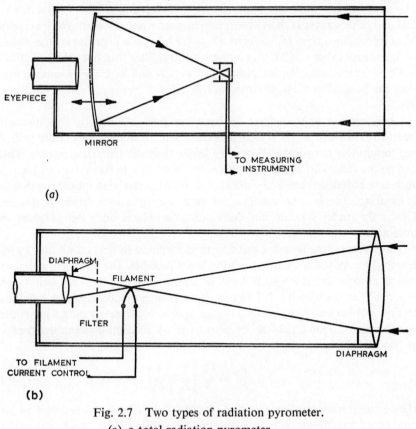

Fig. 2.7 Two types of radiation pyrometer.
   (a) a total radiation pyrometer
   (b) a disappearing filament instrument

39

the receiving unit containing the thermo-electric device—and also mirror for determining exact focusing. The concave mirror must be movable as indicated in order to give the focusing adjustment.

Because the total radiation depends only on T^4 this instrument is liable to give greater errors in determination of T. In addition air absorbs some of the radiation in the infra-red region. This absorption would be less under laboratory conditions with a (partial) vacuum path, but the instrument is not so likely to be used under these conditions since a relatively larger source is necessary.

A disappearing filament type of instrument is shown in Fig. 2.7(b). Here the radiation entering the instrument is compared with that emitted from a filament which is heated by an electric current. The current is adjusted until effective 'equality' of the two radiations received through a red filter is obtained—the filament is not then distinguishable from the background. In practice the true temperature may be found by taking the mean of two adjustments just on either side of the position of least visibility—for in fact the filament may not quite disappear owing to diffraction effects. Errors due to this and other causes such as absorption in glass windows, prisms or lenses can be corrected by a term of the form $\Delta T = CT^2$. In both instruments the higher temperatures (above 1600 °C) require a neutral filter to reduce the radiation.

Other instruments employ photo-electric cells and have the advantage that they can be used to measure temperatures below the range of visible radiation [6, 31].

Consideration of the effects of departure from black body conditions is necessary for most practical measurements. The measured temperature (called the 'brightness temperature') may be lower than the true temperature. This can be the case with say a piece of cooler metal in a hotter furnace which is therefore radiating heat to the metal. The metal is receiving more heat than it is emitting. The reverse would apply to a hot ingot in a cooler enclosure. Obviously under equilibrium conditions the black body temperature is obtained.

The disappearing filament instrument is assumed to depend on one wavelength. This monochromatic condition is not possible. The filter is of red glass which absorbs all the visible light of shorter wavelength—say below 0·6 microns. The eye's limit is 0·7 and so the instrument records a finite band (as in Fig. 2.6) between 0·6 and 0·7. Under such approximately monochromatic conditions equation (2.13) can be used to give a relation between the energy at two temperatures, viz.

$$\log_e \frac{E_2}{E_1} = \frac{C_2}{\lambda} \left[\frac{1}{T_1} - \frac{1}{T_2} \right] \qquad (2.20)$$

Hence calibration can be obtained directly if T_1 and E_1 correspond to an established fixed point such as the melting temperature of gold. The extension of the scale by a neutral filter is simply equivalent to assuming E_1 is radiated at T_1 and the filter then gives an apparent temperature T_2 corresponding to $E_2 = fE_1$, where f = the fraction transmitted; i.e.

$$\frac{1}{T_1} - \frac{1}{T_2} = \frac{\lambda \log_e f}{C_2} = \text{constant} \qquad (2.21)$$

An emissivity ϵ_λ gives a similar apparent shift on the $1/T$ scale (ϵ_λ replaces f). A rotating sector can be used instead of a filter, in which case $\theta/2\pi$ is the fraction of transmitted radiation replacing f in the formula (θ = angular range open to radiation).

For the most accurate temperature measurements either conditions approaching the black body enclosure must be ensured or the emissivity correction found (see for example Bailey [31, p. 460]). For further information on the subject of radiation pyrometry the reader may refer to a book by Harrison [33].

2.2 THERMAL ANALYSIS

2.2.1 Introductory

Thermal analysis uses thermal effects to detect and study the structural changes which a metal, or other material, may undergo. Such changes are solidification, melting or phase changes in the solid state. The rate at which a mass of metal loses heat on cooling, or gains heat, from its surroundings depends on the difference in temperature. In regulated temperature surroundings (as in a furnace) the cooling, or heating, rate can be controlled with a fairly constant temperature difference. When a phase change occurs additional heat is absorbed or evolved by the specimen. This is the basis of the use of heating and cooling curves in determining thermal equilibrium diagrams or for other metallurgical investigations. The minimal information given by such cooling curve experiments is the temperature range or critical temperature of a transition. It is, however, possible to provide a fuller interpretation which is, at least, semi-quantitative in many cases. The specific heat, for example, is an important physical property which is directly related to the curves.

Changes of state or other transitions fall into two fundamental types:

1. Changes involving latent heat.
   (a) At constant temperature—for a pure element, compound, eutectic, etc.
   (b) Over a range of temperature as in the solidification of an alloy with solid solubility. This condition will sometimes be followed or completed by a constant temperature evolution due to a eutectic or peritectic.
2. Changes involving an additional specific heat over a range of temperature rising to a critical point. This is the result of a 'co-operative' process. Examples which are of structural or physical interest to the metallurgist, and are referred to in later chapters, are the loss of ferromagnetism at the Curie

41

temperature, and the order–disorder transformation. Combinations of 1 and 2 may occur.

The methods for the study of heat evolution or absorption in such changes come within the general category of thermal analysis and can be applied to a wide variety of materials as well as metals. In particular, differential thermal analysis, or DTA, is now widely applied to problems of chemistry and mineralogy. It has been pointed out that in the early stages metallurgy was the primary field in which the differential methods found applications. From 1920 onwards the techniques were increasingly developed and exploited in mineralogy and more recently (since about 1945) for chemistry. Also during the last period, the techniques have become more quantitative. In view of the present metallurgical interest the elements of cooling and heating curve analysis and their quantitative interpretation are dealt with and then the determination of specific heat. In section 2.2.5 some further brief details are then given in reference to differential thermal analysis.

2.2.2 Cooling and Heating Curves

2.2.2.1 Furnaces for obtaining cooling curves from alloys undergoing solidification require provision for supporting a crucible to contain liquid metal, and must be provided with a suitable means of controlled heat input in order to give controlled cooling rates. Provision for gas atmosphere or vacuum is often required. Temperature measurement by thermocouple is general but radiation pyrometry may be needed. A number of experimental arrangements have been described in connection with the accurate determination of equilibrium diagrams [8]. Similar or sometimes less elaborate methods may be used for commercial alloys or in circumstances where less precision is required. For the determination of solidus temperatures heating curves may be used, but under certain conditions other techniques may be needed. A conventional method is to make a cylinder of the alloy with a small axial hole for the thermocouple. It will be apparent that because of the relative slowness of diffusion in the solid state, thermal methods are not so readily applied to solid state changes involving diffusion. This objection is not, however, important if the atom movements are rapid or over very short distances as with C or N in steel, the order–disorder transformation or diffusionless (martensitic) transformations.

2.2.2.2 In Fig. 2.8 the principal types of—temperature/time—cooling and heating curves are indicated in relation to the standard types of binary solidification behaviour condition—firstly the ideal curve and then the curves that may be obtained in practice. These curves bring out the following points:

1. A sharp change in direction is not usually obtained because the thermocouple is (necessarily) enclosed in a sheath and so heat must be conducted

42

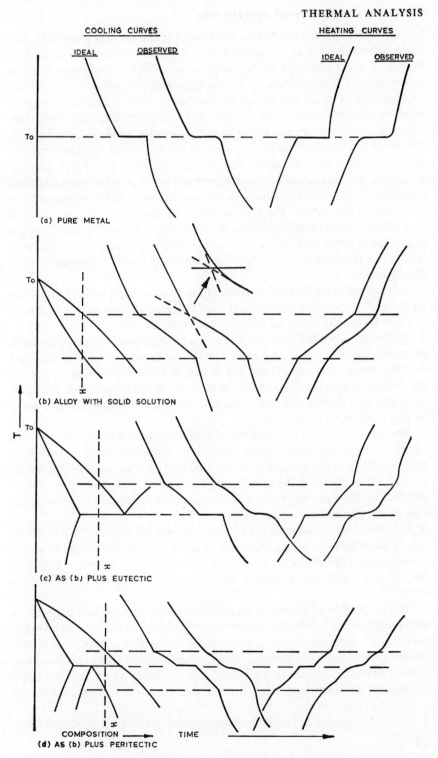

Fig. 2.8 Thermal analysis of solidification and melting.

through it before a change in heat evolution (or absorption) is detected. The thermocouple is thus at a slightly higher temperature than the melt. This leads to a slight rounding at the change point which is less pronounced at slower rates of cooling when the temperature gradient is less.

2. A second cause of this curvature in cooling curve solidification experiments is inadequate stirring of the melt in which case solidification will commence at different times due to the uneven temperature distribution. Towards the end of solidification the mixture of solid and liquid will become so viscous that even with a pure metal the end of solidification is not sharply defined. A reliable estimate of the solidus temperature is not possible from a cooling curve (see 5 below). A method for estimating the liquidus temperature is indicated in Fig. 2.8(b). There is, however, no guarantee that the point of intersection of the tangents is the true temperature. Obviously the curvature begins when some part of the melt away from the thermocouple begins to freeze, but the thermocouple itself is at a slightly higher temperature when this happens.

3. It follows that the thermal capacity of the thermocouple and sheath should be as small as possible relative to that of the melt and the sheath as thin as possible to minimise over all temperature gradients.

4. Extra care is needed when there is supercooling since in this case the melt, thermocouple and sheath have all reached a temperature below the true freezing point. The latent heat which begins to be evolved will raise the temperature again but this takes a finite time and the furnace will have continued to cool during this time. This fact renders extrapolation as indicated in Fig. 2.9(a) open to question.

5. Because of the way in which the equilibrium solidus composition would change with falling temperature continuous diffusion would normally be required in the solid as well as in the liquid. The latter is provided for by stirring, but diffusion in the solid is only very limited. This point gives further emphasis to the conclusion that a cooling curve cannot give a reliable indication of the true solidus temperature.

6. It follows that a heating curve will give the best indication of this temperature if the alloy used is heat treated long enough below the solidus to ensure homogeneity. Melting will then begin simultaneously at many points in the specimen and give a more precise change in direction to the heating curve.

7. The proportion of phases in equilibrium at any temperature depends on the lever principle, whilst the amount of solid phase formed for a given fall in temperature depends on the slope of the liquidus and solidus. Hence it is possible under the more unfavourable combinations of these two factors to have a very slight change in the rate of heat evolution at the liquidus and a consequent tendency to estimate the temperature too low. This condition is indicated in Fig. 2.9(b).

8. Although eutectic or peritectic reactions may give well defined arrests on

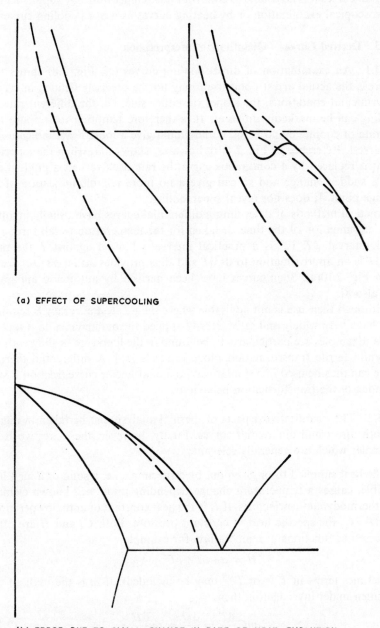

(a) EFFECT OF SUPERCOOLING

(b) ERROR DUE TO SMALL CHANGE IN RATE OF HEAT EVOLUTION
(WIDE FREEZING RANGE, AND FOR STEEP LIQUIDUS SLOPE)

Fig. 2.9 Two potential sources of error in determining freezing temperatures.

thermal curves it is advisable to confirm these temperatures by supplementary microscopical examination or by heating curves as well as cooling curves.

2.2.3 Derived Curves—Quantitative Interpretation

2.2.3.1 An examination of direct cooling curves (cf. Fig. 2.8) shows that whereas the actual arrest point itself may not be precisely defined in certain experimental conditions, the slopes on either side, i.e. the different rates of cooling can be markedly different. It is therefore tempting to suppose that the rate of cooling could be used and would give a more exact indication of the arrest. Reference to Fig. 2.8(b), however, shows that when the inflection point is replaced by a continuous curve the rate itself reveals a gradual and not a sudden change and so can give a no more reliable indication of the change point. It does not reveal supercooling.

Practical methods of determining differential curves have usually involved the determination of the time Δt taken for the temperature to fall through a fixed interval ΔT. (2° is a practical interval.) Plotted against T the ratio $\Delta t/\Delta T$ is an approximation to dt/dT and thus provides an *inverse rate curve* as in Fig. 2.10(a). Such curves have been derived by automatic apparatus [34, also 8].

Although their use is not advisable where the greatest accuracy is required they have been widely and satisfactorily applied to commercial alloys such as steels. Examples are particularly to be found in the literature dealing with the austenite–ferrite transformation range in steels [35]. A differential thermal curve can thus be used in a similar way to a dilatometer curve (section 2.3) as an index of the transformation behaviour.

2.2.3.2 The quantitative aspects of thermal analysis can be taken in regard to both direct and differential curves. Firstly, however, there are two basic principles which are generally relevant:

1. The heat supplied to or given out by a specimen, i.e. a solid or a melt in a crucible, causes a temperature change depending upon well-known elementary thermodynamic principles. If H is the heat constant or enthalpy per mole, $dH/dT = C_p$ the specific heat at constant pressure. Both C_p and H are often expressed as functions of temperature, for example

$$H = a + bT + cT^2 + \cdots$$

(sometimes terms in T^{-1} or $T^{1/2}$ may be included). If m is the mass of the specimen under investigation then

$$mC\frac{dT}{dt} = \frac{dQ}{dt} = m\frac{dH}{dt} \tag{2.22}$$

the rate of heating or cooling ($Q = mH$) (the subscript p can be omitted). In this equation the variation of H and C with T can be included, but for experiments with cooling or heating curves to determine solidification or

46

phase changes C may be taken as *approximately constant*. In alloys of varying composition it may also vary with the composition.

2. The specimen (with or without crucible) is enclosed in a furnace and supported on a refractory of low thermal conductivity. The heat is therefore chiefly transmitted by radiation. It is supposed that the furnace is closed and cold air currents cannot enter from outside. There could still be heat transfer by convection but it should be a relatively small proportion at typical temperatures—and a vacuum, which is often employed, would reduce this factor. The rate of heat transfer depends in the following ways on the difference in temperature, $\Delta T = T - T_e$, between the furnace enclosure at T_e and the specimen at T.

1. Conduction $\propto \Delta T$ (see section 2.4)
2. Convection $\propto \Delta T^{5/4}$ (see for example Ref. [36])
3. Radiation $\propto T^4 - T_e^4$.

The last expression is seen to derive from equation (2.15) since, apart from possible effects of different emissivities which may be ignored under approximate black body conditions, the enclosure is radiating heat $\propto T_e^4$ and the specimen heat $\propto T^4$. The difference is therefore transferred. Interpretation of thermal curves assuming the fourth power radiation law is cumbersome and Newton's 'Law' has been taken as an approximation, namely that heat transfer $\propto \Delta T$. This relationship is closer in fact to transfer by convection and conduction. For what is predominantly radiative transfer therefore the 'Law' must be accepted with some reservation and used with care. It is apparently satisfactory for practical cases where the difference in temperature is not too great. This point has been considered by Woodhead [7] and formed the basis of an analysis of thermal curves by Russell [37].

It is seen that $T - T_e$ is a factor of $T^4 - T_e^4$ and the remaining factor contains *third* powers of the temperatures. A simple calculation from equation (2.15) shows that radiation alone between 1000 and 990 K (i.e. 727 and 717 °C) would equal $3 \cdot 94\sigma \times 10^{10}$. At a temperature 100 °C lower, i.e. between 900 and 890 K, the factor is $2 \cdot 85\sigma \times 10^{10}$. This is a large difference. If, however, a phase change occurs the difference in temperature increases and then decreases again often over a comparatively small temperature range.

2.2.3.3 This last point can be explored in order to introduce a more quantitative interpretation. For this purpose only two simple cases will be considered and the approach used by Russell [37] will be generally followed but there are some minor differences in notation.

Fig. 2.10(a) represents a transformation at constant temperature and we first consider a specific case. Let the temperature difference be 10 °C at the beginning of the transformation B and suppose it increases to 30 °C before the end C, with the transformation at 730 °C (1003 ° abs.), and the value of σ previously given and converted to calories (1 cal = $4 \cdot 184 \times 10^7$ ergs). The heat

47

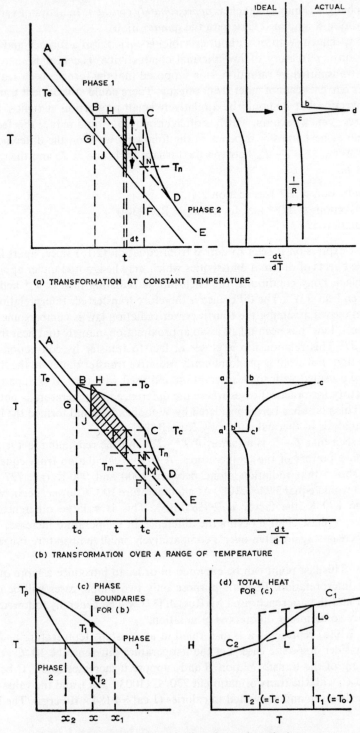

(a) TRANSFORMATION AT CONSTANT TEMPERATURE

(b) TRANSFORMATION OVER A RANGE OF TEMPERATURE

(c) PHASE BOUNDARIES FOR (b)

(d) TOTAL HEAT FOR (c)

Fig. 2.10 Quantitative thermal analysis.

transfer is calculated from the radiation formula and then compared with what would be obtained from a linear law thus:

Specimen temp (°C)	Furnace temp (°C)	Radiation energy[a]	Linear law	Percentage difference
730	720	0·054 27	0·054 27	0
730	710	0·106 93	0·108 54	1·52
730	700	0·158 01	0·162 81	3·04

[a] cal cm^{-2} s^{-1} deg^{-4}

The difference is not excessive having regard to the necessarily approximate nature of the analysis. It is the transformation temperature ranges which are of metallurgical interest and although the total range may be considerable the temperature difference will generally be small enough.

Reference to the cooling curve in Fig. 2.10(a) shows that, providing the heat transferred may be taken as proportional to the temperature difference, i.e. $T - T_e = \Delta T$, then the heat transferred in the interval dt is given by $K \Delta T \, dt$. If the specimen has mass m any latent or specific heats per gram atom or mole must be multiplied by m. On the other hand the heat transfer constant K is necessarily in terms of surface area. The best way to overcome the difficulty of having quantities multiplied by mass on one side of an equation and area on the other is to assume that there is a constant K for the conditions in the individual experiment under consideration. Secondly it can be understood that specific heats C and latent heats L refer to specimens of *unit mass* unless otherwise stated. Let the cooling rate of the furnace be R (this need not be confused with the gas constant), i.e. $R = (-dT_e/dt)$: assume R is linear.

From equation (2.22), for the specimen

$$C\left(\frac{-dT}{dt}\right) = \frac{-dQ}{dt} = K \Delta T \tag{2.23}$$

Along the part AB, Phase 1 will cool at an approximately constant temperature interval above that of the furnace, i.e.

$$C_1 R = K \Delta T_1 \tag{2.24}$$

The second phase may not necessarily have the same specific heat and so the specimen temperature after transformation will eventually approach a parallel line with a new value of ΔT:

$$C_2 R = K \Delta T_2 \tag{2.25}$$

Let T_0 be the temperature of BC and suppose $t = 0$ at B and t_c at C. Between these times only latent heat is transferred. The temperature of G is $T_0 - \Delta T_1$

49

and between G and F the furnace temperature continues to fall, and after time t is $T_0 - \Delta T_1 - Rt$, i.e. ΔT increases from ΔT_1 to $\Delta T_1 + Rt$:

$$\frac{dL}{dt} = K(\Delta T_1 + Rt)$$

(We omit consideration of the sign of L.)

After time t the total latent heat transfer is

$$L = K \int_0^t (\Delta T_1 + Rt)\, dt = K\left(\Delta T_1 t + \frac{Rt^2}{2}\right)$$

$$= K\left(\frac{RC_1 t}{K} + \frac{Rt^2}{2}\right) \tag{2.26}$$

The total latent heat is thus:

$$K\left(\Delta T_1 t_c + \frac{Rt_c^2}{2}\right)$$

Hence L is not a linear function of time as is sometimes stated, and the total latent heat is not proportional to the (time) length of BC. For example, in the case of a eutectic arrest the amount of eutectic is not proportional to this length. The integral in equation (2.26) is clearly equal to the area $BCFG$ in Fig. 2.10(a). This area $=L/K$ and so is independent of C_1 and R. These quantities occur in (2.26) but obviously at faster rates, (RC_1/K) is also larger and so t_c is correspondingly smaller. Furthermore (2.26) can be treated as a quadratic in t, i.e.

$$t = \frac{\sqrt{(2LKR + R^2 C_1^2)}}{KR} - \frac{C_1}{K} \tag{2.27}$$

(At C, $t = t_c$ and the evolution of L is complete.)

At the end of the transformation Phase 2 begins to cool and approach the line DE. For convenience, we take the furnace temperature at F as $T_F = (T_0 - \Delta T_1 - Rt_c)$ and recommence calculation with $t = 0$ at C. Along CD:

$$C_2\left(\frac{-dT}{dt}\right) = K(T - T_e) = K(T - T_F + Rt) \tag{2.28}$$

The solution of this equation must include an exponential term which would apply with $R = 0$ and a function of t, e.g. $f(t)$.

$$\frac{dT}{dt} = -\frac{K}{C_2} A \exp\left[-\frac{Kt}{C_2}\right] + \frac{df}{dt} = -\frac{K}{C_2}[T - f(t)] + \frac{df}{dt}$$

$$= -\frac{K}{C_2}(T - T_e) \quad \text{(from (2.28))}$$

Therefore
$$-\frac{K}{C_2}\left[-f(t)-\frac{C_2}{K}\frac{df}{dt}\right] = -\frac{K}{C_2}(T_e)$$

$$T_e = T_F - Rt = f(t) + \frac{C_2}{K}\frac{df}{dt}$$

i.e.
$$f(t) = T_F - Rt + \frac{RC_2}{K}$$

and when $t=0$, $T=T_0$. The solution is

$$T = \left(T_0 - T_F - \frac{RC_2}{K}\right)\exp\left[-\frac{Kt}{C_2}\right] + \left(T_F - Rt + \frac{RC_2}{K}\right) \qquad (2.29)$$

Note that T_F is not independent of R.

The first, exponential, term represents the decrease in rate along CDE which thus eventually settles down to a line parallel to GF—since the second term is linear in Rt.

Further observations may be made about the areas involved in Fig. 2.10(a). If ED is extrapolated back to H it intersects the ordinate CF in N (temperature T_N) and BC in H. It follows that:

(a) The area $HJNF$ represents the heat which would be evolved if the Phase 2 had cooled from T_0 to T_N. Hence this area must equal the area

$$CND = \frac{C_2}{K}(T_0 - T_N)$$

(b) The same area could therefore be used to find C_2/K and so $(RC_2)/K = NF$, and thus to locate the furnace cooling line GF.

(c) The area $BHJG$ (a trapezium)

$$= \tfrac{1}{2}BH(HJ+BJ) = \frac{1}{2}\frac{\Delta T_2 - \Delta T_1}{R}(\Delta T_2 + \Delta T_1)$$

$$= R\frac{C_2{}^2 - C_1{}^2}{2K^2} \qquad (2.30)$$

and vanishes when $C_2 = C_1$.

(d) The area $BCFG = L/K$ has two parts, i.e. the part given by (30) due to $C_2 \neq C_1$ and the remainder which is given by areas

$$HCFJ = HCN + HJNF = HCN + CDN = HCD \qquad (2.31)$$

Hence the whole area above the specific heat line HDE represents this part of the latent heat.

(e) When $C_1 = C_2$, AB extrapolates to DE and the single enclosed area then $= L/K$, and is independent of R and can so be estimated (even if the cooling curve is not prolonged far enough to justify the extrapolation back to H and the determination of the furnace cooling curve by the steps (a) and (b)).

These conclusions are also of value in connection with the interpretation of the inverse rate curves. The area between such a curve and the temperature axis is obviously a time (i.e. $\int (-dt/dT)\, dT$). In the 'ideal' case, at a constant temperature arrest $dT/dt=0$ and $dt/dT=\infty$. The 'area' here, i.e. the time t_c, is thus within or along this line. In practice the time to fall a constant temperature interval, say $1\,°C$, is taken and the curve has the actual form shown. The length $ab=1/R$ and the length $bd=$ the time taken to fall through $1\,°C$ in the region of $T_0=t_c$. The ratio bd/ab is of interest. From (2.27) for t_c this ratio is

$$\frac{bd}{ab} = Rt_c = \frac{\sqrt{(2LKR+R^2C_1{}^2)}-C_1R}{K} \qquad (2.32)$$

This increases as R increases. Now ac, a very short interval on the diagram, is the inverse rate of cooling at C when the specimen is just beginning to fall along CE.

$$\frac{1}{ac} = \frac{-dT}{dt} = \frac{K}{C_2}(T_0-T_t) = \frac{K}{C_2}\left(\frac{RC_1}{K}+Rt_c\right)$$

$$= \frac{\sqrt{(2LKR+R^2C_1{}^2)}}{C_2} \qquad \text{from (2.32))}$$

Therefore $$\frac{bc}{ab} = \frac{cd-bd}{ab}$$

or $$\frac{ab-ac}{ab} = 1-\frac{C_2}{\left(\dfrac{2LK}{R}+C_1\right)} \qquad (2.33)$$

The expressions (2.32) and (2.33) thus indicate how the shape of an inverse rate curve will alter in the region of a transformation—with different cooling rates or different values of the latent or specific heats. In particular, providing the rate of cooling is not fast enough to cause supercooling or suppression of the transformation, the peak becomes more pronounced with faster cooling rates and the return towards the temperature axis becomes shorter. The treatment of the case represented by Fig. 2.10(a) will be most reliable in an alloy of eutectic or eutectoid composition where atomic diffusion over large distances is not necessary. It will also be applicable to a eutectic at the end of a solidification or a diffusionless transformation in the solid state. Less accuracy would be expected with other types of transformation including the peritectic case.

2.3.3.4 The same approach can be used (with less accuracy) in considering a transformation taking place over a range of temperature. In this case the second phase is progressively replacing the first with an evolution of latent heat whilst the mixture of phases is cooling down. Simultaneously each phase will separately evolve its own specific heat in an amount depending on the mass present at any given time. Again a linear furnace cooling rate can be assumed.

52

The conditions are indicated in Fig. 2.10(b). As in Fig. 2.10(a) we have an effect—exaggerated here—due to the difference in specific heats. If this difference is negligible, extrapolation from AB to DE will, as before, exclude an area equivalent to the latent heat only. From the point C onwards cooling along CD will be analogous to that in the same range as in Fig. 2.10(a) except that the temperature T_c at C is now lower than T_0. It is not possible to determine the relative amounts of two phases accurately since equilibrium is not generally reached. An equilibrium diagram can, however, be used to indicate ideal proportions. Woodhead [7] has provided a treatment of cooling under radiative conditions. The mass transfer from one phase to another is given by a formula due to Masing [38]. These methods and Russell's treatment are modified and adapted here.

Let the masses of the two phases be m_1 and m_2 and $m = m_1 + m_2$. In a weight per cent phase diagram:

$$m_1 = \frac{m(x - x_2)}{x_1 - x_2}; \qquad m_2 = \frac{m(x_1 - x)}{x_1 - x_2}$$

This is the 'lever law'. Latent heat L is liberated as m_2 increases.

$$\frac{-\mathrm{d}m_2}{\mathrm{d}T} = \frac{\mathrm{d}m_1}{\mathrm{d}T} = \frac{m[(x_2 - x)(\mathrm{d}x_1/\mathrm{d}T) + (x - x_1)(\mathrm{d}x_2/\mathrm{d}T)]}{(x_1 - x_2)^2} \qquad (2.34)$$

The heat transfer at T is given by

$$\mathrm{d}Q = \left[m_1 C_1 + m_2 C_2 + L\left(\frac{-\mathrm{d}m_2}{\mathrm{d}T}\right) \right] \mathrm{d}T \qquad (2.35)$$

Note that L itself generally varies with T and $-\mathrm{d}m_2/\mathrm{d}T$ is algebraically positive. If the specific heats are equal or not greatly different

$$\frac{\mathrm{d}Q}{\mathrm{d}T} = mC + L\frac{\mathrm{d}m_1}{\mathrm{d}T} = m\left(C + \frac{L}{m}\frac{\mathrm{d}m_1}{\mathrm{d}T}\right) \qquad (2.36)$$

The quantity in parentheses is a modified specific heat. $\mathrm{d}m_1/\mathrm{d}T$ is a complicated function of T unless simplifying assumptions are made. We consider two cases.

Case 1

The phase boundaries are approximately straight lines passing through T_p, which represents a pure component, or the intersection of tangents, in which case the zero of x must be moved to this point. Then

$$x_1 = S_1(T_p - T) \qquad (2.37)$$
$$x_2 = S_2(T_p - T)$$
$$S_1, S_2 \text{ are the slopes, i.e. } \mathrm{d}x_1/\mathrm{d}T = -S_1$$

If $S_2 > S_1$ the boundaries are as illustrated. If $S_1 > S_2$ they rise up with increasing x. (Case not considered by Russell.)

$$m_1 = m\left[\frac{x - S_2(T_p - T)}{(S_1 - S_2)(T_p - T)}\right]$$

$$m_2 = m\left[\frac{S_1(T_p - T) - x}{(S_1 - S_2)(T_p - T)}\right] \tag{2.38}$$

From (2.34)

$$\frac{dm_1}{dT}\left(= \frac{-dm_2}{dT}\right) = \frac{mx}{(S_1 - S_2)(T_p - T)^2} \tag{2.39}$$

The variation of L with T is given (for $C_2 > C_1$) by

$$L = L_0 + (C_2 - C_1)(T_0 - T) \tag{2.40}$$

where L_0 = the value at T_0 in Fig. 2.10(b). This equation can easily be deduced from a diagram of heat (H) versus temperature—Fig. 2.10(d). L is the heat which would be evolved if all transformed completely at T, but we need to take account of the fact that it is progressively given up between T_0 and T_c as in equation (2.35). From (2.36) with (2.38), (2.39), (2.40) inserted we have:

$$\frac{dQ}{dT} = m\left[\frac{x(C_1 - C_2)}{(S_1 - S_2)(T_p - T)} + \frac{C_2 S_1 - C_1 S_2}{S_1 - S_2}\right.$$

$$\left. + \frac{x\{L_0 + (C_2 - C_1)(T_0 - T)\}}{(S_1 - S_2)(T_p - T)^2}\right]$$

$$= mF(T) = m\{G(T) + L(T)\} \tag{2.41}$$

i.e. 'specific heat terms' plus 'latent heat terms'. Therefore

$$\frac{dQ}{dt} = \frac{dQ}{dT}\frac{dT}{dt} = mF(T)\frac{dT}{dt} = -K\left[T - \left(T_0 - \frac{mRC_1}{K} - Rt\right)\right] \tag{2.42}$$

(For unit mass m is omitted.)

This equation ($C_1 \neq C_2$) cannot be used to give an explicit function of t or T in terms of the other. This means that the precise shape of BC is not known and hence relationships for the inverse rate curve cannot be deduced, but $F(T)$ can be integrated and deductions made about certain areas:

(i) The temperature T_c (end of transformation), T_N and T_m (Fig. 2.9(c)) can be found by construction providing the curve has continued far enough.

$$BG = \frac{RC_1}{K} \left(\text{or } \frac{mRC_1}{K}\right), \qquad NF = \frac{RC_2}{K} \left(\text{or } \frac{mRC_2}{K}\right)$$

Thus area $BGFM$ (time $(t_c - t_0)$) $= (C_1/K)(T_0 - T_m)$; area $HJFN = (C_2/K)(T_0 - T_N)$, and similarly reduced areas for cooling either phase over a reduced range of temperature, e.g. $(T_0 - T_c)$.

(ii) Hence to cool Phase 2 from C to DE the heat evolution is

54

$(C_2/K)(T_c - T_N) = $ area CND. In this case the shape of CD is given by equation (2.19) and hence the shape of $-dt/dT$ from b' downwards. This area can be used to find C_2/K and so the length NF. Hence even if the furnace cooling curve is not obtained it can be approximately located at this point and then extended parallel to AB and DE (as for Fig. 2.10(a)).

(iii) The total area $BCFG \times K$ is given by $\int_{T_c}^{T_0} F(T)\,dT$ since this is the total heat actually transferred between T_0 and T_c as determined by (2.41). The specific heat terms give (unit mass):

$$\int_{T_c}^{T_0} G(T)\,dT = \left[\left(\frac{C_2 S_1 - C_1 S_2}{S_1 - S_2} \right)(T_0 - T_c) - \frac{x(C_2 - C_1)}{(S_1 - S_2)} \log\left(\frac{T_p - T_c}{T_p - T_0} \right) \right] \quad (2.43)$$

The latent heat terms give:

$$\int_{T_c}^{T_0} L(T)\,dT = \left[\frac{xL_0(T_0 - T_c)}{(S_1 - S_2)(T_p - T_0)(T_p - T_c)} \right.$$

$$\left. + \frac{x(C_2 - C_1)}{(S_1 - S_2)} \log\left(\frac{T_p - T_c}{T_p - T_0} \right) - x\left(\frac{C_2 - C_1}{S_1 - S_2} \right)\left(\frac{T_0 - T_c}{T_p - T_0} \right) \right]$$

Simplification follows:

From Fig. 2.9(c) and *if equilibrium is attained*

$$T_0 = T_1 \quad \text{and} \quad T_c = T_2 \quad (2.44)$$

where $x = S_1(T_p - T_1) = S_2(T_p - T_2)$ from which

$$\frac{x(T_0 - T_c)}{(S_1 - S_2)(T_p - T_0)(T_p - T_c)} = 1$$

i.e. the coefficient of L_0. This result is in any case demanded by the fact that all the mass is transferred from one phase to the other over this interval. The second terms in each integral are numerically equal but opposite in sign.

The first term in $\int G(T)\,dT$ is given by:

$$\frac{C_2 S_1 - C_1 S_2}{S_1 - S_2}(T_0 - T_c) = \left[C_2 + \frac{S_2(C_2 - C_1)}{S_1 - S_2} \right](T_0 - T_c)$$

$$= \left[C_1 + \frac{S_1(C_2 - C_1)}{S_1 - S_2} \right](T_0 - T_c) \quad (2.45)$$

But the third term in $\int L(T)\,dT$ is *negative*

$$x\left(\frac{C_2 - C_1}{S_1 - S_2} \right)\left(\frac{T_0 - T_c}{T_p - T_0} \right) = x\left(\frac{C_2 - C_1}{S_1 - S_2} \right)\left(\frac{T_0 - T_c}{T_p - T_2} \right)$$

$$= \frac{S_2(C_2 - C_1)(T_0 - T_c)}{S_1 - S_2} \quad (2.46)$$

The integral $\int F(T)\,dT$ therefore reduces to

$$[C_2(T_0 - T_c) + L_0] \quad (2.47)$$

which is equivalent to the condition that the same resultant heat content in Phase 2 would be achieved by evolving all the latent heat L_0 at T_1 and then cooling at the rate C_2 ($=dH/dT$) as in Fig. 2.10(d).

(iv) It is apparent that the total heat (Fig. 2.10(d)) follows an intermediate path in practice and at any point there is a mean specific heat and a latent heat given by equations (2.41) and (2.40). The fraction of heat transferred between any two temperatures can be deduced from the integrals (2.43), (2.44) with the appropriate limits, e.g. $T-\frac{1}{2}\Delta T$, $T+\frac{1}{2}\Delta T$. This would require a knowledge of the quantities S_1, S_2, T_p, etc., but would give the height of a narrow strip at T. Hence a theoretical cooling curve could be built up by this method and used to check selected experimental curves.

(v) Assuming equilibrium the total area:

$$BCFG = \frac{1}{K}[C_2(T_0 - T_o) + L_0]$$

Deduction of the shaded area $(C_2/K)(T_0 - T_o)$ leaves L_0/K. Since $CND =$ remainder of area to NF, i.e. $(C_2/K)(T_c - T_N)$, $L_0/K =$ area above the line $HD +$ the small area (actually dependent on $C_2 - C_1$) slightly less than $BHGJ$ as in Fig. 2.10(a). If $C_1 = C_2$, extrapolation from AB to DE then encloses the single area $= L_0/K$. Equations (2.40), (2.41), (2.43), (2.44), (2.45), simplify considerably since all terms in $C_2 - C_1$ vanish. This analysis clearly gives C_1, C_2, L_0 as relative magnitudes so that a knowledge of one enables a determination of the other two (approximate).

Case 2

The slopes of the phase boundaries are equal, i.e. $S_1 = S_2$ but $T_p \rightarrow \infty$, so

$$x_1 - x = -S(T_1 - T)$$
$$x - x_2 = -S(T - T_2)$$
$$\frac{dx_1}{dT} = \frac{dx_2}{dT} = -S \tag{2.48}$$

Hence:

$$\frac{x_1 - x}{x - x_2} = \frac{T_1 - T}{T - T_2}$$

so that the 'lever law' applies equally well over the temperature interval $T_1 - T_2$ and compositions can be eliminated from equation (2.35). (This is the case treated by Russell.) Corresponding to the previous equations:

$$m_1 = m\left(\frac{x - x_2}{x_1 - x_2}\right) = m\left(\frac{T - T_2}{T_1 - T_2}\right), \qquad m_2 = m\left(\frac{T_1 - T}{T_1 - T_2}\right) \tag{2.49}$$

$$\frac{dm_1}{dT} = -\frac{dm_2}{dT} = \frac{1}{T_1 - T_2} \tag{2.50}$$

$$L = L_0 + (C_2 - C_1)(T_0 - T) = L_0 + (C_2 - C_1)(T_1 - T) \tag{2.51}$$

(i.e. assuming equilibrium is continuously established).

56

$$\frac{dQ}{dT} = m\left[\frac{T-T_2}{T_1-T_2}C_1 + \frac{T_1-T}{T_1-T_2}C_2 + \frac{L_0+(C_2-C_1)(T_1-T)}{T_1-T_2}\right]$$

$$= m\left[\frac{T_1C_2-T_2C_1}{T_1-T_2} - \frac{(C_2-C_1)T}{T_1-T_2} + \frac{L_0+(C_2-C_1)T_1}{T_1-T_2} - \frac{(C_2-C_1)T}{T_1-T_2}\right] \quad (2.52)$$

$$\int_{T_2=T_c}^{T_1=T_0} G(T)\,dT = \left[(T_1C_2-T_2C_1)-(C_2-C_1)\frac{T_1+T_2}{2}\right] \quad (2.53)$$

(cf. equation (2.43))

$$= \tfrac{1}{2}(C_2+C_1)(T_1-T_2) = C_2(T_1-T_2)-\tfrac{1}{2}(C_2-C_1)(T_1-T_2)$$

$$\int_{T_2}^{T_1} L(T)\,dT = \left[L_0+(C_2-C_1)T_1-(C_2-C_1)\frac{T_1+T_2}{2}\right] \quad (2.54)$$

$$= [L_0+\tfrac{1}{2}(C_2-C_1)(T_1-T_2)]$$

(cf. equation (2.44))

$$\text{The sum} = \int F(T)\,dT = [C_2(T_1-T_2)+L_0]$$

Thus, in this case which is identical with (2.47) the specific heat term in (2.52) is a linear function of T and the part of the integral which represents the difference in specific heat is transferred from $G(T)$ to $L(T)$ as before. The analysis of areas in Fig. 2.10(b) follows as for Case 1. When $C_2=C_1$ the same simplifications result. Furthermore, in this case, from (2.52) and analogously with (2.42) the heat transfer equation (for unit mass) is:

$$\left[C+\frac{L_0}{T_1-T_2}\right]\frac{dT}{dt} = -K\left[T-\left(T_0-\frac{RC}{K}-Rt\right)\right] \quad (2.55)$$

Apart from different constants this equation is integrable exactly as (2.28) giving an equation similar to (2.29). The constant in front of the exponential term is found by putting $T=T_1$ when $t=0$.

$$T = \frac{-RL_0}{K(T_1-T_2)}\exp\left[\frac{-Kt(T_1-T_2)}{C(T_1-T_2)+L_0}\right]$$

$$+\left(T_1-\frac{RC}{K}-Rt\right)+\frac{R}{K}\left[C+\frac{L_0}{T_1-T_2}\right] \quad (2.56)$$

The first term in this expression represents a curve. The second, in parentheses, is a linear function of t approached asymptotically. The third (square brackets) is simply a constant.

$$\frac{dT}{dt} = R\left[\left\{\frac{L_0}{C(T_1-T_2)+L_0}\right\}\exp\left\{\frac{-Kt(T_1-T_2)}{C(T_1-T_2)+L_0}\right\}-1\right] \quad (2.57)$$

The following deductions can be made.

(a) When $t=0$ this expression becomes:

$$\left(\frac{dT}{dt}\right)_{t=0} = -R\left[\frac{C(T_1-T_2)}{C(T_1-T_2)+L_0}\right] \tag{2.58}$$

This expansion shows that the initial slope cannot be zero and must be negative. Its reciprocal expression is therefore the finite distance ac in the inverse rate curve.

(b) ab is $1/R$, therefore

$$\frac{ac}{ab} = \frac{C+[L_0/(T_1-T_2)]}{C} = \frac{\text{Heat per degree below } T_1}{\text{Heat per degree above } T_1}$$

(c) Because of the diminishing exponential term the rate increases numerically but in a negative direction. Hence the inverse curve falls along cc' according to (2.58) and the equation (2.56) for BC (in Fig. 2.10(b)) approaches a line $T=\text{constant}-Rt$, i.e. parallel to the furnace cooling line GF. (This is a consequence of $C_1=C_2$ and $L=L_0=\text{constant}$.)

(d) The diagram implies there should be a marked change in the rate of cooling at C or C'. This point of inflection may not be easy to recognise if the conditions are far from equilibrium.

(e) From b' onwards the rate of cooling could be determined if necessary from equation (2.29) with suitable change of constants. T_t in the initial difference T_c-T_t will depend on the rate of cooling. Otherwise from (2.28)

$$a'b' = -\frac{dt}{dT} = \frac{C}{K(T_c-T_t)}$$

which is generally smaller with increasing R, since $K\,\Delta T=CR$, and $a'c'=1/R=ab$. The rate of cooling represented by (2.58) increases numerically with R so ac decreases but ac/ab is constant. Further analysis and consideration of the properties of thermal curves has been given by Russell and Woodhead in the papers referred to.

It must be emphasised that the methods cannot in practice give accurate values of latent and specific heats but in favourable cases (especially for the transformation at constant temperature) the analysis should be of value in giving a reasonable semi-quantitative basis of interpretation. The general features of differential cooling curves deduced here are in fact matched by curves obtained from steels and Russell gives interesting examples.

2.2.4 Direct Methods of Specific Heat Determination

2.2.4.1 The last section points naturally to the possibility of determining the specific heat by actually controlling and measuring the heat transferred to (or from) a specimen at a known rate or in a known temperature interval. This procedure gives a direct measure of specific heat and, whilst still limited in regard to transformations over a range of temperature (when diffusion between the phases is required for equilibrium), otherwise provides a very useful technique.

The specific heats of metals and other materials can obviously also be determined by the conventional physical methods such as the method of mixtures. These methods will be accepted and not discussed here. The student should consult suitable textbooks, for example Kubachewski and Evans [39], and some of the more recent papers dealing with applications and methods [40–42].

2.2.4.2 The basis of the direct methods is equation (2.22) provision being made for the measurement of the rate of heat supply to the specimen alone. Obviously this provision can most easily be met by using electrical heat. The supply of heat can be arranged to give continuous heating or alternately to provide a known fixed quantity at a time so that the heating produced represents a specific heat contribution over a small range of temperature. The principal methods are based on the work of Sykes and collaborators [43, 44], Awberry and Griffiths [45, 46] and C. S. Smith [47]. (See also Ref. [7] and Ref. [8], Chap. 14.)

The Sykes method is summarised in Fig. 2.11. The specimen is supported inside a copper block from which it is thermally insulated. Inside this, hollow, specimen there is a small nichrome heater. The outer specimen surface and inner block surface are highly polished whilst the outer block surface and inner surface of its containing tube are blackened. The whole is operated *in vacuo*. Fluctuations in furnace supply and other sources of thermal instability are smoothed out as a result of these provisions. The thermocouple arrangement ensures a measure of block temperature by A_1B_1 and the difference by A_2A_3. The wires can be firmly fixed into the copper blocks and contact resistances avoided. (Other details can be obtained from the original papers, e.g. Ref. [44].)

The experimental procedure depended on a continuous external rate of heating applied to the system through the furnace. Normally the specimen temperature would lag behind that of the block. The small internal heater is therefore used to control the difference and by suitable manipulation the specimen temperature T_s can be made to follow a curve which crosses the curve for the block T_B as shown in Fig. 2.11(b). At any point where the two temperatures are exactly equal, the specimen is receiving no heat from the block with which it is exactly in thermal equilibrium. When T_s is rising through such a point the rate of rise must clearly depend on the heat being supplied by the interior heater alone, and equation (2.22) will apply, that is

$$\frac{dQ}{dt} = mC\frac{dT_s}{dt} \qquad (2.59)$$

Equations (2.59) and (2.22) are equivalent.

In the original papers, Q was used for the actual *power*, but it is probably better to keep to the convention that Q is a quantity of heat and then power is represented by dQ/dt. This must be equated to VI/J where V is the voltage supplied to and I is the current across the heater. J is the electrical equivalent of heat.

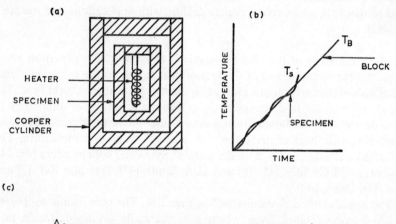

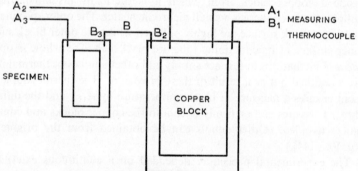

Fig. 2.11   The Sykes method of determining specific heat.
(a) experimental arrangement
(b) heating curves
(c) thermocouples arranged for differential measurement

By applying equation (2.59) at a succession of points and suitably adjusting V to control the slope it is possible to determine the instantaneous specific heat C at each point. The possible difficulty in measuring dT_s/dt accurately enough will be immediately apparent from Fig. 2.11(b) and from possible effects of the (discontinuous) adjustments necessary in voltage. The time–temperature curve of the copper block is likely to be very smooth since *in vacuo*, and with the heat capacities involved, small fluctuations in the supply will tend to be smoothed out. The differential thermocouple arrangement permits the separation of this temperature rate from the difference $T_s - T_B$. Thus

$$\frac{dT_s}{dt} = \frac{dT_B}{dt} + \frac{d}{dt}(T_s - T_B) \tag{2.60}$$

The first term is then measured by the thermocouple A_1B_1, and the second by the differential couple whose connections are A_2A_3 (Fig. 2.11(c)). The authors point out that when $T_s \neq T_B$ the method can still be used since a radiation correction can be applied but in view of variation of emissivity with temperature experimental determination of this correction is necessary.

An essentially similar procedure employing a furnace of special design and continuous reading was used in the work of Awberry and Griffiths [45, 46]. An automatic system was used by Jellinghaus [48]. Smith's procedure and analysis [47] does not, however, require such elaborate equipment and can be used for specific and latent heat evaluation. (In some respects it stands between Russell's and Syke's methods.) A refractory container is arranged so that the specimen is completely enclosed within it. It has a low thermal conductivity and is set up in a furnace with thermocouples arranged to give specimen temperature and the temperature on the outside of the container. By opposing these two e.m.f.'s, the difference can be made to control the furnace heating supply, so that there is *constant temperature difference* through the refractory specimen enclosure. Let this difference be Δ, then apart from a 'shape factor' (S) for the container, we can equate the heat flow at any temperature (into or out of the specimen) to $k\Delta$ where k is the thermal conductivity. Relatively small but significant variations of k, and the differential e.m.f., i.e. Δ, with temperature necessitate the use of a standard specimen for calibration. Otherwise the equation is:

$$\frac{dQ}{dt} = kS\,\Delta \tag{2.61}$$

Hence by keeping Δ constant for two or more experiments we can eliminate the constant rate of heat transfer dQ/dt—the quantity which is clearly difficult to determine experimentally by any simple procedure. Furthermore the heat conducted is partly used to heat the refractory container but this contribution can be eliminated by carrying out an experiment with the container empty.

Although somewhat different notations have been used in accounts of this method, there need be no confusion since the actual heat transferred in a temperature interval dT is always represented by

dQ = [sum of specific heat terms + any heat of transformation] dT

as in equations (2.35) and (2.36). For generality we replace L/m by ΔH = heat

61

of transformation unit mass. Then in a small time interval Δt_1, the heat transferred is, in such a case,

$$dQ_1 = \frac{dQ}{dt} \Delta t_1 = [m_1 C_1 + m_2 C_2 + m_r C_r + \Delta H \, dm/dT] \, \Delta T_1$$

where $m_r C_r$ is added to represent the refractory contributions, and we have assumed two phases of different specific heat in process of transformation. We consider first the case in which there is no transformation:

$$\frac{dQ}{dt} \Delta t_1 = [m_1 C_1 + m_r C_r] \, \Delta T_1 \tag{2.62}$$

and with the standard (no transformation):

$$\frac{dQ}{dt} \Delta t_s = [m_s C_s + m_r C_r] \, \Delta T_s \tag{2.63}$$

and with the refractory alone:

$$\frac{dQ}{dt} \Delta t_r = [m_r C_r] \, \Delta T_r \tag{2.64}$$

From 2.62

$$\frac{dQ}{dt} \left(\frac{\Delta t_1}{\Delta T_1} \right) = m_1 C_1 + m_r C_r$$

and so by subtraction of the corresponding expressions from (2.63) and (2.64) and then dividing to eliminate the constant quantity dQ/dt we obtain:

$$\frac{m_1 C_1}{m_s C_s} = \frac{[(\Delta t/\Delta T)_1 - (\Delta t/\Delta T)_r]}{[(\Delta t/\Delta T)_s - (\Delta t/\Delta T)_r]} \tag{2.65}$$

The ratio quantities in parentheses are all determined experimentally—directly in inverse rate curves or derived from heating or cooling curves. The standard of known mass (m_s) and specific heat (C_s) should preferably have similar properties to the test specimen (for example copper with copper base alloys, iron with steels, etc.). When a transformation occurs the additional heat terms appear in the numerator on the left-hand side of equation (2.65). Obviously if a mass m_1 undergoes a transformation at constant temperature in a time $\Delta t'$ there is no refractory or specific heat contribution in this interval and so

$$\Delta H \frac{dm}{dT} \, dT_1 = \Delta H \, m_1$$

whence

$$\frac{m_1 \Delta H}{m_s C_s} = \frac{\Delta t'}{[(\Delta t/\Delta T)_s - (\Delta t/\Delta T)_r]} \tag{2.66}$$

In the case of a transformation over a range of temperature it is clear that extrapolation of specific heat contributions could be used to eliminate the ΔH contribution at any temperature, but if the transformation involved solid state diffusion the accuracy would generally be reduced except where rapid diffusion of interstitial atoms is involved or the transformation involves excess specific heat.

2.2.4.3   The phenomena involving excess specific heat are those physical or metallurgical changes for which the methods of Sykes *et al.* were primarily designed, and for which the accurate direct determination of specific heat is most useful. These are the changes of the second kind referred to in section 2.2.1, viz.

(a) The order–disorder transformation.
(b) The loss of ferromagnetism at the Curie temperature. Representative curves are shown in Fig. 2.12.

In Fig. 2.12(a) for the order–disorder change in β-brass the specific heat temperature curve is shown for an alloy near to the CuZn composition based on a diagram of Sykes and Wilkinson [49]. This illustrates some quite general principles. The broken line AE represents the specific heat temperature curve which would probably be followed by the alloy without transformation. This can be assumed to follow a normal relationship which corresponds approximately to the weighted mean of the specific heats of the two pure components. The actual specific heat follows the curve AB and then suddenly drops to C, and falls off along CD to a line parallel to but slightly higher than AE. Since $C = \mathrm{d}H/\mathrm{d}T$ and $\int C\,\mathrm{d}T = H$ the area under such a curve is the total heat content and the area *between* the two curves thus represents the excess heat involved in the transformation. The rise along AB is typical of a co-operative phenomenon—the transformation proceeds with greater intensity as the amount transformed increases. The crystallographic aspects of the order–disorder transformation are referred to in Chapter 4 and the effect on electrical resistance in Chapter 3, Vol 2. It is noted that the atoms are supposed to be virtually completely ordered at low temperatures—in the CuZn case this would mean that the distribution of atoms is Cu atoms at the corners of the unit cells and Zn atoms at their centres (caesium chloride structure). Increasing disorder along AB results in a random distribution of atoms on either site (body-centred cubic structure), but some of the tendency to order still remains, i.e. atoms of the same kind have a tendency to keep further apart. This is residual short-range order and associated with it is the small remaining area between the curves above D.

Fig. 2.12(b) shows the same type of curve for the alloy Cu_3Au. In this case the process of ordering (on cooling) is not as simple as in alloys where equal numbers of atoms are involved. Two features should be noted. The curve is likely to be less regular in shape and the final portion is much steeper—so much so that it appears to involve an emission of latent heat at the critical point. It would be true to say that the co-operative type of transformation passes into a critical temperature type more closely resembling a phase change.

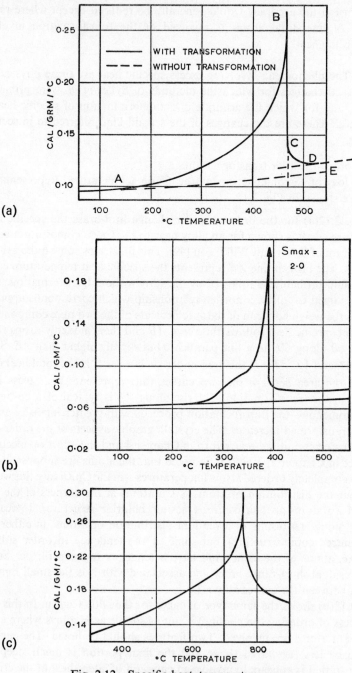

Fig. 2.12   Specific heat–temperature curves.

(a) Cu–Zn alloy (after Sykes and Wilkinson [49])
(b) Cu₃Au alloy (after Sykes and Jones [50])
(c) α-iron (after Awberry and Griffiths [45])

Fig. 2.12(c) shows the specific heat of α-iron and here the peak is associated with the loss of ferromagnetism, the peak being the Curie temperature (see Chapter 3, Vol. 2).

The methods have been applied to other alloy systems such as copper palladium (Ref. [51]) and can be used to study more complex alloys including steels (Ref. [46]) whose interest attaches to the heat evolved in the eutectoid transformation and over the range into the austenite region ($A_1 - A_3$ temperature range). A simplified method was used by Griffiths and Pallister [52] to study the specific heat of metastable austenite and martensite in subcritical transformation regions.

2.2.5 Differential Thermal Analysis (DTA)

The essential principle used in differential thermal analysis is that heat evolution or absorption is measured by a difference in temperature ΔT which develops over the critical range between the sample under examination and an inert reference standard. The experimental arrangements are thus basically similar to those used for the method of specific heat determination indicated in Fig. 2.11(c). It is possible to heat or cool a specimen over a wide range of temperature and record the variation of ΔT as a function of T. A number of commercial instruments are available for automatic recording.

The heat may be evolved from phase changes of one kind or another. In the case of some minerals heat changes also accompany the evolution of water vapour, carbon dioxide or other components. The technique has been widely applied. For a full and comprehensive account of methods and applications reference may be made to Mackenzie [52].

The more sophisticated modern methods of DTA have originated in simpler metallurgical cooling and heating curve investigations. The quantitative interpretation naturally uses similar principles. The use of the DTA equipment to investigate metals and alloys is therefore recognised as an appropriate extension of earlier thermal techniques and is useful for making overall thermal surveys of materials, including estimates of the specific, or latent heats, determination of transformation ranges and of phase boundaries.

2.3 THERMAL EXPANSION: DILATOMETRY AND PHASE CHANGES

2.3.1 General Observations

2.3.1.1 The thermal expansion of a metal or alloy is of interest for three principal reasons:

(i) It is related to the structure and other physical properties and is therefore fundamentally significant in the theory of metals and alloys.

65

(ii) It influences the practical application of materials as in the choice of alloys for service in steam generating plant, turbines and other engines. The requirement of dimensional stability in precision measuring scales and instruments including clocks, involves another aspect—the development of special low expansion alloys.

(iii) Since a phase change usually involves difference in specific volume, the thermal expansion shows a discontinuous change in length or volume with temperature when the phase change occurs. Hence this general metallurgical application of dilatometry.

It is interesting to note a fourth reason which has not been extensively explored but may be more important than has generally been recognised, viz.

(iv) The existence in an alloy of two or more phases which have cooled from a high temperature may give rise to internal stresses due to the different thermal expansion (see Chapter 5 of Vol. 2).

2.3.1.2   In the most elementary determinations of thermal expansion it has been usual to define a mean coefficient over a range of temperature as

$$\alpha = \frac{L_2 - L_1}{L_1(T_2 - T_1)} \tag{2.67}$$

where L_1 is the length at T_1 (°C) and L_2 is the length at T_2. The curve of L against T is not usually linear and although mean coefficients may be useful for technical calculations, a more exact definition is the instantaneous value of α at any temperature, i.e.

$$\alpha = \underset{\Delta T \to 0}{\text{Lt}} \left(\frac{\Delta L}{L} \frac{1}{\Delta T} \right) = \frac{1}{L} \frac{dL}{dT}$$

L is generally a function of T of the form

$$L = L_0(1 + aT + bT^2 + \cdots) \tag{2.68}$$

The coefficient of volume expansion is likewise

$$\beta(= 3\alpha) = \frac{1}{V} \frac{dV}{dT} \tag{2.69}$$

2.3.2   Experimental Methods

2.3.2.1   The experimental methods are essentially the same whether intended to provide physical data or for metallurgical dilatometry. Greater accuracy is obtained with a longer length L, but it then becomes more difficult to ensure uniform temperature along the whole length. Classical work in this field was done by Holborn et al. [53] and by Hidnert [54].

2.3.2.2 There are several methods for amplifying the increase in length and recording it. Some of these methods are more clearly appropriate for a general dilatometer record where less absolute accuracy is required.

1. The simplest type of dilatometer uses a dial gauge. If the specimen is enclosed in a silica furnace tube and a silica rod is used to transmit the expansion then the instrument actually indicates the difference between the very low expansion of silica and that of the specimen. If necessary a correction can be applied. (Refs. [55, 56].) Another way of constructing a dilatometer with a gauge is shown in Fig. 2.13(a). Here three silica rods form a frame and the fourth down the centre transmits the expansion. This arrangement has been used in transformation studies where rapid thermal contact can be established with a quenching bath or other medium at a chosen transformation temperature.

2. Another method which uses mechanical means to transmit and magnify the expansion is that of Chevenard [57]. Further sensitivity can be obtained by using a mirror to reflect a light spot, the movement of which is recorded on photographic paper. Fig. 2.13(b) shows how mechanical levers transmit the expansion to the corners of a triangle. One of the movements is provided by a rod of a standard alloy 'pyros' which is regular, reversible and stable. This movement causes the point p_2 to rotate about the pivot point p_3 on amounts proportional to a temperature change. Similarly the expansion of the rod under examination moves the point p_1 about p_3. A needle fixed to the centre of the triangle then moves in two co-ordinate directions corresponding to temperature and specimen expansion. (The reflecting mirror would replace the needle at the centroid of the triangular plate.) If a time–temperature–expansion relation is required the pen can be used in conjunction with a moving chart.

An alternative arrangement is shown in Fig. 2.13(c)—by re-arranging the points of the triangle a differential record is obtained. A limitation of this method is the uncertainty whether both rods are at the same temperature exactly. There is, however, no reason why supplementary thermocouple measurement should not be used, and in any case instruments of this kind are capable of high accuracy.

3. The third is an optical method and is illustrated in Fig. 2.13(d). It is appropriate for accurate thermal expansion coefficient determination especially when only small specimens are available, and would not be used in ordinary metallurgical dilatometry. It is more suitable for ceramics and non-metals. The specimens are rings or small pieces shaped like a flattened tripod, and are placed between two optically flat discs of quartz (E and T) which are actually at a very small angle to each other ($\sim \frac{1}{3}°$). Monochromatic light is used to illuminate the discs and they are viewed as shown. Interference fringes are seen and these fringes move as the separation of the plates changes with the expansion of the specimen. Over the area of the circular disc E the alternate black and light bands measure the expansion whilst bands over part of the

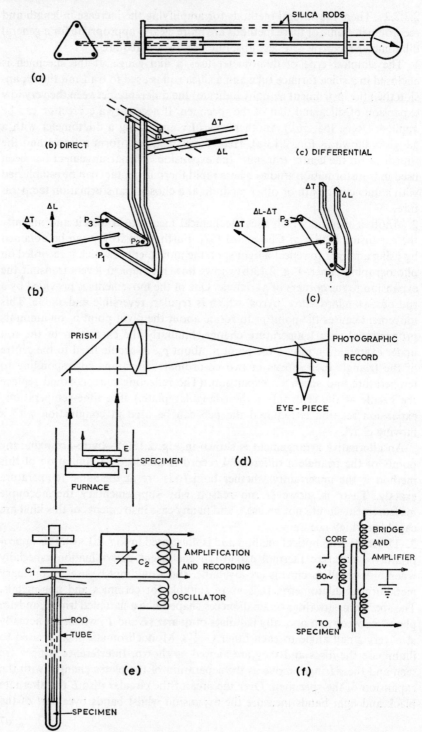

(a)

(b) DIRECT

(c) DIFFERENTIAL

ΔT

ΔL

ΔL

ΔT

ΔL

ΔT

ΔL-ΔT

ΔT

P₃

P₂

P₁

P₃

P₂

P₁

(b)

(c)

PRISM

PHOTOGRAPHIC
RECORD

EYE-PIECE

E

SPECIMEN

T

FURNACE

(d)

C₁

C₂

L

AMPLIFICATION
AND RECORDING

OSCILLATOR

ROD

TUBE

SPECIMEN

(e)

CORE

4v
50~

TO
SPECIMEN

BRIDGE
AND
AMPLIFIER

(f)

SILICA RODS

lower disc T can be used to indicate temperature. An apparatus of this type was described by Saunders [58, 59].

4. There are obvious difficulties and potential sources of error with methods involving mechanical levers for amplification, and electrical methods have therefore been devised. One of these has a plate of a condenser attached to the end of the specimen. The position of the other plate can be adjusted suitably. The condenser forms part of a tuned circuit as in the simple arrangement of Fig. 2.13(e). The rod connecting the specimen must be free to move yet supported centrally, e.g. by light tension springs or wires. Instruments of this kind have been described by Haughton and Adcock [60] by Stanton [61] and others. Improved means of measuring changes of capacity [62, 63] have led to sufficient sensitivity for the instruments to be classed as suitable for accurate work (~ 1 in 10^{-6}).

5. Another electrical method uses the movement of an iron core in a coil or differential transformer (Fig. 2.13(f)). High accuracies should again be attainable with the best arrangement. A number of circuits have been used (e.g. Ref. [64]) including a fully automatic instrument by Werner [65].

6. The measurement of phase transformations at very high rates of cooling requires an instrument which is capable of very rapid response. The above method can be arranged to meet this requirement. Another such method was developed by Christenson, Nelson and Jackson [66] and advanced by Martin and Raring [67]. The dilatation is measured by an 'electrical micrometer'. In this device a filament is placed between two plates which are coupled mechanically through a lever to the specimen. Displacement of the plates relative to the filament is recorded through a bridge circuit and amplifier to an oscillograph. Obviously in all dilatometry the thermocouple should be in close contact with the specimen and under these conditions it should be welded on. The maximum cooling rate claimed with the instrument was 5000 °F (2760 °C) per second and average rate 2900 °F (1593 °C). The instrument is particularly suitable for studying transformations in low alloy steels.

For work where intermediate to moderately fast cooling rates are required the electrical instruments using the capacity or inductance methods are clearly the most satisfactory. Typical applications have been described by Cottrell [68, 69].

7. In a somewhat different category from all the above is the X-ray method (Chapters 4 *et seq.*). Changes in lattice parameter are measured as a function

---

Fig. 2.13   Principles of instruments for thermal expansion and dilatometry.

    (a)   dial gauge with specimen frame holder
    (b), (c) the Chevenard dilatometer
    (d)   interferometric method
    (e)   recording by condenser
    (f)   differential transformer

of temperature by means of a high temperature powder camera. The method is referred to in some detail in Chapter 7. In effect it gives the actual change in the average distance between the atoms and is largely independent of the presence of defects, porosity, voids, grain boundaries, etc., which might affect measurements of bulk expansion. Another advantage is that if the material is anisotropic, having different coefficients of thermal expansion in different crystallographic directions, these coefficients can be determined from one experiment. Examples include the accurate determination of the expansion and transformations of pure iron [70], the expansion of hexagonal metals such as Be [71] or rare metals (e.g. Ru [72]). The different thermal expansions of different phases in an alloy (of importance in relation to their influence on internal stresses) has been studied by Goldschmidt [73] in relation to high speed steel. (See also Chapter 5, Vol. 2.)

2.4 THERMAL CONDUCTIVITY AND DIFFUSIVITY

2.4.1 Introduction: Definitions and Equations

2.4.1.1 The thermal conductivity is an index of the rate of flow of heat through a material and is important for several reasons.
1. It is again related to the structure and to its other physical properties. Of special interest is the relation to electrical conductivity (section 2.4.3).
2. It influences the choice of metals or materials for particular applications. At one extreme one may instance the use of a high conductivity metal in a heat exchanger, and at the other a refractory insulator used in furnace walls to reduce heat losses.
3. It is involved in the calculation of heating and soaking times for ingots and other large masses of metal, and therefore makes a necessary contribution to heat treatment technology and economics. (Refs. [36] (Chap. 5), [74], [75], [7] (pp. 202–14).)

2.4.1.2 The simplest definition of thermal conductivity is the heat conducted through unit area per unit temperature gradient in unit time.

Thus, if a 'flat slab' has parallel faces of area A, thickness x, and there are temperatures of T_1, T_2, at the respective surfaces, then the quantity of heat which flows in time t is given by

$$Q = kAt \frac{T_1 - T_2}{x} \tag{2.70}$$

Since the flow of heat may not be uniform in time, and the temperature gradient is not necessarily linear we should consider the equation in terms of small quantities, i.e.

$$\Delta Q = -k \frac{\Delta T}{\Delta x} A \, \Delta t$$

and so
$$\frac{\mathrm{d}Q}{\mathrm{d}t} = -kA\frac{\mathrm{d}T}{\mathrm{d}x} \tag{2.71}$$

The negative sign is required by the fact that heat flows down the gradient. It will now be appreciated that the 'flat slab' can be a thin slice within a material which in terms of (2.70) becomes an interior surface, at which there is specific value of the gradient and through which heat flows normally to the surface. The following further relationships apply:

(a) Q/A or $\mathrm{d}Q/A$ is heat per unit area, hence the flux of heat is given by

$$w = \frac{1}{A}\frac{\mathrm{d}Q}{\mathrm{d}t} = -k\frac{\mathrm{d}T}{\mathrm{d}x} \tag{2.72}$$

(b) Since $C\,\mathrm{d}T$ represents heat per unit weight, $C\rho\,\mathrm{d}T$ represents heat per unit volume. Hence the 'flow' of temperature, rather than heat, is given by

$$\frac{1}{A}\frac{\mathrm{d}T}{\mathrm{d}t} = -\frac{k}{C\rho}\frac{\mathrm{d}T}{\mathrm{d}x} = -\alpha\frac{\mathrm{d}T}{\mathrm{d}x} \tag{2.73}$$

where α is the thermal diffusivity.

(c) The above equations do not take account of the possibility that k or α might vary with temperature. They are particularly suitable for experimental methods of determining the conductivity (or diffusivity) where a *steady state* is established, i.e. the temperature and thermal gradient at a given point x have settled down to an equilibrium value. In practical applications, however, it is often necessary to deal with conditions during which a steady state is being approached.

The most general equations for conductive heat transfer can be derived thus:

We consider a small rectangular volume of dimensions, δx, δy, δz, at a point xyz. The heat that flows in or out of this region is divisible into three components parallel to the co-ordinate axes, e.g. parallel to Ox, heat flows through an area $\delta y\,\delta z$. Thus at one end of the volume heat flows *out* in time δt equal to:

$$k\,\delta y\,\delta z\left(\frac{\partial T}{\partial x}\right)\delta t$$

and at the opposite end heat flows *in* equal to:

$$\delta y\,\delta z\left[\left(k\,\frac{\partial T}{\partial x}\right) + \frac{\partial}{\partial x}\left(k\,\frac{\partial T}{\partial x}\right)\delta x\right]\delta t$$

The heat gained is the difference, viz.

$$\delta x\,\delta y\,\delta z\left[\frac{\partial}{\partial x}\left(k\,\frac{\partial T}{\partial x}\right)\right]\delta t$$

71

and correspondingly for the other two directions. The resultant net gain is also given by

$$\delta x \, \delta y \, \delta z \, C\rho \left[\frac{\partial T}{\partial t} \, \delta t \right]$$

and on equating and cancelling out we obtain:

$$\frac{\partial}{\partial x} \left(k \frac{\partial T}{\partial x} \right) + \frac{\partial}{\partial y} \left(k \frac{\partial T}{\partial y} \right) + \frac{\partial}{\partial z} \left(k \frac{\partial T}{\partial z} \right) = C\rho \frac{\partial T}{\partial t} \qquad (2.74)$$

or if k can be regarded as constant

$$k \left[\frac{\partial^2 T}{\partial x^2} + \frac{\partial^2 T}{\partial y^2} + \frac{\partial^2 T}{\partial z^2} \right] = C\rho \frac{\partial T}{\partial t} \qquad (2.75)$$

This equation can also be written in terms of the diffusivity, viz.

$$\alpha \left[\frac{\partial^2 T}{\partial x^2} + \frac{\partial^2 T}{\partial y^2} + \frac{\partial^2 T}{\partial z^2} \right] = \frac{\partial T}{\partial t} \qquad (2.76)$$

It is recognised that k, C and ρ vary with temperature and α less markedly, C, the specific heat/unit mass, should be known. Alternatively C/ρ (the specific heat/unit volume) can be considered as one variable, or one may use the relation $\rho \propto 1/[(1+\beta)T]$, where β is the coefficient of volume expansion. Since $C = \partial H/\partial T$, and H is discontinuous at phase changes which may also involve a change in specific volume and a discontinuous change in k, all these changes should be taken into account in using the equations—even if it is only possible to do this by approximate numerical methods. Such circumstances apply in the heat treatment and cooling of steel masses. It is also necessary to recognise the existence of change points when steady state or any other physical methods are being used to determine α or k. The use of these properties in furnace and heat transfer problems is covered, for example, by Dusinberre [74].

2.4.2 Standard Methods

2.4.2.1 The methods that may now be regarded as 'standard' and are of practical metallurgical importance, generally involve heat flow in one direction. They may be broadly divided into two procedures, one for materials of relatively high conductivity (metals), and the other for materials of low conductivity (refractories and ceramics). The difference is simply understood from equation (2.70)—if k is large an adequate heat flow will occur if the area A is small and x is large, whereas the same heat flow with much smaller k requires a larger A and smaller x.

2.4.2.2 For a metal, the arrangement involves the provision of a source of heat at one end of a bar and a calorimeter (usually a water flow calorimeter) at the opposite end. It is necessary to measure the temperature at accurately located points along the bar and so to obtain the (mean) temperature gradient between these points.

Loss of heat laterally should be reduced to a minimum—ideally to zero. For this purpose lagging or insulation are not adequate. The guard-ring principle is therefore generally used. A tube, usually of the same material, surrounds the bar, and thermocouples are placed along it at positions corresponding to those on the bar. By the use of small subsidiary heaters it is possible to arrange that this outer cylinder has the same temperature distribution within very close limits. In view of the possible variation of k with temperature it is desirable to keep the temperature interval between measurement points as small as possible. This is difficult if the conductivity is required over a wide range of temperature since the water flow calorimeter fixes (within narrow limits) the lower end of the scale and the gradient increases unduly as the temperature of the hot end is raised. Hence methods have been developed in which the heat is supplied to the centre and conducted away symmetrically [76, 77]. The heater is placed in a cavity within the specimen—two halves are joined at the centre. Providing all the heat supplied is conducted along the two arms of the specimen and the gradients are identical it is possible to establish the conductivity without measuring the heat which leaves the ends. The calorimeter is not needed and so the temperature at the ends can be much nearer that of the centre.

A method employing heating at, and symmetrical conduction away from, the centre has been used for measurements over a small range from room temperature upwards. In this case there is no objection to having calorimeters to measure the quantity of heat. An apparatus of this kind has been used to determine the conductivity of nickel below 100 °C [78]. An alternative procedure is to use a material of known conductivity in series with the specimen under examination [76, 79]. Lead and iron [80] have been used for this purpose.

2.4.2.3 For materials of low conductivity such as refractories, including insulating materials and other oxides and ceramics the most suitable specimen is the flat plate (A large relative to x). This is the principle of Lee's disc method and subsequent variants and refinements. In the original method a flat heating coil was placed between two equal copper discs and the substance under examination sandwiched between one of these discs and another also of copper. The total heat supplied is emitted from the end and cylindrical surfaces according to how much is conducted on either side and the gradients in the respective components, the temperatures of the copper blocks being measured by thermocouples. The analysis is given by Roberts (Ref. [81], p. 290).

In another procedure a flat heater is surrounded by a guard ring and both are embodied in the material under examination. Heat flow perpendicular to the surface is then easy to arrange and the amount from the area A is that provided for the main heater as determined by the electrical input. In this group of methods there are a number which have been developed specially for refractory materials. These include the elaborate apparatus and procedure of Patton and Norton [82] and other simpler experimental arrangements [83, 84].

2.4.3 Other Procedures

2.4.3.1 Other methods include those in which heat is supplied by an electric current passed through the specimen. This is usually in the form of a rod and the ends are kept at the same temperature. The heat produced at any point by the current plus heat reaching it by conduction is necessarily equal to the heat leaving by conduction plus that lost to the surroundings. A discussion of the theory of the method has been provided by Woodhead (Ref. [7], pp. 207–10).

There are also procedures involving periodic heating to one end of a bar (see for example Ref. [85]) and radial heat flow [80].

2.4.3.2 Powell has described a 'thermal comparator' [86] which could have uses as a non-destructive test. Two metal balls are mounted in a block of insulating material (balsa wood), one ball being below the surface and the other projects, so that when it is placed in contact with a material the latter ball which makes the contact, cools more rapidly than the other. The thermal conductivity depends on the difference in cooling rates and calibration graphs can be constructed.

2.4.3.3 There is a relationship between the thermal conductivity k and electrical conductivity σ. The Wiedemann–Franz (–Lorenz) 'law' states that k/σ varies as the absolute temperature, i.e. $k/\sigma = LT$ or $k = L\sigma T$, there L should be a constant. In practice it is found that L is not constant, although it can be shown that two constants fairly well represent the relationship for a given metal or alloy base. It appears that the thermal conductivity may have a temperature independent part so

$$k = L'\sigma T + k' \qquad (2.77)$$

and L' and k' can be used to calculate k from σ. The method has been used for steels [80], copper [87], aluminium [88] and magnesium alloys [89].

Since the electrical conductivity is much easier to measure it could be used generally to determine the thermal conductivity if this relationship could be more accurately established. At present accuracies of the order of ± 10 per cent can be claimed and may be sufficient for approximate calculations.

74

References

1. TATE, J. T. *et al.* (eds.), *Temperature, Its Measurement and Control in Science and Industry*, vol. I, Reinhold, New York, 1941. For American Institute of Physics.
2. WOLFE, H. C. (ed.), ibid., vol. II, 1955.
3. HERZFELD, C. M. *et al.* (eds.), ibid., vol. III (Parts 1 and 2), 1962/3.
4. GRIFFITHS, E., *Methods of Measuring Temperature*. Griffin and Co. Ltd, London, 1947.
5. HALL, J. A., (a) *Fundamentals of Thermometry*, (b) *Practical Thermometry*. The Institute of Physics and The Physical Society Monographs for Students, London, 1953.
6. National Physical Laboratory, *Calibration of Temperature Measuring Instruments*, Notes on Applied Science, No. 12. HMSO, London, 1955.
7. WOODHEAD, J. H., Chapter III of *The Physical Examination of Metals*, ed. B. Chalmers and A. G. Quarrell. Arnold, London, 1960.
8. HUME-ROTHERY, W., CHRISTIAN, J. W. and PEARSON, W. B., *Metallurgical Equilibrium Diagrams*. The Institute of Physics, London, 1952.
9. MCLAREN, E. H., Ref. [3], Part 1, **23**, 185.
10. DAUPHINEE, T. M., ibid., **30**, 269.
11. ROBERTSON, D. and WALCH, K. A., ibid., **32**, 291.
12. PROSSER, L. E., *Engineering*, 1939, **139**, 95.
13. GARTSIDE, F., *Metallurgia*, 1950, **42**, 211.
14. Principles incorporated in commercially available instruments.
15. CHAMPION, F. C. *University Physics* Blackie, London and Glasgow, 1960.
16. STARLING S. C. and WOODALL A. J. *Physics*, Longmans Green, London, 1952.
17. Anon, *The Metallurgist*, 1962, **2**, 121.
18. HAUGHTON, J. L., *J. Inst. Metals*, 1920, **23**, 499.
19. GALIBOURG, J., *Rev. Metall.*, 1925, **22**, 400.
20. WAGNER, H. J. and STEWART, J. C., Ref. [3(1)], 27, 245.
21. HUNTER, M. A. and JONES, A., Ref. [1], 1227.
22. LOHR, J. M., HOPKINS, C. H. and ANDREWS, C. L., Ref. [1], 1232.
23. CRUSSARD, C., *Conference on Strength of Solids*, 119. The Physical Society, London, 1948.
24. BRINDLEY, G. W., ibid., 95.
25. WERT, C., *Thermodynamics in Physical Metallurgy*, 178, Amer. Soc. Met., Cleveland, Ohio, 1950.
26. WHITE, W. P., Ref. [1], 265.
27. STANSFIELD, *Phil. Mag.*, 1899, **46**, 59; CARPENTER, H. C. H. and KEALING, B. P., *J. Iron Steel Inst.*, 1904 (i), **65**, 224.
28. BROOKS, H. B. and SPINKS, W. B., *J. Res. Nat. Bur. Stds*, 1932, **9**, 781.
29. DITCHBURN, R. W., *Light*, 17.27–17.26, Blackie, London, 1952.
30. RICHTMYER, F. K. and KENNARD, E. H., *Introduction to Modern Physics*, 4th edn., Chap. 5, McGraw-Hill, New York, London, Toronto, 1947.
31. BAILEY, G. L. J., Chapter IX. *The Physical Examination of Metals*, ed. B. Chalmers and A. G. Quarrell. Arnold, London, 1960.
32. VAN DUSEN, M. S. and DAHL, I. A., *J. Res. Nat. Bur. Stds.*, 1947, **39**, 291.
33. HARRISON, T. R., *Radiation Pyrometry and Its Underlying Principles of Radiant Heat Transfer*. Wiley, New York and London, 1960.
34. ROSENHAIN, W., *J. Inst. Metals*, 1915, **13**, 160.
35. JENKINS, C. H. M., BUCKNALL, E. H. and JENKINS, G. C. H., *Second Report of the Alloy Steels Research Committee*, Iron and Steel Inst. Special Report No. 24, 1939, 215.
36. ZEMANSKY, M. W., *Heat and Thermodynamics*, 91–3, McGraw-Hill, New York and London, 1957.
37. RUSSELL, T. F., *J. Iron Steel Inst.*, 1939, **139**, 147.
38. MASING, G., *Z. Metallkunde*, 1941, **33**, 36.
39. KUBACHEWSKI, O. and EVANS, E. LL., *Metallurgical Thermochemistry*, 3rd edition. Pergamon Press, London, 1958.

40. LEAKE, L. E. and TURKDOGAN, E. T., *J. Sci. Instrum.*, 1954, **31**, 447.
41. GINNINGS, D. C., DOUGLAS, A. B. and BALL, A. F., *J. Res. Nat. Bur. Stds*, 1950, **45**, 23.
42. OELSEN, W., RIESKAMP, K. H. and OELSEN, O., *Arch. Eisenhüttenwesen*, 1955, **26**, 253.
43. SYKES, C., *Proc. Roy. Soc. A*, 1935, **148**, 422.
44. SYKES, C. and JONES, F. W., *J. Inst. Met.*, 1936, **59**, 257.
45. AWBERRY, J. H. and GRIFFITHS, E., *Proc. Roy. Soc. A*, 1940, **174**, 1.
46. AWBERRY, J. H., *Second Report of the Alloy Steels Research Committee*, Iron and Steel Inst. Special Report No. 24, 1939, 216.
47. SMITH, C. S., *Trans. Am. Inst. Min. Met. Eng.*, 1940, **137**, 236.
48. JELLINGHAUS, W., *Arch. Eisenhüttenwesen*, 1951, **22**, 67.
49. SYKES, C. and WILKINSON, H., *J. Inst. Metals*, 1937, **61**, 223.
50. (a) SYKES, C. and JONES, F. W., *Proc. Roy. Soc. A*, 1936, **157**, 213; (b) JONES, F. W. and SYKES, C., *J. Inst. Metals*, 1940, **65**, 419.
51. GRIFFITHS, E. and PALLISTER, P. R., *J. Iron Steel Inst.*, 1953, **175**, 30.
52. MACKENZIE, R. C. (ed.), *Differential Thermal Analysis*, vols. 1 and 2, Academic Press, London and New York, 1970/72.
53. HOLBORN, L. and VALENTINER, S., *Ann. Phys.*, 1907, **22**, 12.
54. HIDNERT, P., *Nat. Bur. Stds Sci. Paper* No. 17, 1922, 91.
55. GALE, R. C., *J. Sci. Inst.*, 1930, **17**, 131.
56. WALTERS, F. M. and GENSAMER, M., *Trans. Am. Soc. Steel Treatment*, 1939, **19**, 608.
57. CHEVENARD, P., *Rev. de Metallurgie*, 1950, **47** (11), 805. (Also: ibid., 1926, **23**, 92.)
58. SAUNDERS, J. B., *J. Res. Nat. Bur. Stds*, 1939, **23** (1), 179.
59. SAUNDERS, J. B., *J. Res. Nat. Bur. Stds*, 1945, **35** (3), 157.
60. HAUGHTON, J. L. and ADCOCK, F., *J. Sci. Instrum.*, 1933, **10**, 178.
61. STANTON, L. R., *J. Iron Steel Inst.*, 1943, **147**, 95P.
62. BRADSHAW, E., *J. Sci. Instrum.*, 1945, **22**, 112.
63. DAYTON, R. W. and FOLEY, G. M., *Electronics*, 1946, **19**, 106.
64. EWELES, J. and CURRY, C., *J. Sci. Instrum.*, 1947, **24**, 261.
65. WERNER, O., *Z. Metallkunde*, 1956, **47**, 28.
66. CHRISTENSON, A. L., NELSON, E. C. and JACKSON, C. E., *Trans. A.I.M.*, 1945, **162**, 606.
67. MARTIN, F. E. and RARING, R. H., *Trans. A.I.M.E.*, 1956, **206**, 191.
68. COTTRELL, C. L. M., *J. Iron Steel Inst.*, 1953, **174**, 17.
69. COTTRELL, C. L. M., *J. Iron Steel Inst.*, 1954, **176**, 273.
70. BASINSKI, Z. S., HUME-ROTHERY, W. and SUTTON, A. L., *Proc. Roy. Soc. A*, 1955, **229**, 459.
71. MARTIN, A. J. and MOORE, A., *J. Less Common Metals*, 1959, **1**, 85.
72. HALL, E. O. and CRANGLE, J., *Acta Crystall*, 1957, **10**, 240.
73. GOLDSCHMIDT, H. J., *J. Iron Steel Inst.*, 1957, **186**, 68.
74. DUSINBERRE, G. M., *Heat Transfer Calculations by Finite Differences*. International Text Book Co., Scranton, Penna U.S.A., 1961.
75. JAKOB, M., *Heat Transfer*. Chapman and Hall, London; Wiley, New York, 1949.
76. POWELL, R. W., *Proc. Phys. Soc.*, 1934, **46**, 659.
77. BENEDICKS, C., BACKSTRÖM, H. and SEDERHOLM, P., *J. Iron Steel Inst.*, 1926, **114**, 127.
78. SPYRA, W., D.E.W. *Tech. Berichte*, 1962, **4** (2), 166.
79. VAN DUSEN, M. S. and SHELTON, S. M., *J. Res. Nat. Bur. Stds*, 1934, **12**, 429.
80. POWELL, R. W. and HICKMAN, M. J., *J. Iron Steel Inst.*, 1946, **154**, 112.
81. ROBERTS, J. K. and MILLER, A. R., *Heat and Thermodynamics*, Blackie & Son Ltd, London and Glasgow, 1954.
82. PATTON, T. C. and NORTON, C. L., *J. Amer. Ceram. Soc.*, 1943, **26**, 350.
83. CLEMENTS, J. F. and VYSE, J., *Trans. Br. Ceram. Soc.*, 1954, **53**, 134.
84. BLAKELEY, T. H. and COBB, J. W., *J. Soc. Chem. Ind.*, 1932, **51**, 237.
85. BOSANQUET, C. H. and ARIS, R., *Br. J. Appl. Phys.*, 1954, **5**, 252.
86. POWELL, R. W., *J. Sci. Instrum.*, 1957, **34**, 485.
87. SMITH, C. S. and PALMER, E. W., *Trans. A.I.M.E.*, 1935, **117**, 225.
88. KEMPF, L. W., SMITH, C. S. and TAYLOR, C. S., ibid., 1937, **124**, 287.
89. POWELL, R. W., *Phil. Mag.*, 1939, **27**, 677.

Chapter 3

SOLIDIFICATION TECHNIQUES (Including Growth of Single Crystals)

3.1 GENERAL LABORATORY TECHNIQUES

3.1.1 Conventional Methods

Conventional methods of melting, casting—and heat treatment—on a laboratory scale generally employ the same principles as their larger industrial counterparts. In this category also come vacuum melting and casting. Refinements and developments on an industrial scale have usually followed the corresponding laboratory scale developments. The former techniques are dealt with elsewhere [1].

3.1.2 Copper Hearth Arc Melting

A useful type of furnace for laboratory vacuum melting of small quantities employs a water-cooled hemispherical copper hearth. The charge of metal or alloy to be melted is placed in the hearth and it is melted by an arc which is struck between it and an electrode. The electrode is positioned by hand and the melt usually takes the form of a button into which shape it solidifies. Furnaces of this kind have been used for the production of alloys for constitutional and structural studies. It is doubtful if there are many industrial production uses but these furnaces have been produced in a variety of sizes.

3.2 GROWTH OF SINGLE CRYSTALS

3.2.1 Introductory

The remainder of this chapter is concerned with special laboratory techniques which may nevertheless have some applications in production. The subject covered in this section—the growth of single crystals—involves solidification as the basis for one group of procedures and so falls into place here. There are, however, certain other methods which involve growth in the solid state. All methods do in fact presuppose the existence of the crystalline solid which otherwise forms the basis of subsequent chapters. (General Refs. [2]–[5].)

The methods to be described apply to both metals and alloys. The

77

techniques of zone melting and refining described in the next section also have applications to single crystal growth.

3.2.2 Growth from the Liquid State

3.2.2.1 Andrade [6] first grew single metal crystals by cooling slowly from the melt. Historically the next step forward was the method devised by Czochralski [7] and this has formed the basis of several adaptations since. The basic principles are shown in Fig. 3.1(a). A seed crystal is held in a small chuck and brought into contact with the surface of the molten metal and then slowly withdrawn vertically and rotated. The speeds of removal and rotation and the temperature must be carefully controlled in order to maintain contact and ensure a suitable size of crystal. The orientation of the crystal is determined by that of the seed. Rotation is intended to ensure a uniform cross-section and simultaneously stirs the melt, thus producing a more uniform distribution of impurities. The method has been applied successfully to a number of metals including silicon and germanium [8, 9]. A modern version of this method is described by Burton, Prim and Slichter [10].

3.2.2.2 A second important group of solidification methods for growing single crystals are developed from that devised by Bridgman [31]. The melt is contained in a mould or crucible which is slowly lowered through a furnace in which there is a marked temperature gradient so that solidification starts at the bottom end of the mould and progresses slowly upwards—see Fig. 3.1(b). Alternatively the furnace is moved upwards or the mould and furnace are stationary and the temperature of the furnace is slowly reduced through a suitable variable transformer. In this last procedure the melt can be completely free of any vibration due to mechanical movement but the maximum temperature reached by some of the melt would be higher because the melt is entirely in the furnace.

Both the methods illustrated in Fig. 3.1 have a vertical mould or crucible to contain the melt. If impurities are present or the growth of an alloy crystal is being attempted, the solute elements will tend in normal circumstances (i.e. lowering of freezing range by solute) to be enriched in the liquid. In the Bridgman method the whole ingot solidifies progressively from bottom to top hence the top will tend to contain more of the solutes or impurities. Evidently the problem of segregation is less serious, the narrower the temperature and composition differences through the freezing range.

3.2.2.3 An alternative arrangement to the vertical furnace in Fig. 3.1(b) is to have a horizontal furnace. The tube containing the melt in a suitable mould is drawn slowly through the furnace—or more often the furnace moved along the, comparatively long, inner tube. Moulds for this purpose have been cut from solid graphite—usually boat-like with a small recess extending from the

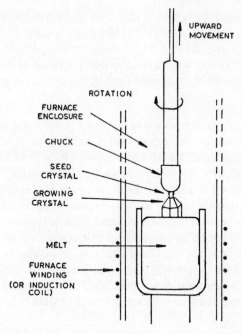

UPWARD
MOVEMENT

ROTATION

FURNACE
ENCLOSURE

CHUCK

SEED
CRYSTAL

GROWING
CRYSTAL

MELT

FURNACE
WINDING
(OR INDUCTION
COIL)

(a)

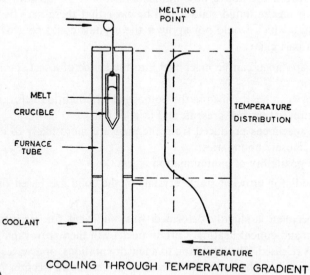

MELTING
POINT

MELT

CRUCIBLE

FURNACE
TUBE

TEMPERATURE
DISTRIBUTION

COOLANT

TEMPERATURE

COOLING THROUGH TEMPERATURE GRADIENT

(b)

Fig. 3.1   Methods for growing single crystals from a melt.
     (a) drawing from melt [7]
     (b) moving crucible method [11]

end which passes down the gradient first. The seed crystal of known orientation can be placed here. Chalmers [11] has described the production of bi- and tri-crystals by the same method using two separated seed crystals. Crystals grown in this way will necessarily take the shape of the mould except for the open surface which may need further preparation (e.g. chemical cutting or polishing).

3.2.2.4 In all the above methods it is necessary to take account of the usual factors such as furnace atmosphere and suitability of mould or crucible material for the particular metal being grown. It is possible to use heat-resisting glass or silica for low melting point metals. These may be coated with graphite which is itself a useful material for many applications. The higher melting point metals may require oxide refractories such as alumina. A hard mould naturally constrains a solidified metal mass to its own shape. Moulds for producing single crystal tensile, or other test, specimens with the necessary shoulder ends may be made from hard materials, but owing to stresses which tend to arise on cooling especially after solidification soft moulds have been suggested [12]. Alumina powder has been used for shaping such moulds and the resultant (unfired) shape is sufficiently friable to yield before the specimen itself.

3.2.3 Growth in the Solid State

3.2.3.1 From the above description of methods for growing single metal crystals from the liquid state it will be clear that there may be certain disadvantages which would not apply if the crystals could be grown from the solid. In particular:

(a) The specimen can be machined into the shape of a test specimen before treatment.
(b) Alloys can be obtained in the homogeneous condition before growth and segregation is thus not present, and in addition.
(c) The specimens produced from the melt are more likely to contain subgrain or mosaic boundaries.
(d) The possibility of contamination is less.

Methods for growing single crystals in the solid are based on two principles.

1. A specimen is slightly deformed—just sufficient for recrystallisation to begin on subsequent heating. Certain nuclei will then grow rapidly, as strain energy is released. The limitation to a few orientations, or even a single one, is often assisted by the presence of an orientation texture (section 7.6).
2. Alternatively if a metal is annealed, one grain will often grow into the surrounding matrix and absorb its neighbours as it grows. Growth is again favoured by existing orientations and in this case there is a tendency for those grains to grow most rapidly which have a particular orientation relationship

80

to the grains they absorb. In any case growth is limited by the use of a temperature gradient.

3.2.3.2   In applications of the strain–anneal method it is important to provide the right amount of critical strain. Slow heating to a suitable temperature may follow. Alternatively a tapered specimen is sometimes used and this provides a stress gradient so that some grains will be at the critical stress. This method can thus be used to determine the critical strain (see Ref. [4]). In some procedures the strained specimen is moved slowly through a furnace with a temperature gradient. Growth then starts at one end and nucleation does not occur where the temperature is too low. The known preferred orientation of some materials can be used to control the orientation developed by suitable bending or twisting—a technique which has been applied to aluminium [13], to silicon iron [14] and to copper [15].

3.2.3.3   In the grain growth methods a moving temperature gradient is employed. A small furnace is moved along a wire or rod and a single grain then grows at the expense of others. The methods have been applied to iron [16, 17], to molybdenum [18], niobium [19] and uranium [20]. In the last example a dispersion of very small inclusions was used to provide a fine-grained but oriented structure for the low temperature α phase. In this way one or a small number of grains would grow—probably those particularly favourably oriented—the growth of the remainder being prevented by the inclusions.

Analogous with this method is one which takes specific advantage of the preferred orientation in a deformed metal. In some metals this can be the initial orientation or it can be one formed on annealing or a secondary recrystallisation texture. The latter may give rise to a few strongly oriented large grains as for copper [15], austenitic iron nickel alloys [16] and silicon iron [14, 17].

An interesting possibility arises when a phase change occurs. If gradual cooling takes place down a gradient the phase interface can be made to move slowly along a specimen such as a rod or wire. As the new phase replaces the old (high temperature) phase it is sometimes possible to grow it with a single orientation.

3.3   ZONE MELTING AND REFINING (INCLUDING NORMAL FREEZING)

3.3.1   Introductory

The techniques which come under this general classification depend primarily on the different concentrations of solute in the liquid and solid phases in equilibrium at any temperature within the solidification range. This circumstance has already been taken account of in connection with thermal analysis

81

of alloy freezing (Fig. 2.10(c), p. 48) and is well known as the cause of segregation in castings and ingots and as the basis of long established processes for metal refining by fractional crystallisation. The distinctive new feature, introduced by Pfann [21] in 1952, is the melting of a small quantity of the metal or alloy at a time by moving a narrow hot zone along a rod or wire. As a refining process for example, it is necessary for the liquidus and solidus to decrease in temperature with increasing solute content. The solid that reforms as the zone moves along will be purer than is represented by the initial mean composition whilst impurity will concentrate in the liquid phase and so be transferred towards the end of the rod. Repetition of the process provides successively greater purification. As the material becomes purer fewer nuclei are generally present and successive zone refining tends to produce single crystals and so the method can be operated with this object in view.

The zone melting methods are important because of their application on a technological scale to the production of semi-conductors as in the purification of germanium and other materials for transistors. They are valuable on a laboratory scale for the general purification and production of single or poly-crystalline specimens. There are several variants and developments of the general principle which are of interest to physical metallurgy. (General Refs. [22, 23].) It is important to consider a more general type of solidification first.

3.3.2 Basic Principles (a) Normal Freezing

3.3.2.1 The elementary principles of normal freezing and zone refining have been elaborated in considerable mathematical detail. A simplified account will

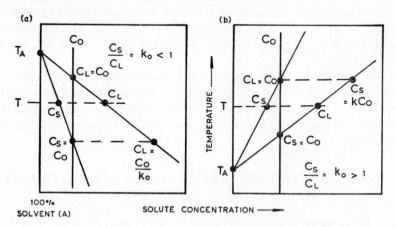

Fig. 3.2 Definition of distribution coefficient k_0.

be given here and Pfann's notation will be generally followed, apart from one or two minor changes.

Figs. 3.2(a) and (b) show the two possibilities for the solidification of an alloy. The boundaries are not necessarily straight lines but if they are it is easy to establish (from equation (2.37)) that the ratio of solidus composition to liquidus composition is constant, i.e.

$$\frac{C_\text{S}}{C_\text{L}} = k_0 \qquad (3.1)$$

In Fig. 3.2(a) $k_0 < 1$ and in Fig. 3.2(b) $k_0 > 1$. k_0 will generally depend on composition and can be obtained for any temperature from the phase diagram if it is known. k_0 is the *distribution coefficient* for *equilibrium* freezing.

In general equilibrium is not followed because insufficient time is allowed for diffusion in the solid. In so-called *normal freezing* there is no diffusion in the solid, but diffusion or mixing must occur in the liquid—there are three cases:

(1) Concentration in liquid is always uniform, i.e. complete mixing.
(2) Incomplete mixing in liquid phase.
(3) No stirring of liquid—minimum movement of atoms by diffusion only.

If (1) is fulfilled it is either because the solidification rate is large relative to diffusion in solid but small relative to diffusion in liquid or because the process is assisted by stirring. For (1) the distribution coefficient is $k = k_0$. For (2) and (3), it is $k \neq k_0$. Conditions giving control over the variables affecting the three cases are obtained if a solidification interface moves in a horizontal direction, e.g. along a cylinder (Fig. 3.5(a)).

The actual composition of the solid in equilibrium with the liquid may be found as follows:

C_0 = mean volume composition of alloy
C_L = mean composition of liquid
C = local composition of solid phase at interface
g = volume fraction solidified
$C = kC_\text{L}$ at any time

Also let $s = $ *actual amount* of solute in liquid (e.g. gram atoms if C is in atomic units). Then $1 - g = $ fraction of liquid (volume) and $s/(1-g) = $ amount/unit volume $= C_\text{L}$ (for total unit volume).

If $(-\delta s)$ is transferred and the volume transferred is δg then the concentration (C) in the thin slice of new solid must be

$$-\frac{\delta s}{\delta g} = C = kC_\text{L} = \frac{ks}{1-g}$$

$$\int_{s_0}^{s} \frac{\text{d}s}{s} = -\int_{0}^{g} \frac{k}{1-g}\,\text{d}g$$

from which $s = s_0(1-g)^k$. But when $g = 0$, $s_0 = C_0 (= C_\text{L} = $ mean composition of alloy at point of commencement of solidification—Fig. 3.2.) Thus

$$s = C_\text{L}(1-g) = C_0(1-g)^k$$

83

Therefore

$$C = kC_{\mathrm{L}} = kC_0(1-g)^{k-1}$$

i.e. $\qquad\qquad\qquad C = kC_0(1-g)^{k-1} \qquad\qquad\qquad (3.2)$

The normal freezing equation (3.2) was derived by Pfann [21] and in a different form and context by Hayes and Chipman [24] (and by others).

The equation can only be regarded as approximate because it assumes:

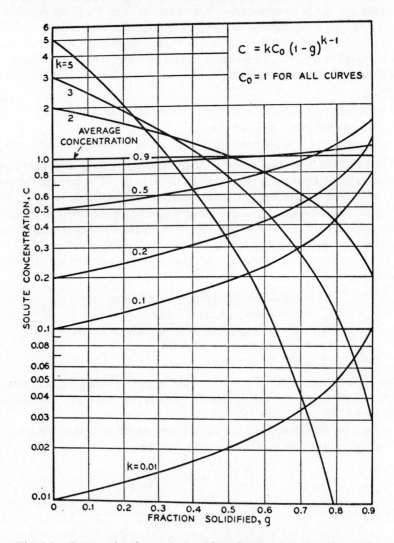

Fig. 3.3 Curves showing concentration C of solute during normal freezing (as function of volume fraction solidified, g, and for different values of k). Reproduced by courtesy of author[21] and publisher (see Fig. 3.5).

(a) No change or difference in density between liquid and solid.

(b) k is a constant ($=k_0$ for case (1)).

This cannot be true for all concentrations since either a eutectic or a peritectic will be reached or the liquidus and solidus meet again. Furthermore:

(c) The equation fails as $g \to 1$ and is probably only correct up to ~ 0.8.

C/C_0 can be plotted against g for different constant values of k as in Fig. 3.3 which is of general application since actual C values are obtained by multiplying by C_0. The values of $k < 1$ are of interest since they generally apply to solidification as in Fig. 3.2(a) where solute is concentrated in the liquid. In a typical case, $k = 0.2$, it is seen that even when 0.7 (70%) has solidified the solute concentration in the solid has only risen from the minimum value of 0.2–0.5 of the mean. Hence the remainder of the solute is concentrated in the liquid. The removal of the solid before the concentration has begun to rise would therefore provide a purified alloy in which the concentration nowhere exceeds, say, half the initial concentration. It is thus possible to see how fractional crystallisation can be based on this principle and that repetition of the process can give a substantial purification at each stage.

3.3.2.2   Under the conditions of solidification postulated (see Fig. 3.5(a)) and in case (1) complete mixing in the liquid gives uniform concentration in this phase as shown in Fig. 3.4(a) (for $k < 1$). This condition might be taken to correspond to an infinitely slow growth rate and $k = k_0$. If the growth rate is faster the liquid will tend to circulate in such a way that the movement in the immediate vicinity of the interface is vertically upwards. This creates a region in which the horizontal movement of atoms is by diffusion only. The effective thickness of this region is δ as in Fig. 3.4(b). If the diffusivity is D and growth rate f, then the effective distribution coefficient (cases (2) and (3)) is given by

$$k = \frac{k_0}{k_0 + (1 - k_0)\exp(-f\,\delta/D)} \tag{3.3}$$

an equation derived by Burton et al. [10]. Beyond the region in which there is a concentration rise, the circulation of the fluid itself promotes uniformity. k must necessarily refer to the relationship between the solid in immediate contact with the liquid and C_0 as in (3.2).

If $f\,\delta/D \to 0$, $k \to k_0$ as in case (1), $f\,\delta/D$ is called the normalised growth velocity. For a given alloy system and value of k_0 it is possible to arrange considerable differences in the effective k by altering the growth rate f. It is not always easy to determine δ and D separately but (δ/D) can be determined by measuring k in crystals grown at different rates but with stirring conditions identical. δ has, however, also been calculated [10].

Fig. 3.4(c) gives an indication of how the effective k approaches unity as the growth rate increases thus reducing the effectiveness of refining if that is

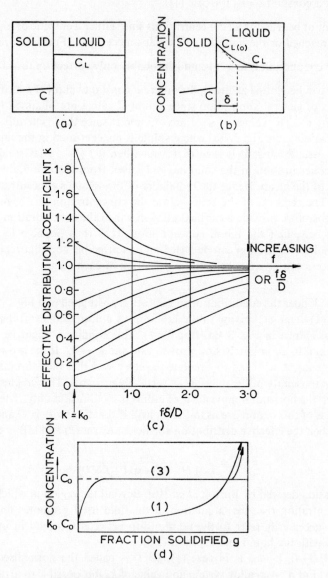

Fig. 3.4 Effective distribution coefficient.
(a) equilibrium—very slow growth rate
(b) steady state—with finite growth rate
(c) variation of k with growth rate
(d) distribution of solute after normal freezing—limiting cases (1) and (3)

required, but on the other hand if $k\to 1$ uniformity of composition can be obtained over most of the length of the solidifying ingot. The two possible extremes (for a given k_0) are represented schematically in Fig. 3.4(d) and correspond to cases (1) (cf. Fig. 3.3) and (3). It should be appreciated that $C_{L(0)}$ in Fig. 3.4(b) will tend to rise with increasing rate. C_L for the bulk of the liquid will tend to fall, and for this reason k tends to approach 1 and $C_S \simeq C_L \simeq C_0$ *apart from the narrow region of high concentration* which then appears at the end of solidification as in Fig. 3.4(d)—for case (3).

To establish the freezing conditions and segregation effects as are represented for example by equation (3.2) (or under other conditions referred to below) it is thus desirable to know or be able to determine k. If k_0 is not available from the equilibrium diagram it may sometimes be estimated thermodynamically [24, 25]. The value of k is then found from equation (3.3) with or without the aid of diagrams such as Fig. 3.4(c) providing the growth conditions are sufficiently known. Alternatively k can be determined experimentally for constant growth conditions—such as may be used later in the actual zone melting process. For example a cylinder would be solidified by normal freezing and the composition C determined as a function of g by an analytical method. Thus at any stage the liquid could be poured off and the surface layer analysed. A minimum of two points but preferably more enable k to be found from equation (3.2) using a straight line logarithmic plot, since (from (3.2))

$$\log \left(\frac{C}{C_0} \right) = \log k + (k-1) \log (1-g)$$

The method is not considered reliable for high values of k (> 0.9)—see Pfann [23].

3.3.3 Basic Principles (b) Extended to Zone Melting

3.3.3.1 The fundamental difference between the conditions to which equation (3.2) applies and the zone-melting processes are now indicated in Fig. 3.5(a). In the zone melting case it is better to consider the actual length x because the length l of the molten zone is also necessarily involved. Otherwise the only difference is that the concentration of solute in the liquid is replenished, as the heater moves along, with fresh material of base composition C_0. Using the same notation as before, although the quantity of solute s is now confined always to a fixed volume of liquid, the derivation is similar to that for (3.2). Taking unit area of cross-section and assuming the heater and liquid zone move a distance dx, then the volume solidified is dx and the *quantity* of solute transferred is the actual concentration (kC_L) times this volume. Also

$$C_L = \frac{\text{amount in liquid}}{\text{volume}} = \frac{s}{l}$$

87

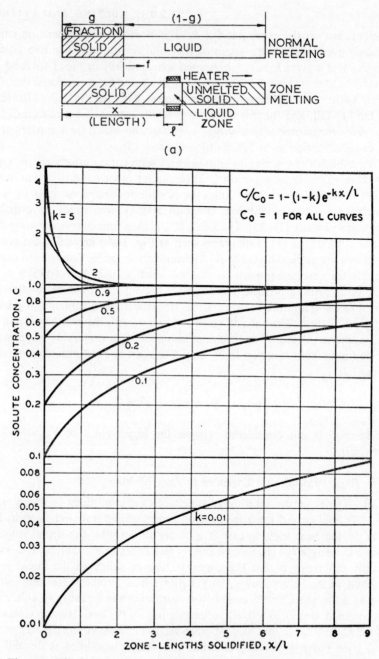

Fig. 3.5 Single-pass zone melting. From paper by Pfann, W. G., *Trans. A.I.M.E.*, 1952, **194**, 747. Published by the Metallurgical Soc. of A.I.M.E. Reproduced with permission of the Society.

(a) conditions for normal freezing and the zone melting process
(b) curves showing solute concentration in terms of number of zone lengths along an ingot

i.e. $(ks/l)\,\mathrm{d}x = C\,\mathrm{d}x$, is taken out of the zone, but $C_0\,\mathrm{d}x$ is added from the newly melted material of concentration C_0. Hence:

$$\mathrm{d}s = \left(C_0 - \frac{ks}{l}\right)\mathrm{d}x$$

i.e.

$$\int \frac{ds}{\left(C_0 - \frac{ks}{l}\right)} = \int \mathrm{d}x$$

$$\left(-\frac{l}{k}\right)\log\left(C_0 - \frac{ks}{l}\right) = x + \text{constant}$$

$$= x + \left(-\frac{l}{k}\right)\log C_0(1-k)$$

since when $x=0$, $s_0 = C_0 l$.

Replacing ks/l by C and re-arranging

$$\log\frac{C_0 - C}{C_0(1-k)} = -\frac{kx}{l}$$

or

$$\frac{C}{C_0} = 1 - (1-k)\,\mathrm{e}^{-kx/l} \tag{3.4}$$

This is the equation for *single-pass zone melting*. It is noted that the ratio of the two lengths, i.e. x/l appears in the exponent and represents the *number* of zone lengths in the distance x. The function (3.4) is thus best represented in terms of this number and typical curves are shown in Fig. 3.5(b)—these particularly represent the distribution over the early part of a long horizontal ingot. These curves are analogous to those in Fig. 3.3 but rise to higher concentrations more rapidly (convex towards the C axis). If they are continued to higher values of x/l the concentration becomes more nearly constant approaching C_0 except that at the very end it rises when the (impure) zone solidifies (under normal freezing conditions). Hence the curve for a single pass of zone melting for solute distribution along the whole length of a charge is rather like that in Fig. 3.4(d) case (3)—the limiting case for normal freezing without mixing (other than by diffusion). If the process is to be employed as a refining process then a *single pass* is thus not as advantageous as normal freezing with complete mixing (case (1) in Fig. 3.4(d)). The removal of impurity during the initial stage depends on the initial value of k (when $k=k_0$) and the impurity then removed in this region may be regarded as transmitted along the specimen and included in the last portion to solidify. Two conclusions follow, and provide the basis of the two principal uses of zone melting:

1. Although a single pass of zone melting and freezing has an apparent disadvantage it is clear that a repetition of the process one or more times would provide a progressive advantage and this is the basis of *multi-pass zone refining*.

2. Alternatively the level portion can be used in another way—to even out the composition. If desired a controlled amount of solute can be introduced and so an alloy of fixed uniform composition obtained. This is the basis of *zone levelling*.

These processes are now considered.

3.3.3.2   The effects of repeated zone melting can be appreciated in a general way by noting that the point indicated by k_0C_0 in Fig. 3.4(d) would be replaced by a starting composition approaching but slightly greater than $k_0(k_0C_0)$ for the second pass and $k_0{}^3C_0$ for the third (and so on). There will be a corresponding improvement on moving along the ingot but the impurity abstracted in this region will necessarily add to the impurity concentration at the other end. The distribution curve will become lower and lower at one end but higher at the other. The mathematical formulation is more difficult since no general equation has been derived but computer techniques have been used to provide working data. The mathematical treatments available have been discussed by Pfann [23], and their general significance is as follows:

The problem is to find the distribution for n passes, i.e. $C_n(x)$ or $C_n(a)$, where a is the dimensionless ratio $x/l =$ number of zone lengths, as a function of the distribution C_{n-1} for the previous pass. The differential equation is obtained exactly as for the derivation of (3.4). Thus, the amount of solute in the molten zone (for unit cross-section) is $[lC_n(x)]/k$ and for movement dx, $C_n(x)\,dx$ is frozen out and $C_{n-1}(x+l)\,dx$ is taken in. Therefore

$$d\left(\frac{lC_n(x)}{k}\right) = \frac{l}{k}\,dC_n(x) = [C_{n-1}(x+l) - C_n(x)]\,dx \qquad (3.5)$$

Equation (3.4) follows if $n=1$, $C_{n-1}=C_0=$ constant. Otherwise the solution must deal with $C_{n-1}(x+l)$ as a variable (i.e. for $n=1$ the solution of (3.5) is (3.4)). If this function is now inserted into (3.5) a second integration gives the equation for $C_2(x)$ and so on. The solution obtained by Lord [26] is of the form

$$\frac{C_n(a)}{C_0} = 1 - (1-k)[Z]\exp - ka \qquad (a = x/l) \qquad (3.6)$$

which is very similar to (3.4) apart from the factor $[Z]$ which is a complicated function of n. This solution necessarily excludes the end of the ingot where the last molten zone is itself solidifying (in this region the behaviour is obviously governed by the same conditions as normal freezing). A variant of this mathematical solution was provided by Milliken [27] and Reiss [28] used an alternative approach. These and other treatments are considered in Pfann's book. The important result is that providing $k < 1$ there is a progressive purification with each successive pass. A few passes give the effects indicated in Fig. 3.6(a) for $k = 0.25$. For larger k, with less refinement at each pass, the theoretical curves show that larger numbers of passes should eventually produce the same result. With increasing n the curve takes a shape like the lower curve in Fig. 3.4(d) and then becomes more concave towards the ordinate (composition) finally approaching a limit for $n \to \infty$. This condition is indicated in Fig. 3.6(b) (for a value of k nearer to 1). It is important to note:

(a) There is a very great purification indicated over a considerable length of an ingot although some of the mathematical treatments assume a 'semi infinite' ingot.

(b) The impure part of the ingot can be cut off and discarded.

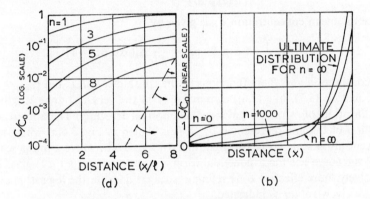
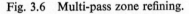

Fig. 3.6   Multi-pass zone refining.
(a) curves for $k \simeq 0.25$—arrows show effect of increasing k
(b) curves for k nearer 1 (e.g. 0.95)

(c) In practice the efficiency of purification decreases with successive passes and the *ultimate distribution* for $n \to \infty$ is not approached as closely as it would otherwise be. The ultimate distribution is, however, a useful criterion and its form can be simply derived as follows:

This is the concentration distribution $C(x)$ for which a further pass makes no difference. Since $C(x)$ is the concentration in the solid at the point x where it is in contact with the liquid zone of length l, the concentration in this liquid must be $C(x)/k$. On the other hand, this liquid concentration must be the mean of that in the solid from which the liquid fraction has come. This is:

$$\frac{1}{l} \int_{x}^{x+l} C(x)\, dx$$

(since the liquid extends from the point x to $x+l$). Therefore

$$C(x) = \frac{k}{l} \int_{x}^{x+l} C(x)\, dx$$

91

which must have an exponential solution. Let this be

$$C(x) = A \exp Bx \qquad (3.7)$$

Then substituting:

$$A \exp Bx = \frac{k}{lB} [A \exp B(x+1) - A \exp Bx]$$

$$= \frac{k}{lB} [(\exp Bl) - 1] A \exp Bx$$

Therefore

$$k = \frac{lB}{(\exp Bl) - 1}$$

Since the mean concentration C_0 is given by

$$C_0 = \frac{1}{L} \int_0^L A \exp Bx \, dx = \frac{1}{BL} A[(\exp BL) - 1]$$

the constants A and B are functions of the constants, k, l, L and C_0.

(Equation (3.7), like (3.6), cannot apply to the very last zone length the solidification of which is governed by normal freezing.) The approximate location of the line represented by (3.7) is shown in Fig. 3.6(a) where a log concentration scale gives a straight line. In Fig. 3.6(b) where a linear scale is used the curve has the exponential form. Since decreasing k (<1) gives progressively more effective zone refining, straight lines on the logarithmic plot progressively rotate as indicated.

3.3.3.3   Zone levelling depends primarily on the existence of the level region resulting from a single pass as represented by equation (3.4) and Fig. 3.5(b) (using the appropriate value of k which derives from k_0 according to equation (3.3) or Fig. 3.4(c)). Since, however, it is desired to level the composition the excess concentration at the end of the first pass can be reduced by beginning the second pass at this end. At the end of this second pass the initially lower concentration is now to some extent counterbalanced by the excess concentration which has been transferred back from the other end.

By repeated passes in alternate directions the composition can be completely levelled out except in the last position to solidify, which can therefore be removed and discarded at the end of the process. This portion has the length l of the molten zone and if L is the total length then $(L-l)$ has concentration C_f, say,

$$l \text{ has concentration } \frac{C_f}{k}$$

which is the concentration for liquid in equilibrium with C_f. Hence total amount of solute $= C_0 L = C_f (L-l) + C_f l / k$, or

$$\frac{C_f}{C_0} = \frac{kL}{kL + (1-k)l} \qquad (3.8)$$

An alternative method naturally suggests itself—the initial region where the concentration is small and below, but increasing towards, C_0 can be allowed for by adding solute separately in the right quantity. The amount to be added is deduced as follows. When the initial zone melts it contains C_0 and this precipitates solid of concentration kC_0. The liquid therefore has a higher concentration than C_0 if $k < 1$ (or lower if $k > 1$). It will go on accumulating (or losing) solute until it reaches a composition C_0/k which is the concentration in equilibrium with the rest of the ingot. Naturally it cannot accumulate any more because it will then throw out solid on either side of concentration C_0. It therefore reaches equilibrium with the ingot of mean composition thus giving the horizontal portion. Hence the initial region of low solute concentration is eliminated by adding enough solute to the first molten zone to raise the composition over a length from C_0 to C_0/k. The methods for doing this are quite simple, e.g. a short (wafer-like) piece of rod of pure solute can be placed on the end before commencing.

In cases where k is less than unity and small (e.g. 0·1 or less) and an ingot of uniform solute concentration is required this result can be achieved by starting with pure solute and solvent. The starting charge of solute melts into the first zone length (l) giving a concentration C_1. This liquid is in contact with pure solvent and as the zone moves along it deposits solid of decreasing solute concentration whilst the liquid zone becomes progressively more dilute. Using the same notation as for the derivation of (3.2) and (3.4) we have $s_0 =$ solute added, so $C_1 = s_0/l$ at $x = 0$; elsewhere $ds = (0 - (ks/l))\, dx$ (since no more solute enters the progressing zone).

$$\int_0^s \frac{ds}{s} = -\int_0^x \frac{k}{l}\, dx$$

$s = s_0 \exp(-kx/l)$ and $C =$ concentration of solute in solid at x, i.e. $-ds/dx$. Therefore

$$C = kC_1 \exp(-kx/l) \tag{3.9}$$

This indicates that the concentration decreases exponentially along the ingot. A logarithmic plot gives straight lines of slope k. Therefore for low values of k (< 1) only a slight concentration gradient results after one pass, e.g. with $k = 0·01$ the concentration decreases by 10 per cent in ten zone lengths.

This gradient can be eliminated by altering the length of the zone l, or the cross-sectional area, as the zone travels along, so as to keep C constant. In either case the volume of the liquid zone, v, is adjusted to meet the requirement. Using the same notation as for unit cross-sectional area but now with (i) area a constant and the function $l(x)$ variable, or (ii) area $a(x)$ variable and l constant, in either case:

$$C = -\frac{1}{a}\frac{ds}{dx} = \frac{ks}{al} = \frac{ks}{v} = \text{constant} \tag{3.10}$$

93

In case (i)

$$dC = \frac{k}{a}\left[\frac{ds}{l(x)} - s\frac{dl}{l^2(x)}\right] = 0$$

or

$$\frac{dl}{l(x)} = \frac{ds}{s} = -\frac{k\,dx}{l(x)}$$

so

$$dl = -k\,dx \quad \text{and} \quad l(x) = l_0 - kx$$

Similarly in case (ii)

$$\frac{dv}{v} = \frac{ds}{s} = -\frac{ka(x)\,dx}{v}$$

so

$$dv = -ka(x)\,dx$$

But in this case the volume v is also given by

$$v = \int_x^{x+l} a(x)\,dx$$

so that

$$dv = [a(x+l) - a(x)]\,dx$$

Hence

$$a(x+l) - a(x) = -ka(x)$$

and so

$$\frac{a(x+l)}{a(x)} = (1-k)$$

which is easily seen to require a to be a function of the form

$$a(= a_0 b^x) = a_0 \exp x/D \tag{3.11}$$

where D is a constant (of linear dimensions). Thus

$$\frac{a(x+l)}{a(x)} = \frac{a_0 \exp x/D \exp l/D}{a_0 \exp x/D} = \exp l/D = (1-k)$$

which gives

$$D = \frac{l}{\log(1-k)}$$

Equations (3.9), (3.10), (3.11) are valid for $k < 1$ and do not refer to the last zone to solidify. It is also noted that whereas (3.11) refers to exponential tapering which could extend to infinity, (3.10) indicates that l diminishes to zero when $x = l_0/k$ and the process would then stop.

3.3.4 Techniques

3.3.4.1 From the above account of principles it follows that a practical process must take proper account of the various dimensions and other parameters in order to obtain the best results. Thus:

(a) It is seen from (3.3) that a higher value of f reduces the efficiency of purification with $k_0 < 1$ by increasing k, but for practical reasons higher f is required for speed.

(b) Therefore δ should also be as low as possible and can be reduced by circulation of the liquid in the zone. In many cases f must be low in order for k to be near to k_0.

(c) The value of k_0 is a limiting factor. When several impurities are present the system is necessarily determined by the member with k_0 nearest to 1.

(d) The distance i between zones in multi-pass refining can be as short as reasonably possible providing a solid barrier exists between two liquid zones.

(e) The length of zone, l, should be small because it makes for better separation.

(f) The total length L should be as long as reasonably possible.

(g) The ratio L/l indicates the number of zone lengths. For levelling in particular this should be as large as possible (equation (3.4) and Fig. 3.5(b)).

3.3.4.2 In multi-pass zone-refining a number of hot zones can follow along an ingot at close intervals. Single-pass melting is already indicated in Fig. 3.5(a). In typical arrangements for moving heaters, three ring heaters are suitably spaced and move along the outside of a furnace tube containing the metal. This requirement is illustrated in Fig. 3.7(a). If d is the separation between heaters the desired result can be achieved by moving the three heaters slowly in the required direction, a distance d, and then rapidly reversing so that the third heater takes over from the second and the second from the first. The reversing can be effected by a reversible two-speed motor and switching device using either screw or cable to transmit movement. Alternatively a cam device can be used to transmit the oscillatory movement—using a sloping furnace for gravity assisted return or weight over pulley.

Although a straight cylindrical or nearly cylindrical rod type ingot is often used other arrangements are possible. These include ring-shaped specimens as in Fig. 3.7(b). A rotating turntable with the charge in a circular recess round the circumference is also shown. Other variants naturally suggest themselves. Thus a variety of spiral forms with radial heaters as in Fig. 3.7(c) will automatically move zones along the spiral. Other forms have a spiral heater in which case the zones travel outwards radially along suitably separated sectors.

Yet another device is to use a spiral specimen (in a tube). Some arrangements of this kind have used a longitudinal heater as in Fig. 3.7(d) and the spiral itself has been partly immersed in a cooling liquid. It is seen that as the spiral slowly rotates successive zones will move along the helix in the direction indicated.

Some metals which react with common refractory materials may be refined by the floating zone method, Fig. 3.7(e). The specimen is held vertically and the zone is produced by a ring-shaped heater. The liquid metal is held in place by surface tension.

95

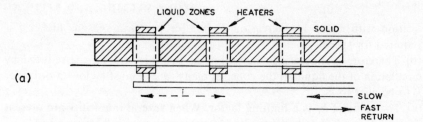

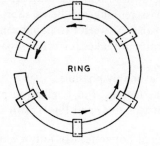

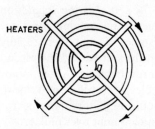

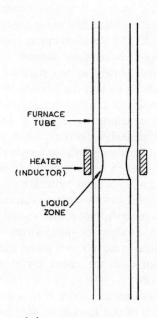

(a)

(b)

(c)

(d)

(e)

Fig. 3.7   Representative arrangements for zone refining (based on diagrams due to Pfann).

   (a) multi-pass zone melting (schematic)
   (b) circular or ring-shaped specimens
   (c) spiral charge with radial heaters
   (d) rotating spiral tube method
   (e) floating zone

3.3.4.3   For heating, electricity is perhaps the best and most common means, and resistance coil heating, induction or arc melting are obviously suitable. Gas heaters and focused radiation have also been used. (Devices of this kind are described by Pfann.)

One interesting procedure is to use the Joule heat from an electric current passing through the specimen by cooling the ends with suitably attached (water) cooling devices. By progressively adjusting the cooling at the two ends a hot zone can be created at one end and moved along. A development of this method employs the addition of a heat reflector at the zone position and cylinders on either side which are cooled and act as absorbers. In this way the Joule heat provides a uniform temperature along the ingot somewhat below the melting point but there is a small region where it rises above the melting point. This method is most useful when radiation is the main cause of heat transfer, i.e. for the higher melting point metals.

The advantages of stirring the melt or contriving to make the most use of circulation by convection have been indicated and it is worth noting that by reducing δ in this way it is possible to speed up the process by as much as ten times. A good conductor can clearly be circulated by electromagnetic means. Induction heating provides circulation in many cases but may not always be adequate. Magnetic stirring devices are available.

3.3.4.4   Two other aspects require brief consideration.

1. *Continuous zone refining* is possible under certain conditions. It has the advantage that once flow is established the process can go on indefinitely and the amount of material wastage is relatively very small. The process is in some respects similar to the separation of crystals from melt by counter current methods, i.e. when liquid and crystals move in opposite directions.

The movement of a molten zone in one direction can be used to assist such a process. In principle it is supposed that enough passes have been made to create a condition approaching the ultimate distribution. Suppose that new material is then introduced into a liquid zone somewhere near the middle of the specimen thus increasing its length in two directions. The length is then reduced again by removing an equivalent amount of purified metal from one end and waste material from the other. The continual movement of the liquid zones maintains the refinement so that the product material is of uniform composition.

In one of the simplest arrangements the feed enters at the top of an inverted U-tube in which purified products proceed down one arm and waste material down the other. The feed zone represents a large static completely molten zone into which impurities move from the product arm. The movement of material can be assisted by the introduction of voids which naturally rise to the top of the liquid zones. On the product side where the heaters move upwards, material drops through the voids as it melts. The voids on the waste side, however, move upwards intermittently. The zone heaters move downwards on this side. There are several other procedures for achieving continuous zone refining.

2. *Temperature gradient zone melting* [29, 30], can be understood by considering Fig. 3.2. Supposing a specimen is in a temperature gradient, then the boundaries of a liquid zone will correspond to two different points on the liquidus and solidus (curves), and the liquid zone tends to develop a corresponding concentration gradient. Advantage is taken of the effect to create a type of transistor known as an n–p–n junction, for growing single crystals, for joining and for measurements of diffusion. The starting point in a typical process is to take two blocks of pure solvent A and to insert between them a thin layer of the second metal (solute) B. This combination is placed in a temperature gradient the extremes of which correspond to temperatures between the melting point of A and that of B (or any eutectic or peritectic reached by the liquidus) as in Fig. 3.2. Therefore:

(a) B will melt and form a liquid zone.

(b) This zone will dissolve A but less A dissolves at the lower temperature and more A at the higher (liquidus nearer pure A).

(c) The concentration difference will tend to equalise by diffusion.

(d) But this will create supersaturated solution of A in liquid at the lower temperature and cause solidification of an alloy of composition determined by k (< 1) and this solid is richer in A and contains less B than the liquid.

(e) This leaves the liquid enriched in B, unsaturated, and so able to dissolve more A at the upper surface temperature.

(f) Consequently the liquid zone creeps up the temperature gradient and leaves a solid alloy of nearly pure A (especially if the solubility of B in A is small and k considerably less than 1).

It is thus easy to appreciate how this process could, for example, be used to produce a completely firm welded junction between two pieces of metal (or alloy) A, or to provide a system A–(alloy of A and B)–A such as is used in the transistor n–p–n junction.

References

1. GILCHRIST, J. D., *Furnaces*. Pergamon, Oxford, 1963.
2. BUCKLEY, H. E., *Crystal Growth*. Wiley, New York; Chapman & Hall, London, 1951.
3. CHALMERS, B., *Modern Research Techniques in Physical Metallurgy*, 170, Amer. Soc. Metals, 1953.
4. HONEYCOMBE, R. W. K., *Metal. Rev.*, 1959, **4**, 1.
5. GILMAN, J. J. (ed.), *The Art and Science of Growing Crystals*. Wiley, New York, London, 1963.
6. ANDRADE, E. N. DA C., *Phil. Mag.*, 1914 (vi), **27**, 869.
7. CZOCHRALSKI, J., *Z. Phys. Chem.*, 1917, **92**, 219.
8. TEAL, G. K. and LITTLE, J. B., *Phys, Rev.*, 1950, **78**, 647.

9. HORN, F. H. and NEUBAUER, R. L., *Rev. Sci. Instrum.*, 1953, **24**, 1154.
10. BURTON, J. A., PRIM, R. C. and SLICHTER, W. P., *J. Chem. Phys.*, 1953, **21**, 1987.
11. CHALMERS, B., *Can. J. Phys.*, 1953, **31**, 132.
12. NOGGLE, T. S., *Rev. Sci. Instrum.*, 1953, **24**, 184.
13. TIEDEMA, T. J., *Acta Cryst.*, 1949, **2**, 261.
14. DUNN, C. G. and LIONETTI, F., *Trans. A.I.M.E.*, 1949. **185**, 72, 125.
15. BURGERS, W. G., MEIJS, J. C. and TIEDEMA, T. J., *Acta Met.*, 1953, **1**, 75.
16. CUSTERS, J. F. and RATHENAU, G. W., *Philips Res. Rept.*, 1949, **4**, 241.
17. DUNN, C. G., *The Cold Working of Metals*, 113 Amer. Soc. Metals, 1949.
18. CHEN, N. K., MADDIN, R. and POND, R. B., *Trans. A.I.M.E.*, 1951, **191**, 461.
19. MADDIN, R. and CHEN, N. K., *Trans. A.I.M.E.*, 1953, **197**, 1131.
20. FISHER, E. S., *Trans. A.I.M.E.*, 1957, **209**, 882.
21. PFANN, W. G., *Trans. A.I.M.E.*, 1952, **194**, 747.
22. PFANN, W. G., *Metal. Rev.*, 1957, **2**, 29.
23. PFANN, W. G., *Zone Melting*. Wiley, New York; Chapman and Hall, London, 1958.
24. HAYES, A. and CHIPMAN, J., *Trans. A.I.M.E.*, 1939, **135**, 85.
25. DARKEN, L. S. and GURRY, R. W., *Physical Chemistry of Metals* esp. pp. 324–25. McGraw-Hill, New York, London, 1953.
26. LORD, N. W., *Trans. A.I.M.E.*, 1953, **197**, 1531.
27. MILLIKEN, K. W., *Trans. A.I.M.E.* (*J. Metals*, 1955, **7**, 838).
28. REISS, H., *Trans. A.I.M.E.*, 1954, **200**, 1053.
29. PFANN, W. G., *Trans. A.I.M.E.*, 1955, **203**, 961.
30. WERNICK, J. H., *J. Chem. Phys.*, 1956, **25**, 47.
31. BRIDGMAN, P. W., *Proc. Amer. Acad. Arts Sci.*, 1925, **60**, 305.

Chapter 4

OUTLINE OF CRYSTALLOGRAPHIC PRINCIPLES

4.1 INTRODUCTORY

This chapter provides an account of some of the basic principles of crystallography which are required in the subsequent chapters or which are otherwise of most direct interest to the metallurgist. It must be emphasised that this cannot be any more than an outline.

The first steps in the development of a crystal science were concerned with observations and measurements in connection with the outward forms and shapes and the regular faces of crystals—particularly those found in nature. The existence of regular facets and their arrangement leads naturally to the supposition that, as a crystal grows, atoms or molecules are added or stacked in a regular manner—perhaps even building up the faces in successive layers.

The essential characteristic of the crystalline state of a solid is thus that the atoms or molecules of which it is composed are arranged in a regular repetitive pattern in space. Amorphous materials and glass do not have a regular arrangement in the same sense and are not therefore regarded as crystalline. In these materials and in liquids—glass can be described as a supercooled liquid—there is indeed a kind of order. It is perhaps useful to regard the states of complete order with perfect crystallinity on the one hand, and complete disorder as opposite extremes which are only closely approached in rare instances. For example, the metallurgist soon becomes aware of the existence of a variety of ways in which a metal crystal can depart from perfection. Furthermore, most metals do not show regular faces because they exist in a polycrystalline mass with irregular grain boundaries. Metallurgical materials do, however, include a number of minerals and oxides and other compounds which may develop crystal facets. There are a number of ways in which the crystallinity of a metal in a polycrystalline mass may be revealed under visual observation:

1. The occurrence of polygonal (straight) grain boundaries in some metals.
2. The existence of twin bands.
3. The shape of some precipitated phases, inclusions, or phases which have developed a shape in the melt.
4. The patterns formed by slip lines.

5. Patterns formed by precipitated particles.
6. The facets revealed by fractographic techniques.
7. The shapes of etch pits.

It is necessary to point out that crystalline regularity can exist in one dimension (a line) or two dimensions (a plane). Although this observation is of less direct interest in physical metallurgy there are circumstances in which a metallurgical constituent can be considered in one of these ways—more often the two-dimensional form. (General Refs. [1, 2, 3].)

4.2 CRYSTAL SYMMETRY

4.2.1 External Symmetry of Crystals

4.2.1.1   Before the discovery of X-ray diffraction which made possible the study of the internal arrangements of atoms, crystallographers had established several laws and principles relating to the geometry of crystals. For example, measurements of the inter-facial angles showed that faces of one type would make the same angle with faces of a second type. Often faces were disposed round an axis or 'zone' in a regular manner, even though the actual width of a face would depend on the (accidental) circumstances under which it grew. Since a crystal exists in three spatial dimensions its surfaces can be completely described in relation to three axes which are not co-planar and can be regarded as co-ordinate axes. Because, in the present context, the internal arrangement of atoms is of most interest these concepts will be developed in that connection. It is, however, useful to approach the symmetry of crystals briefly from the more classical viewpoint first.

4.2.1.2   A symmetry operation could be defined as any process by which a regularly faceted crystal with corresponding faces, equally developed, could be rotated, reflected or otherwise translated so that its position appeared to be unaltered. The operator is called a *symmetry element*. The operation of a symmetry element on the face of a crystal will translate it into another face of the same type. By a study of the external form of crystals in this way it is possible to show that they fall into:

Seven crystal systems
each divided according to its symmetry into a number of classes giving altogether,

Thirty-two crystal classes
Each of these classes can be shown to arise from one of the symmetry elements alone or from combinations of two or more elements called (in either case),

TABLE 4.1 *Summary of Crystal Systems,*

Crystal systems	Crystal classes (Point group symmetry)			Space groups (as divided between the 7 systems and 32	
	Number in each system	Full symbol	Abbreviated symbol	Numbers isomorphous with point group	Symbols of space groups
(1) Triclinic	2	1	1	1	$P1$
		$\bar{1}$	$\bar{1}$	1	$P\bar{1}$
(2) Monoclinic					(P) (C)
	3	m	m	4	Pm Cm
		2	2	3	$P2$ $C2$
		$2/m$	$2/m$	6	$P2/m$ $C2/m$
(3) Ortho-rhombic					(P) (C) (I) (F)
	3	222	222	9	$P222$ $C222$ $I222$ $F222$
		$2mm$	mm	22	$Pmm2$ $Cmm2$ $Imm2$ $Fmm2$
		$2/m\ 2/m\ 2/m$	mmm	28	$Pmmm$ $Cmmm$ $Immm$ $Fmmm$
(4) Hexagonal	7	6	6	6	$P6$
		$\bar{6}$	$\bar{6}$	1	$P\bar{6}$
		$6/m$	$6/m$	2	$P6/m$
		622	62	6	$P622$
		$6mm$	$6mm$	4	$P6mm$
		$\bar{6}m2$	$\bar{6}m2$	4	$P\bar{6}m2$
		$6/m\ 2/m\ 2/m$	$6/mmm$	4	$P6/mmm$
(5) Trigonal (including rhombohedral)	5	3	3	3+1	$P3$
		$\bar{3}$	$\bar{3}$	1+1	$P\bar{3}$
		$3m$	$3m$	4+2	$P3m$ $P31m$
		32	32	6+1	$P312$ $P321$
		$\bar{3}2/m$	$\bar{3}m$	4+2	$P\bar{3}1m$ $P\bar{3}m1$
(6) Tetragonal					(P) (I)
	7	4	4	6	$P4$ $I4$
		$\bar{4}$	$\bar{4}$	2	$P\bar{4}$ $I\bar{4}$
		$4/m$	$4/m$	6	$P4/m$ $I4/m$
		422	422	10	$P422$ $I422$
		$4mm$	$4mm$	12	$P4mm$ $I4mm$
		$\bar{4}2m$	$\bar{4}2m$	12	$P\bar{4}2m$ $I\bar{4}2m$ $P\bar{4}m2$ $I\bar{4}m2$
		$4/m\ 2/m\ 2/m$	$4/mmm$	20	$P4/mmm$ $I4/mmm$
(7) Cubic					(P) (I) (F)
	5	23	23	5	$P23$ $I23$ $F23$
		$2/m\bar{3}$	$m3$	7	$Pm3$ $Im3$ $Fm3$
		432	432	8	$P432$ $I432$ $F432$
		$\bar{4}3m$	$\bar{4}3m$	6	$P\bar{4}3m$ $I\bar{4}3m$ $F\bar{4}3m$
		$4/m\ \bar{3}2/m$	$m3m$	10	$Pm3m$ $Im3m$ $Fm3m$
Total	32			230	

Note: The hexagonal lattices are alternatively indicated by C. The trigonal lattices are referred to hexagonal axes. The rhombohedral sub-division of the trigonal lattices can be referred to either the hexagonal axes or to rhombohedral. Hence (4) and (5) are sometimes regarded as one system.

Classes, Lattices and some Space Groups

classes)		Space lattices	
isomorphous with the point groups	Numbers with each space lattice	Symbols	Numbers
	2	P	1
Pc Cc $P2_1$ $P2_1/m$ $P2/c$ $P2_1/c$ $C2/c$	8 5	P C	2
$P222_1$ $P2_12_12$ $P2_12_12_1$ $I2_12_12_1$ $C222_1$ $Pmc2_1$ $Pba2$ $Cm2m$ $F2mm$ $I2_1na$ (+13 others) $Pmma$ $Pmna$ $Cmca$ $Fddd$ $Ibca$ (+19 others)	15 30 9 5	$C(A)$ P I F	4
$P6_1$ $P6_5$ $P6_4$ $P6_2$ $P6_3$			
$P6_3/m$ $P6_122$ $P6_522$ $P6_422$ $P6_222$ $P6_322$ $P6cc$ $P6_3cm$ $P6_3mc$ $P\bar{6}c2$ $P\bar{6}2m$ $P\bar{6}2c$ $P6/mcc$ $P6_3/mcm$ $P6_3/mmc$	27	P (C)	1
$P3_1$ $P3_2$ $R3$ $R\bar{3}$ $P3c1$ $P31c$ $R3m$ $R3c$ $P3_112$ $P3_212$ $P3_121$ $P3_221$ $R32$ $P\bar{3}1c$ $P\bar{3}c1$ $R\bar{3}m$ $R\bar{3}c$	18 7	P or R	1
$P4_1$ $P4_3$ $P4_2$ $I4_1$			
$P4_2/m$ $P4/n$ $P4_2/n$ $I4_1/a$ $P4_122$ $P4_322$ $P4_222$ $P4_12_12$ $P4_32_12$ $P4_22_12$ $P42_12$ $I4_12_2$ $P4bm$ $P4nc$ $P4cc$ $P4_2nm$ $P4_2bc$ $I4cm$ $I4_1md$ (+3 others) $P\bar{4}2c$ $P\bar{4}2_1m$ $P\bar{4}b2$ $P\bar{4}n2$ $I\bar{4}2d$ (+3 others) $P4/mbm$ $P4/ncc$ $P4_2/mmc$ $P4_2/nbc$ $P4_2/ncm$ $I4/mcm$ $I4_1/acd$ (+11 others)	49 19	P I	2
$P2_13$ $I2_13$ $Pn3$ $Pa3$ $Ia3$ $Fd3$ $P4_232$ $P4_332$ $P4_132$ $I4_132$ $F4_132$ $P\bar{4}3n$ $I\bar{4}3d$ $F\bar{4}3c$ $Pn3n$ $Pm3n$ $Pn3m$ $Ia3d$ $Fm3c$ $Fd3m$ $Fd3c$	15 10 11	P I F	3
	230		14

103

Point groups

because they describe the symmetry operations about a point.

The symmetry elements are:

1. The plane of (reflection) symmetry *m* (mirror).
2. Rotation axes $X = 1, 2, 3, 4, 6$.
3. Inversion axes $\bar{X} = \bar{1}, \bar{2}, \bar{3}, \bar{4}, \bar{6}$, or axes of rotary inversion.

The symbols are a convenient shorthand and will be followed here in preference to an earlier system which is less descriptive. The plane *m* acts like a mirror and reflects one half of the crystal into the other half. Rotation axes translate faces into each other by rotation about a fixed axis through an angle equal to 2π divided by the number describing the axis. Thus in the case of 4, a four-fold or tetrad axis, the steps are 90°. An inversion axis follows the same process but at each stage the face is transposed through an origin on the axis. In the case of $\bar{1} = 360°$ rotation + inversion, we have a special element, viz.

4. A centre of symmetry although it is also represented as an axis.

Table 4.1 records the seven systems and the crystal classes in terms of these symmetry elements. For the present we note that *Xm* signifies that the plane *m* passes through the axis *X* (or $\bar{X}$) and *X/m* implies that *X* is normal to *m*. (Other information in this table will be referred to later.)

As an elementary exercise one may investigate the symmetry of simple crystals by the following tests.

1. A piece of plain hexagonal pencil is cut off so as to have flat ends. This has a six-fold (hexad) axis down the centre. It has two-fold axes (diads) perpendicular to the centres of the faces and another set through the edges. Mirror planes could also be placed in these positions. Point group symmetry $6/m\ 2/m\ 2/m = 6/mmm$ (see table).
2. A cube is examined and found to have the following types of symmetry elements:

   (a) Four-fold axes perpendicular to middle of faces.
   (b) Planes containing these axes parallel to cube faces.
   (c) A second set of planes also through these axes and in addition cutting opposite edges.
   (d) Each of these planes contains cube diagonals which form three-fold (triad) axes.
   (e) They also contain two-fold axes (where?).
   (f) There is a centre of symmetry at the centre of the cube.

It is useful to tabulate the number of each. It will be noticed that in the table this symmetry is given by $4/m\ \bar{3}2/m$ but, as in the hexagonal case, this can be abbreviated—to $m3m$—because such a combination of symmetry elements requires the presence of the others.

3. Obtain a cube of suitable material (plasticine or wax) and cut off material from the eight corners progressively and by cutting equally along the three edges. This gives a regular octahedron which still has the same symmetry elements as the cube.

4. Repeat, but only cut alternate corners. This gives a regular tetrahedron. Some of the symmetry elements are now absent (which remain?).

5. Obtain a rectangular block with two edges equal and write down the symmetry elements noting that some of those present in the cube have now vanished. Find the point group symmetry in the table.

6. Cut a tetrahedron from this—the faces of which are isosceles triangles (not equilateral)—and examine as in (4). A cardboard model can be used.

7. Obtain a rectangular block with the three edges different and note the reduction in symmetry from (4). Write the elements down.

If these exercises are done *carefully* they will be sufficient to give an appreciation of the principles. The reader will note that:

1. Any one symmetry element translates any other into one of its own kind. For this reason when m occurs every element on one side must be reflected into identical elements on the other. Similarly we can have a single three-fold axis in one direction, but if we add another, not parallel to it, the first axis requires that we add two more so that each axis operates on the other three and rotates them into each other every 120°. This is how the symmetry elements of the cube arise. It can be shown that if these are the only elements we demand we have actually also introduced two-fold (diad) axes. The minimum point group symmetry in the cubic system is therefore 23.

2. Hence one or more symmetry elements may generate or equate to one or more others. Other examples are—two m planes at right angles have a diad along their intersection, and $\bar{2}$ produces the same effect as m (why?).

4.2.2 Internal Symmetry and Lattices

4.2.2.1 Figure 4.1 shows how four typical symmetry elements relate to the external shape of four crystals and secondly how the same elements would operate on points—an aspect developed in the next section. If it is accepted that the atoms or molecules inside the crystal should have essentially the same symmetry we might expect to find them disposed like these points about axes or planes which are parallel to the external symmetry elements. This principle is found to be essentially correct but additional information is required.

The concept of a regular arrangement in space requires firstly the kind of symmetry, represented by the point group, and secondly that the pattern should repeat itself at regular intervals in three dimensions. The centres of the groups of symmetry elements must be placed on a framework—*the space lattice*.

Two-dimensional lattices or nets can be found in a repeating pattern or design on say wallpaper or cloth. It is possible to draw a parallelogram which contains the smallest unit of pattern and the pattern then can be built up by placing the parallelograms together. The network formed by their edges is the lattice. The smallest unit is *the unit cell*.

Fig. 4.2 should now be studied very carefully because it brings out important aspects of such nets (including their inherent symmetry) which are also

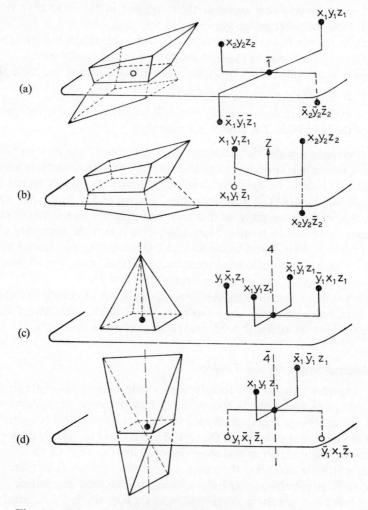

Fig. 4.1 External and internal symmetry of crystals about a point
(after a diagram due to de Jong [4]).

        (a) centre of symmetry $\bar{1}$
        (b) plane of symmetry m
        (c) four-fold rotation axis 4
        (d) four-fold inversion axis $\bar{4}$

found in three-dimensional lattices. In (a) the two edges are different and the angle $\gamma \neq 90°$. There are centres of symmetry or two-fold axes (equal effects in a plane) at the intersections and the middle of each parallelogram. In (b) $\gamma = 90°$. This lattice now has planes of symmetry perpendicular to the plane of the

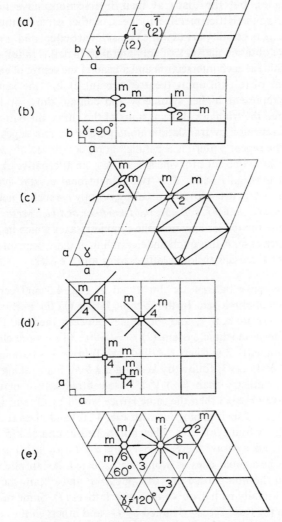

Fig. 4.2   The five two-dimensional lattices.

diagram containing the cell edges and half-way in between. (Two-fold axes at their intersections.)

If two edges are made equal as in (c) but $\gamma \neq 90°$ the parallelogram is still a possible unit cell but we can now draw a rectangle of double the area but with the higher symmetry elements found in (b). In fact equalising the edges ($a=b$) has had the same effect as putting a lattice point at the centre of the rectangle in (b). We thus have the concepts of *a primitive lattice* and *a centred lattice*. The latter is preferred because it reveals the (true) higher symmetry.

If the angle $\gamma=90°$ and $a=b$ we obtain a square lattice (d) with the same symmetry planes, but the diads at their intersections have now become tetrads. Centring the lattice merely produces another square lattice (how?).

An interesting possibility arises if the two sides are equal and $\gamma=120$ or $60°$. In this case we obtain a network of equilateral triangles as in (e), and there is a hexad (six-fold) at each intersection and a triad at the centre of each triangle. The conditions of (c) still apply (centred rectangle) but the symmetry elements of (e) represent higher symmetry and contain those of Fig. 4.2 (c). If we considered the triangles alone as units of the lattice the pattern within it would not be repeated by translation alone parallel to two edges. Hence for this purpose the repeat pattern is a parallelogram comprised of two triangles. Even so if we chose one edge for one axis there is an alternative choice for the second which is equally valid. If a 'two-dimensional crystal' had only the three-fold symmetry, it could fill 'space' by occupying such a network. In a sense *the hexagonal network contains the triangular and vice versa*.

As an exercise the reader may note the symmetry axes which lie *in the plane* of the net in each case. A thorough understanding of these principles makes an appreciation of three-dimensional lattices very much easier.

4.2.2.2 These space lattices are illustrated in Fig. 4.3 and certain useful symbols are again employed. In the triclinic lattice (1) the unit cell edges or *lattice parameters* are a, b, c, and the angles between them, α, β, γ, are all unequal. In the monoclinic system (2) $\alpha=\gamma=90°$. (The monoclinic lattices (2) have the symmetry $2/m$ where the diad (two-fold) axis is parallel to b.) The orthorhombic lattices (3) follow by keeping $a\neq b\neq c$ but making $\alpha=\beta=\gamma=90°$. (Higher symmetry than (2).) If a base-centred lattice in this system happens to have $b=a\sqrt{3}$ then the same lattice would be obtained by putting $a=b$ and $\gamma=120°$. This is analogous with Fig. 4.2(e) and gives the hexagonal lattice (4) with six-fold symmetry. The rhombohedral unit cell (5) has only a three-fold axis and, as drawn here, $a=b=c$ and $\alpha=\beta=\gamma\neq90°$. This lattice can be referred to hexagonal axes as would be expected. As an alternative step from the orthorhombic lattice one can put $a=b$ and obtain the tetragonal lattices (6) and finally with $a=b=c$ the cubic lattices (7). Some of the lattices are centred on the three faces—denoted by F—and others in the centre of the cell—denoted by I. In certain cases, viz. monoclinic and orthorhombic, it is possible to have one face-centred C. Other apparent possibilities involve duplication. Thus a face-centred tetragonal lattice can be referred to a smaller body-centred tetragonal lattice. In the cubic case one could not have centring on one or two faces (A or B) and not the third (C) because this would

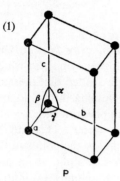
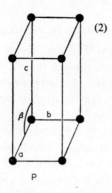
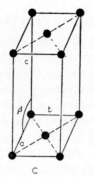

(1)

(2)

P

P

C

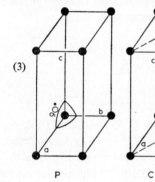

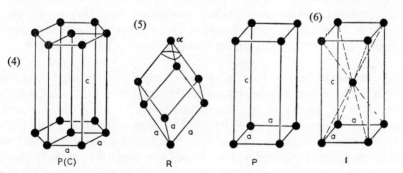

(3)

P

C

I

F

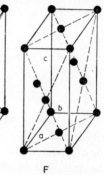

(4)

(5)

(6)

P(C)

R

P

I

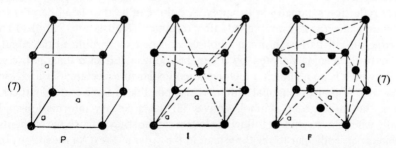

(7)

(7)

P

I

F

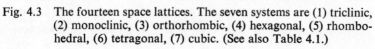

Fig. 4.3 The fourteen space lattices. The seven systems are (1) triclinic,
(2) monoclinic, (3) orthorhombic, (4) hexagonal, (5) rhombohedral, (6) tetragonal, (7) cubic. (See also Table 4.1.)

destroy the cubic symmetry. The final result is *fourteen space lattices*. One can regard Fig. 4.3 as representing a progression to higher symmetry which branches at (3) when there is a choice (4) or (6). It is also noted that the cube has the symmetry elements previously discussed—including a three-fold axis along a diagonal. Thus a simple cube is merely a rhombohedron with $\alpha = 90°$. A face-centred cube contains a rhombohedron with $\alpha = 60°$. This can be seen by joining the points at opposite ends of a diagonal to the points in the centres of the adjacent faces. (One can confirm this by a careful drawing and may similarly confirm that a body-centred cubic lattice contains a primitive rhombohedral cell with the angle $\alpha = 109° \, 28'$.) Hence cubic lattices can also be derived from rhombohedral lattices. We can summarise this as

$$
\begin{array}{c}
\rightarrow (6) \longrightarrow \\
\left\lceil \qquad\qquad \downarrow \right. \\
(1) \rightarrow (2) \rightarrow (3) \quad (5) \leftrightarrow (7) \\
\left\lfloor \quad \updownarrow \right. \\
\rightarrow (4)
\end{array}
$$

The lattices can be denoted by the letters P, I, F or C. Reference to Table 4.1 again will show how the lattices of Fig. 4.3 belong to the seven crystal systems. The description of one matches the other and as seen from both Figs. 4.2 and 4.3 the lattices themselves have point group symmetry. In fact the lattice in each system has the *highest point group symmetry* for that system, although it can accommodate point groups which contain *no more than an essential minimum* of its own point group elements. This observation has already been foreshadowed in connection with the experiments on external symmetry.

4.2.2.3   It follows that the internal symmetry might be described by the lattice plus the point group symmetry, so that by associating the groups in each system with its lattices we produce a number of so-called *space groups*. These do not represent all the possibilities and other space groups arise from the following considerations. Just as the *external* symmetry does not disclose whether the lattice is centred or not, it does not reveal the full facts about symmetry axes or planes *within* the lattice. Thus it is not necessary for the simple reflection symmetry *m* to apply. A point can be translated parallel to the 'glide plane' (as it is now called) after reflection. Translations parallel to a cell edge of half its length are symbolised *a*, *b* or *c* (according to the edge). Alternatively a glide denoted by *n* can be along a cell face diagonal, e.g. $\frac{1}{2}a + \frac{1}{2}b$ which is an *n* glide in a plane perpendicular to *c* (rectangular axes are necessary). A quarter diagonal shift, *d*, is also possible in some centred lattices. Again, the space group symbols convey all the necessary information and usually when three symbols follow the lattice symbol they refer to the elements in planes perpendicular to the three axes in the order a–b–c. (An exception: In the cubic case when two *m* planes are indicated—these are at 45°.) In the same way an axis can involve translations parallel to the axis simultaneously with

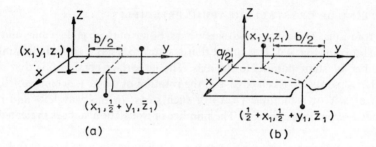

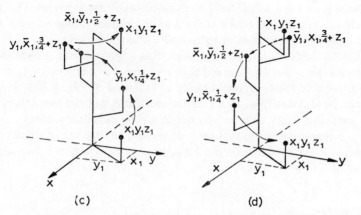

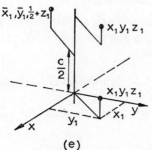

Fig. 4.4 Examples of glide planes and screw axes. (Note: these symmetry elements do not always pass through the origin as shown here.)

(a) glide plane with translation $\frac{1}{2}b$ in plane normal to z. Denoted by b (could also occur in plane yz normal to x)

(b) glide plane with translations $\frac{1}{2}a$ and $\frac{1}{2}b$ (in plane normal to z) denoted by n

(c) 4_1 axis along O_z (R.H. screw)

(d) 4_3 axis along O_z (L.H. screw)

(e) 2_1 axis along O_z

111

the rotations. This gives rise to screw axes. Some of these glide planes and axes are shown in Fig. 4.4. It is noted that a screw axis has a suffix indicating the number of steps parallel to the axis. The important condition is that an X-fold axis parallel to a cell edge, e.g. the rotation of $2\pi/X$, is accompanied by a step of a/X or a multiple ($< X$) of such steps. In this way left- and right-handed screw axes can arise. The number of translation steps is indicated in a suffix.

By considering a given point group in conjunction with the space lattices to which it can be fitted and then replacing the point group elements by the possible equivalent space group elements, the space groups 'isomorphous' with the point groups have been derived. Elimination of certain duplications leads to the numbers—of which some examples are given—in Table 4.1 for each point group and a final total is obtained of *230 space groups*. This number includes certain pairs which are said to be 'enantiomorphous'—one is the mirror image of the other as in right- and left-handed screws.

(Table 4.1 further illustrates the point that trigonal space groups (which only have a three-fold axis) can either be considered as a separate system or as a sub-division of the hexagonal exactly as indicated by Fig. 4.2(e). Seven of these can be also described by a rhombohedral cell, but eighteen others have the trigonal symmetry only and do not fit a rhombohedral lattice.)

For a full account of the symmetry of crystals and the use of space groups reference should be made to the bibliography, especially the *International Tables* [5].

4.2.3 The Crystallographic Axes and Co-ordinates

It is convenient to consider the crystal axes which are denoted by x, y, z and are parallel respectively to the edges of the unit cell, a, b, c. We can use either of two systems according to the problem:

(a) When actual distances are required x, y, z are measured in the same length units as a, b, c (usually in angström units, i.e. Å).
(b) When actual distances are not needed (x, y, z) is short for x/a, y/b, z/c, i.e. fractional distances are involved.

It is seen that, using these co-ordinates, the following positions describe the respective lattices assuming the origin of co-ordinates is at a lattice point.

P $(0, 0, 0)$, (equivalent to $(1, 0, 0)$, (110), etc.)
I $(0, 0, 0)$, $(\frac{1}{2}, \frac{1}{2}, \frac{1}{2})$
F $(0, 0, 0)$, $(\frac{1}{2}, \frac{1}{2}, 0)$, $(\frac{1}{2}, 0, \frac{1}{2})$, $(0, \frac{1}{2}, \frac{1}{2})$
C $(0, 0, 0)$, $(\frac{1}{2}, \frac{1}{2}, 0)$.

Again by referring to Figs. 4.1, 4.2 and 4.4, it can be confirmed that, with the axes of symmetry parallel to z or perpendicular to m, the symmetry elements have the following effects:

4. A point $(x_1 y_1 z_1)$ becomes $(y_1 \bar{x}_1 z_1)(\bar{x}_1 \bar{y}_1 z_1)(\bar{y}_1 x_1 z_1)$ in that order (clockwise) $(\bar{x}_1 = -x_1$, etc.).

$\bar{4}$. The sequence is the same except that the z co-ordinate is alternately $+$ and $-$.

m. Reflects $(x_1 y_1 z_1)$ into $(x_1 y_1 \bar{z}_1)$.

$\bar{1}$. (Centre at (000)) translates $(x_1 y_1 z_1)$ into $(\bar{x}_1 \bar{y}_1 \bar{z}_1)$.

m (z position) replaced by b (see Fig. 4.4(a)).

$(x_1 y_1 z_1)$ translates to $(x_1, \frac{1}{2} + y_1, \bar{z}_1)$; replaced by a. $(x_1 y_1 z_1)$ translates to $(\frac{1}{2} + x_1, y_1, \bar{z}_1)$.

4 replaced by 4_1 gives the sequence (Fig. 4.4(c)): $(x_1 y_1 z_1)$, $(\bar{y}_1, x_1, \frac{1}{4} + z_1)$, $(\bar{x}_1, \bar{y}_1, \frac{1}{2} + z_1)$, $(y_1, \bar{x}_1, \frac{3}{4} + z_1)$ (right-handed).

4_3 gives the same sequence for x and y (Fig. 4.4(d)) but the z co-ordinates are in the order $z_1, \frac{3}{4} + z_1, \frac{1}{2} + z_1, \frac{1}{4} + z_1$. This follows since, if the steps are $\frac{3}{4}$, the sequence $0, \frac{3}{4}, 1\frac{1}{2}, 2\frac{1}{4}$ (in three unit cells) reduces to $1(0), \frac{3}{4}, \frac{1}{2}, \frac{1}{4}$ in one unit cell.

The result is a screw in the opposite sense to 4_1. It can now be appreciated that in a given space group a point in a *general position* (xyz) is operated on by the symmetry operations to give a number of *equivalent points*.

It is not difficult to accept the next step, viz. that a point in a *special position* say on an axis or plane of symmetry is not multiplied in the same way. Thus in an m plane there is no multiplication but in a glide plane the translation occurs without the reflection. These considerations lead to all the possible sets of point positions for each space group.

4.3 CRYSTAL STRUCTURE

4.3.1 From Space Group to Structure

It is important to remember that the term 'group' refers to the collection of one or more symmetry elements including the lattice translations, or to a set of points related by these elements. It is now apparent that a crystal structure can be built up by placing atoms or molecules at each of the points in a set such as those indicated above. The complete arrangement of atoms, the lattice parameters, interatomic distances and any angles involved, together with the space group define the *crystal structure*.

In the determination of crystal structure the chemical composition, density and volume of the unit cell are used to give the number of atoms or molecules per unit cell. This number must generally coincide with one or more of the sets of atomic co-ordinates determined by the symmetry operations. As an example four atoms per unit cell can be placed round a four-fold axis in a crystal of the tetragonal space group P4—as in Fig. 4.1. A low number of atoms per unit cell in a crystal of high symmetry clearly implies that the atoms are at special positions. A notable case arises in connection with the cubic space group $Fm3m$, i.e. point group $4m\bar{3}m2$ abbreviated to $m3m$, with a face-centred lattice and no glide planes or screw axes. The maximum multiplicity of atom positions is when the general position $(x_1 y_1 z_1)$ is operated on to give 48 positions round each point centre of symmetry and so 192 per unit cell. Positions on cell faces or diagonals give a lower multiplicity, and if atoms are put only at the cell corners and face centres they are not multiplied by any of the symmetry operations and hence the *minimum* number per unit cell is 4. One way of arriving at this number is that there are 8 atoms at a cube corner each shared between 8 cells and 6 atoms at the face centres each shared between two cells: $8 \times \frac{1}{8} + 6 \times \frac{1}{2} = 4$.

4.3.2 Simple Metallic and Other Structures

4.3.2.1   The last observation explains one of the simplest metallic structures. There are three such structures:

(a) The face-centred cubic structure with 4 atoms/unit cell (f.c.c.).
(b) The close-packed hexagonal structure with 2 atoms/unit cell (c.p.h.).
(c) The body-centred cubic structure with 2 atoms/unit cell (b.c.c.).

These terms now describe actual *structures* and *not* lattices—a distinction we must carefully preserve. There are many structures with a face-centred *cubic lattice* and some of these belong to the space group $Fm3m$, but only one structure can have this description with 4 atoms/unit cell. Similar remarks apply to the space group $Im3m$ to which the body-centred cubic lattice and structure belong. The first and second of these structures could be taken up by atoms in the form of spheres in contact. One layer of such atoms would give a triangular network. It is easy to draw a network of circles in contact or pack spheres into this arrangement. A second layer could be put over this with atoms in position B or C but not both. A third layer in position A gives the close-packed hexagonal lattice but in position C the face-centred structure is obtained. The face-centred cubic lattice was stated previously to contain a primitive rhombohedral lattice with its three-fold axis along the cube diagonal. Such a rhombohedral lattice can be referred to hexagonal axes. Hence the only apparent difference between the structures (a) and (b) is in the sequences:

(a) $ABCABCABC\ldots$
(b) $ABABAB\ldots$

We notice, however, that if the atoms were spheres, the second structure would have $c/a = 1 \cdot 633$. This is not always the case and sometimes the *axial ratio* is less (as with Be or Mg) and sometimes greater (Zn or Cd). The implication is that some atoms are not spherically symmetrical. The reader may confirm that in this structure the atom co-ordinates can be chosen as $(0, 0, 0)$

and $(\frac{1}{3}, \frac{2}{3}, \frac{1}{2})$. These are special positions of the space group $P6_3mmc$—the highest symmetry represented by these points (see Fig. 4.8(a)).

In certain structures it is found that faults in the sequences may occur. In a face-centred cube for example a *stacking fault* can be represented by

$$ABCABABCABC$$

If the sequence is however

$$ABCAB\underline{C}BACBA$$

the structure is symmetrical about the C plane underlined. This is the sequence for a twinned crystal in which the structure is reflected through the twin plane.

In the hexagonal sequence a stacking fault sequence would be

$$ABABACABABAB\ldots$$
or
$$ABABABACACAC\ldots$$
or
$$ABABABCBCBCBC\ldots$$

In the first case only one layer is out of place, in the second and third only one of the two original positions is followed after the fault. In this case reflection symmetry about a twin plane is not possible.

As a further exercise it is useful to draw a similar diagram of the plane in a body-centred cubic structure which passes through opposite cube edges, and contains two cube diagonals. The arrangement of atoms is very similar to that of the close packed planes but the angle between the diagonals is 70° 32′ instead of 60°. The rows along these directions can still be imagined as spheres in contact—but the spheres do not touch in the third direction. This plane and the corresponding planes in the other two structures are found to be significant in connection with martensitic transformations from one structure to the other. The three structures are shown as arrangements of spheres in Fig. 4.5.

We note that many pure metals have one of these structures and some have two. Other more complicated structures may be found amongst metallic elements. Many other structures arise in the various phases which form in binary and higher order systems. Some of these can, however, be derived from the simpler metallic structures by relatively simple adjustments of, or additions to, the atom positions.

4.3.2.2 The following examples of structures are chosen to illustrate these last observations.

1. The structure of diamond, silicon and germanium. This structure can be derived from the face-centred cubic lattice if additional atoms are placed at positions on cube diagonals half-way between the corner of the unit cell and its centres. One such atom has co-ordinates $(\frac{1}{4}, \frac{1}{4}, \frac{1}{4})$ and the others $(\frac{1}{4}, \frac{3}{4}, \frac{3}{4})$, $(\frac{3}{4}, \frac{1}{4}, \frac{3}{4})$ and $(\frac{3}{4}, \frac{3}{4}, \frac{1}{4})$. One way of regarding this structure is to divide the

115

PACKING
OF SPHERES

LOCATION OF
ATOM CENTRES

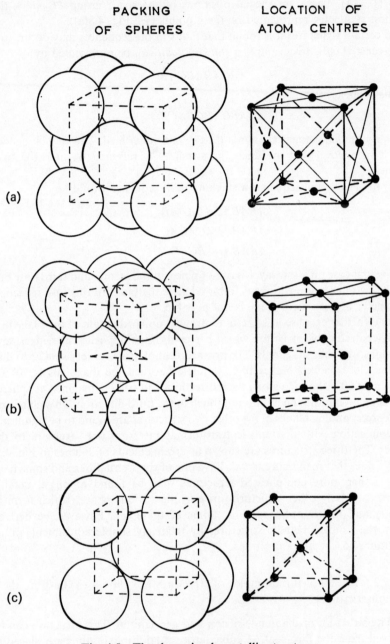

(a)

(b)

(c)

Fig. 4.5 The three simple metallic structures.
    (a) face-centred cubic structure
    (b) close-packed hexagonal structure
    (c) body-centred cubic structure

face-centred cubic unit cell into eight cubes of half the cell edge and to place the extra atoms in centres of *alternate* cubes. The reader may check by means of a drawing that the face-centred atoms occupy alternate corners of these smaller cubes and that these form a tetrahedral group around the one in the centre. Such groupings are very frequent in crystal structures.

The 'extra' atoms also lie on a face-centred cubic lattice with its origin shifted from (000) to $(\frac{1}{4}\frac{1}{4}\frac{1}{4})$. These two inter-penetrating face-centred lattices are connected with each other by the simple replacement of one of the cube face mirror planes m by a glide plane d which involves a quarter translation parallel to it. This would give a shift of $a/4$ in directions parallel to two edges and the three-fold symmetry thus requires a *single* atom to lie on the diagonal at $(\frac{1}{4}\frac{1}{4}\frac{1}{4})$ if one lies at the origin (000). The space group is $Fd3m$ in which the minimum number of atoms per unit cell is 8, as in this structure.

2. The structure of zinc-blende (one modification of ZnS). If instead of the extra atoms being identical they are different as in ZnS (cubic modification) we have a type of compound of general formula AB. The space group is now $F\bar{4}3m$. The tetrahedral symmetry is still there but, because the atoms are different and the two points (000) and $(\frac{1}{4}\frac{1}{4}\frac{1}{4})$ are no longer equivalent, m is not replaced by d but by $\bar{4}$ (the axis implying tetrahedral symmetry).

3. The calcium fluoride structure—CaF_2. In this structure Ca atoms are placed at the face-centred positions and then fluorine atoms at all eight of the centres of the smaller cubes. The fluorine atoms alone would form a simple cubic lattice but the symmetry of the structure as a whole is determined by the Ca atoms which demand the face-centred space group $Fm3m$.

4. The sodium chloride structure—NaCl. This important structure is also found in a number of compounds such as FeO, MnO, CaO, MgO, NbC, VC, TiC, and the corresponding nitrides. The carbides and nitrides are sometimes described as 'interstitial' structures because the metal ions are relatively larger and the smaller ions can be regarded as occupying interstices between the larger close-packed metal ions. In sodium chloride the structure is more properly regarded as ionic—positive and negative ions in equal numbers are placed at the points of two interpenetrating face-centred cubic lattices. The space group is again $Fm3m$. If one set of atoms (ions) is at $(0, 0, 0)$, $(\frac{1}{2}, \frac{1}{2}, 0)$, $(\frac{1}{2}, 0, \frac{1}{2})$, $(0, \frac{1}{2}, \frac{1}{2})$ the others must lie at points half-way between. The reader may check that these positions are $(\frac{1}{2}, 0, 0)$, $(0, \frac{1}{2}, 0)$, $(0, 0, \frac{1}{2})$ and $(\frac{1}{2}, \frac{1}{2}, \frac{1}{2})$ and so that A and B atoms alternate along lines in three directions at right angles. In the case of interstitial structures the close-packed spheres in contact surround points with the second co-ordinates so that they are at corners of a regular octahedron. It can be shown that a smaller sphere can just occupy this space if the ratio of the two radii is 0·59.

5. The caesium chloride structure—CsCl. In this structure the atoms of each kind occupy two interpenetrating simple cubic lattices. Alternatively it resembles a body-centred cube with the atoms at the centres of one kind, and the corner atoms of the other kind. This is no longer a body-centred lattice

because the two points are not equal. An interesting point arises—when an alloy phase of formula AB has a random distribution on a body-centred cubic lattice the points are indistinguishable and the space group is $Im3m$. If the phase then becomes ordered and takes the CsCl structure, the space group is $Pm3m$.

6. Ordered structures arise from face-centred cubic solid solutions in a similar way. If, however, the composition is equiatomic, AB, the ordering cannot occur without destroying the cubic symmetry. The A atoms are found in opposite cube faces and the B atoms in planes half-way between. Because of the different atomic sizes, the structure has a different cell edge perpendicular to the layers and so has become tetragonal. (Face-centred with c/a different from 1·0—more correctly regarded as body-centred with c/a near $\sqrt{2}$.) Example AuCu.

7. The composition AB_3 can undergo ordering and remain cubic. The A atoms occupy the cube corners and the B atoms the mid points of the faces. Again the same point group symmetry elements are present but the lattice points are not all identical and so the space group must now be $Pm3m$. Example AuCu$_3$.

These are only a few simple examples, some of which will be referred to in later chapters. They illustrate a number of important principles.

A note on alloy phase nomenclature:

1. In alloy equilibrium diagrams various phases are frequently denoted by Greek letters. Although the designation is often arbitrary, the sequence α–β–γ–ϵ copper zinc alloys (brasses) has been paralleled by the use of the same symbols to denote *phases of the same structure* in other copper, silver and gold-based alloys. Another example is the so-called sigma phase (σ) which occurs in several alloy systems.

2. In tabulations or lists of structures (e.g. Refs. [6, 7]) a series of letters and numbers have been used. A refers to the elements, B to simple compounds and C to more complicated. The series $A1, A2, \ldots, B1, B2, \ldots$, are quite arbitrary and the order has no significance.

3. The space group symbols are fundamentally useful but should not be put onto phase diagrams because they do not give the actual structure which the metallurgist wants to know.

4. A proposed system [8] which is likely to be adopted, has a logical set of fourteen letters to represent the space lattices. An actual phase is then denoted by the number of atoms per unit cell followed by this letter. An arbitrary sequence of small letters is still required if more than one structure exists with the same number, but this system is probably the shortest conceivable that can reasonably be used in phase diagrams and gives important information which is sometimes completely definitive as when $2B$ means a body-centred cubic structure with two atoms per unit cell, and $4F$ the face-

centred cubic structure. (There need be no confusion with the space group symbols.)

4.4 PLANES AND ZONES

4.4.1 Relation Between Faces, Planes of Atoms and the Symmetry

4.4.1.1 In the discussion of symmetry it was indicated how the connection arose between the external symmetry and the internal symmetry which determined the location of atoms or molecules in the structure. The supposition that as a crystal grows atoms or molecules can be added or stacked in a systematic manner—for example in layers—implies that when crystal faces appear they are parallel to such planes of atoms.

In Fig. 4.6(a) a *two-dimensional lattice* is drawn and a number of straight

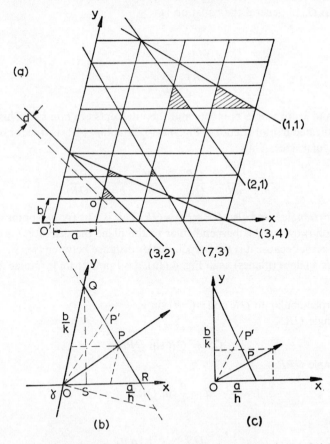

Fig. 4.6   Principles of lattice 'planes' in two dimensions. Origin of indices.

119

lines are shown passing through some of the points. If we consider a simple structure with atoms at the lattice points then these lines correspond to 'planes' perpendicular to the plane of the diagram. Consider the 'plane' marked (3, 2)—this has the equation with origin at O':

$$\frac{x}{2a}+\frac{y}{3b} = 1$$

since it cuts the x axis at $2a$ and the y axis at $3b$. For reasons which will be apparent later, it is useful to shift the origin to O so that the shaded area is the smallest cut off in a single unit cell. The line passes through points whose co-ordinates are $(a/3, O)$ and $(O, b/2)$ so that its equation is now

$$\frac{x}{a/3}+\frac{y}{b/2} = 1$$

One can confirm that this line passes through the points $(a, -b), (-a, 2b)$, referred to O. In general the equation is

$$\frac{x}{a/h}+\frac{y}{k/b} = 1$$

or
$$\frac{hx}{a}+\frac{ky}{b} = 1 \tag{4.1}$$

where h and k are whole numbers and the intercepts are $a/h, b/k$. This point is now indicated in more detail in Fig. 4.6(b). Equation (4.1) is true for oblique or rectangular axes. Thus

$$\frac{hx}{a} = \frac{x}{a/h} = \frac{QP}{QR}, \qquad \frac{ky}{b} = \frac{y}{b/k} = \frac{PR}{QR}$$

by similar triangles, and also $QP+PR=QR$. This is true for *any* point such as P'. When however OP is perpendicular to the plane the distance d is of particular interest because it is the perpendicular distance between two successive parallel (h, k) lines (planes) as in Fig. 4.6(a). d is found from formulae derived as follows:

QS is perpendicular to OR and $=(b/k) \sin \gamma$.
    In triangle QRS,

$$QS = QR \sin QRS$$

    In triangle OPR,

$$d = \frac{a}{h} \sin QRS$$

Therefore

$$QR = \frac{QS}{\sin QRS} = \frac{1}{-} \left(\frac{a}{h}\frac{b}{k}\right) \sin \gamma \tag{4.2}$$

But the cosine formula for triangle QOR gives

$$QR^2 = \left(\frac{a}{h}\right)^2 + \left(\frac{b}{k}\right)^2 - 2\left(\frac{a}{h}\frac{b}{k}\right)\cos\gamma \tag{4.3}$$

By squaring (4.2) and equating to (4.3),

$$\frac{1}{d^2} = \left(\frac{h}{a\sin\gamma}\right)^2 + \left(\frac{k}{b\sin\gamma}\right)^2 - \frac{2hk\cos\gamma}{ab\sin^2\gamma} \tag{4.4}$$

When $\gamma = 90°$ the formula for rectangular axes (Fig. 4.6(c)) is obtained, i.e.

$$\frac{1}{d^2} = \frac{h^2}{a^2} + \frac{k^2}{b^2} \tag{4.5}$$

(A simple direct proof by similar triangles should be found.)
    Further relationships are:

(a) For rectangular axes,

$$\cos P'OX = L$$
$$\cos P'OY = \sin P'OX = M$$

Therefore

$$L^2 + M^2 = 1 \tag{4.6}$$

L and M are called direction cosines.
(b) For OP perpendicular to the plane

$$L = \frac{dh}{a}, \qquad M = \frac{dk}{b}$$

For rectangular axes the equation to the line OP itself is

$$\frac{ax}{h} = \frac{by}{k} \tag{4.7}$$

The integers h and k with l added for the third dimension are called *Miller Indices*.

4.4.1.2 In three dimensions the planar intercepts are thus:

$$\frac{a}{h}, \frac{b}{k}, \frac{c}{l} \text{ as in Fig. 4.7(a)}$$

The equation to a plane is

$$\frac{hx}{a} + \frac{ky}{b} + \frac{lz}{c} = 1 \tag{4.8}$$

or in *fractional co-ordinates*

$$hx + ky + lz = 1 \tag{4.9}$$

121

Other parallel planes which would pass through lattice points (in other cells) are indicated by 2, 3, ... in Fig. 4.7. Their equations would be

$$hx + ky + lz = 2, 3, 4, \dots$$

The number of such planes in one set within one cell is $|h| + |k| + |l|$—including one plane through a corner. In Fig. 4.7(a) the case is illustrated for $h = 3$,

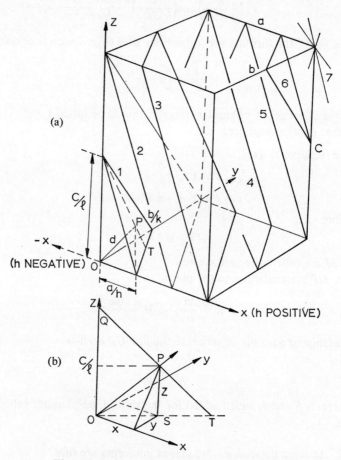

Fig. 4.7 Lattice planes in three dimensions.

$k = 2$, $l = 2$, sum $= 7$. Equations (4.8) or (4.9) are true for oblique axes, and it is also evident that the perpendicular distance OP or d is equal to the distance between any adjacent pair of planes in this direction. As in equation (4.4) this distance is a function of the cell edges and angles. For the triclinic system the derivation is complicated and the rhombohedral case is best derived from it. Otherwise the monoclinic, hexagonal and remaining cases can be dealt with as follows:

The plane containing x and y (angle $xOy = \gamma$) may be regarded as a two-dimensional lattice for which there is a distance d' given by equation (4.4) above as in the two-dimensional case. This distance is OT in Fig. 4.7. The plane OQT is analogous to the condition in Fig. 4.6(c) with a/h replaced by d' and b/k by c/l respectively so that by analogy with equation (4.5)

$$\frac{1}{d^2} = \frac{1}{d'^2} + \left(\frac{l}{c}\right)^2$$

The interplanar spacing formula for the general monoclinic case with z perpendicular to x and y is therefore given by:

$$\frac{1}{d^2} = \frac{h^2}{a^2 \sin^2 \gamma} + \frac{k^2}{b^2 \sin^2 \gamma} + \frac{l^2}{c^2} - \frac{2hk \cos \gamma}{ab \sin^2 \gamma} \tag{4.10}$$

This formula is usually given with the $b(y)$ axis perpendicular to the other two. This interchanges b and c, k and l, γ and β. For hexagonal crystals $\gamma = 120°$ in (10) and so $\sin \gamma = (\sqrt{3}/2)$ and $\cos \gamma = -1/2$. Therefore

$$\frac{1}{d^2} = \frac{4}{3a^2} (h^2 + hk + k^2) + \frac{l^2}{c^2} \tag{4.11}$$

For the orthorhombic case, cf. (5) with $\gamma = 90$

$$\frac{1}{d^2} = \frac{h^2}{a^2} + \frac{k^2}{b^2} + \frac{l^2}{c^2} \tag{4.12}$$

For the tetragonal case $(a = b)$

$$\frac{1}{d^2} = \frac{h^2 + k^2}{a^2} + \frac{l^2}{c^2} \tag{4.13}$$

And for the cubic case $(a = b = c)$

$$\frac{1}{d^2} = \frac{h^2 + k^2 + l^2}{a^2} \tag{4.14}$$

or

$$d = \frac{a}{\sqrt{(h^2 + k^2 + l^2)}} \tag{4.15}$$

In all of these expressions, the integers h, k, l can be positive or negative and this point requires consideration later.

4.4.2 Zone Axes and Symbols

4.4.2.1 Also from Fig. 4.7, for OP perpendicular to the plane, we have, by analogy with (4.6) and (4.7) for rectangular axes:

$$\frac{ax}{h} = \frac{by}{k} = \frac{cz}{l} \tag{4.16}$$

and

$$L^2 + M^2 + N^2 = 1 \tag{4.17}$$

123

(Sometimes small letters l, m, n are used but there need be no confusion, e.g. with the l of hkl.) Since $L = dh/a$, etc. (4.17) leads also directly to equation (4.12) and so to (4.13), (4.14) and (4.15). We now require the more general equation to any straight line through O—(4.16) being a special case. (The significance of (4.16) will be considered in section 4.5.) If this line passes through $(x'y'z')$ its equation is

$$\frac{x}{x'} = \frac{y}{y'} = \frac{z}{z'}$$

Supposing that $(x'y'z')$ is an actual lattice point, then $x' = ua$, $y' = vb$, $z' = wc$, where u, v, w, must be whole numbers. The equation is then

$$\frac{x}{ua} = \frac{y}{vb} = \frac{z}{wc} \qquad (4.18)$$

Comparison with (4.16) shows that for the cubic case with $a = b = c$ the line given by $[uvw]$ is perpendicular to the plane given by (hkl) if $h = u$, $k = v$, $l = w$. In other systems this can only be true for special cases (e.g. tetragonal with $h = u$, $k = v$, $l = w = 0$).

(*Note:* Use of square brackets for lines and parentheses for planes.)

The line (4.18) passes through (000). The planes given by equations (4.8) or (4.9) do not unless 1 is replaced by 0 on the R.H.S. Two such planes have indices $(h_1k_1l_1)$ and $(h_2k_2l_2)$ and equations such as

$$\frac{h_1x}{a} + \frac{k_1y}{b} + \frac{l_1z}{c} = 0$$

Then the line of intersection can be found by elimination, e.g. multiply the $(h_1k_1l_1)$ equation by h_2 and multiply the $(h_2k_2l_2)$ equation by h_1 and subtract:

$$\frac{y}{b}(k_1h_2 - h_1k_2) + \frac{z}{c}(l_1h_2 - h_1l_2) = 0$$

Therefore

$$\frac{y}{b(h_2l_1 - l_2h_1)} = \frac{z}{c(h_1k_2 - h_2k_1)} = \text{(similarly)} \ \frac{x}{a(k_1l_2 - k_2l_1)}$$

Since the numbers in parentheses are integers they can be equated to the indices for a lattice line as (4.18); that is

$$u = (k_1l_2 - k_2k_1), \qquad v = (h_2l_1 - l_2h_1), \qquad w = (h_1k_2 - h_2k_1)$$

which is sometimes written

$$\frac{u}{\begin{vmatrix} k_1 & l_1 \\ k_2 & l_2 \end{vmatrix}} = \frac{v}{\begin{vmatrix} l_1 & h_1 \\ l_2 & h_2 \end{vmatrix}} = \frac{w}{\begin{vmatrix} h_1 & k_1 \\ h_2 & k_2 \end{vmatrix}} \qquad (4.19)$$

$[uvw]$ is the *zone* containing the two planes. Equation (4.19) can be remembered by the cross-multiplication system:

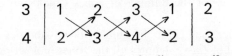

e.g. to find the zone containing (312) and (423)

i.e.

$$u = (3 \times 1 - 2 \times 2), \quad v = (4 \times 2 - 3 \times 3), \quad w = (3 \times 2 - 4 \times 1)$$

$$[uvw] = [-1, -1, 2] \quad \text{usually written} \quad [\bar{1}, \bar{1}, 2]$$

If we equate (4.18) to r then

$$x = rua, \quad y = rvb, \quad z = rwc$$

and any plane which contains these co-ordinates, i.e. the line given by (4.18) is obtained by substitution in

$$\frac{hx}{a} + \frac{ky}{b} + \frac{lz}{c} = 0$$

i.e.

$$hu + kv + lw = 0 \tag{4.20}$$

This is a very important and useful relationship and like (4.19) is independent of a, b, c or the angles between them. Equation (4.20) gives the indices for all possible planes (hkl) contained in a given zone $[uvw]$ or conversely all the possible zone lines contained in a given plane. If two planes are contained in a zone, e.g. $(h_1k_1l_1)$, $(h_2k_2l_2)$ then by addition

$$(h_1 + h_2)u + (k_1 + k_2)v + (l_1 + l_2)w = 0$$

Hence the plane obtained by adding the respective indices is also in the zone or any planes whose indices are given by

$$h = mh_1 + nh_2, \quad k = mk_1 + nk_2, \quad l = ml_1 + nl_2$$

where m and n are integers ($+$ or $-$).

For this reason a large number of lattice planes intersecting along the line $[uvw]$ obey the relation (4.20) and are said to form or lie in the zone $[uvw]$. It also follows that the plane containing any two zones is given by an analogous relation to (4.19) and that any number of zone indices can be fitted to equation (4.20).

4.4.3 Some Other Relationships

4.4.3.1 It will be obvious that the indices of planes or zones can have positive or negative values. An examination of equations (4.14) or (4.15) for

125

the cubic case shows that *hkl* can be interchanged and this makes no difference, but if two or three are equal this reduces the number of permutations. With the $\pm$ change we can have $\pm h$, $\pm k$, $\pm l$, except when an index$=0$. For $h \neq k \neq l \neq 0$ the number of arrangements is $3! = 6$ and the possible sign changes give $2 \times 2 \times 2 = 8$ so the total permutation is 48. If $h = k$ the number is 24, and if $h = k$ with $l = 0$ it reduces to 12, or if $k = l = 0$ only 6 possibilities arise. The *multiplicity factor p* thus denotes the number of planes with the same interplanar spacings, or the number of zones which join equivalent lattice points.

For systems other than cubic the number of possibilities is reduced.

(*Note:* The symbol $\langle uvw \rangle$ denotes all the zones in a set represented by the multiplicity p and $\{hkl\}$ denotes all the planes in a set.)

4.4.3.2   The full multiplicity of a set of planes of indices $\{hkl\}$ all different and non-zero represents the maximum number of such planes that could form external faces on a crystal in this particular system. The class of highest point group symmetry (the symmetry of the lattice itself) can show a fully symmetrical array of all these faces. The classes of lower symmetry show a unique and characteristic selection from the total number. Forms based on planes with two or more indices equal and/or one or more zero can however be found in more than one class.

It is evident that the symmetry axes and planes are generally represented by small integral values of *uvw* or *hkl* respectively, e.g. the cubic triad is [111] and the monoclinic plane of symmetry is a (010) plane perpendicular to $b = [010]$.

4.4.3.3   An important relationship is the angle between two zones. For the cubic case it is given by

$$\cos \rho = \frac{u_1 u_2 + v_1 v_2 + w_1 w_2}{\sqrt{(u_1^2 + v_1^2 + w_1^2)}\sqrt{(u_2^2 + v_2^2 + w_2^2)}} \qquad (4.21)$$

when $\rho = 90°$, $u_1 u_2 + v_1 v_2 + w_1 w_2 = 0$, which since, in this case zone and plane axes are interchangeable, is a variant of equation (4.20). The expression for the angle ϕ between two planes is strictly similar to (4.21). For other systems than cubic the formulae are available in the textbooks or International Tables [9], where tables of calculated angles are provided. Practical tables for cubic, tetragonal and hexagonal systems are also given in Ref. [42] of Chapter 6.

4.4.3.4   It is necessary to take careful notice of certain special features of the hexagonal system, represented in Fig. 4.8(a) (Fig. 4.8(b) is referred to later). Here the plane of projection is (001) and the [001] zone axis is perpendicular to it. It is recalled, e.g. from Fig. 4.2, that with the inter-axial angle of 120° a third axis could have been chosen equally well. The intercepts of planes along

126

this third axis can be shown to be $-(h+k)$ if h and k are the intercepts along the other two axes. Hence hexagonal planes are sometimes denoted by $(hkil)$ where $i = -(h+k)$ or by (hk, l) to indicate the uniqueness of l as in (4.11). The interplanar spacing is the same if h or k is replaced by $-(h+k)$ but not if the sign of either is changed. It is useful to confirm this by some chosen examples.

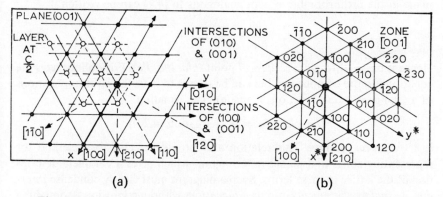

(a)   (b)

Fig. 4.8 A hexagonal lattice showing certain planes and directions (three-index system).

(a) space lattice (N.B. Positions of atoms for c–p hexagonal structure)
(b) reciprocal lattice (N.B. Directions of x^*, y^*, relative to x, y, in (a))

There is a four-index system for zones but it is easier to use the three-index system as in Fig. 4.8(a) and to use the three-index plane system to find the common zone axis by cross multiplication. The positions of atoms in the close-packed hexagonal structure (section 4.3.2) is indicated by the second layer.

4.4.3.5 It must be emphasised that all of the relationships defining planes and zones can be applied on the macroscopic scale to the external shapes of crystals and equally on the sub-microscopic scale to the atomic patterns. Hence they should be apparent or at least implicated in any effects observable on the microscopic scale in between.

4.5 THE RECIPROCAL LATTICE AND PROJECTION METHODS

4.5.1 Introductory

In this concluding section consideration will be given to some concepts which are useful in subsequent applications of crystallography. The expressions for the interplanar spacings in terms of lattice parameters and axial angles are more conveniently represented by equations such as (4.10)–(4.14), except in

127

the cubic system where equation (4.15) is generally used. In the other systems direct expression of d in terms of lattice parameters and angles is more cumbersome but is sometimes used with axial ratios (e.g. c/a in the hexagonal or tetragonal case). The simpler expressions give $1/d^2$ in terms of $1/a^2$, $1/b^2$, $1/ab$, etc., and suggest that use should be made of these *reciprocal* distances for ease of calculation. Furthermore the equation to the perpendicular to a plane in the orthorhombic system analogous to (4.7) can be written

$$\frac{x}{h/a} = \frac{y}{k/b} = \frac{z}{l/c} \tag{4.22}$$

Hence it passes through a *reciprocal* point $(h/a, k/b, l/c)$. For oblique axes the equation is more complicated but can be shown to have the same implications (if a change in axial directions is made). The equation

$$hu + kv + lw = 0 \quad \text{(equation (4.20))}$$

is a *reciprocal relation*. This relation also provides a clue to a method for representing zones and planes in a diagram which gives a complete description of the lattice in these terms. Such a diagram must clearly condense *three-dimensional* measurements into *two-dimensions* without confusion—something like a geographical projection method. Two such methods will be considered and one developed in some detail.

4.5.2 The Reciprocal Lattice

4.5.2.1   The equations (4.5) and (4.12) for $1/d^2$ in terms of lattice parameters for an orthorhombic lattice in two and three dimensions respectively can be seen to correspond to simple Pythagorean expressions for right-angled triangles. The two-dimensional case is represented in Fig. 4.9(a) where a new lattice of points is drawn with axial lengths $1/a$ along OX^* and $1/b$ along OY^*. If any point P^* on this lattice is chosen then

$$OP' = \left(\frac{h}{a}\right)^2 + \left(\frac{k}{b}\right)^2 = \frac{1}{d^2}$$

or
$$(ha^*)^2 + (kb^*)^2 = (d^*)^2 \tag{4.23}$$

This is the *reciprocal lattice* corresponding to the real lattice with cell edges a and b respectively, in which the distance $OP = d$. In this diagram the two lattices are drawn on opposite sides of the origin in order to clarify the relation. It is easy to show by similar triangles that POP' is one straight line. The three-dimensional case is strictly analogous. The tetragonal and cubic systems follow by making $a = b$ and $a = b = c$ respectively.

The derivation of the reciprocal lattice for oblique axes is more complicated. In Fig. 4.9(b) a two-dimensional oblique lattice is shown. If it is made a condition that $OP = d$ and $OP^* = d^* = 1/d$ should lie on a straight line, then we have for $k = 0$, the $(h, 0)$ planes; parallel to OY, and for $(1, 0)$ the

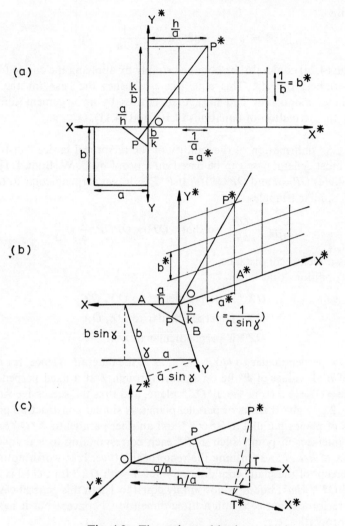

Fig. 4.9 The reciprocal lattice.
(a) two-dimensional lattice $\gamma = 90°$ (illustrated for $h=2$, $k=3$)
(b) two-dimensional lattice $\gamma \neq 90°$ (illustrated for $h=2$, $k=3$)
(c) three-dimensional lattice

separation is $a \sin \gamma$. Thus OX^* is perpendicular to OY and OY^* to OX.
The twin expressions for OP or OP^* become:

$$\frac{1}{d^2} = \left(\frac{h}{a \sin \gamma}\right)^2 + \left(\frac{k}{b \sin \gamma}\right)^2 - \frac{2hk \cos \gamma}{ab \sin^2 \gamma} \qquad \text{(equation (4.4))}$$

or

$$\frac{1}{d^2} = (d^*)^2 = (ha^*)^2 + (kb^*)^2 + 2(ha^*)(kb^*) \cos \gamma^* \qquad (4.24)$$

129

by setting

$$\gamma^* = \pi - \gamma \quad \text{and} \quad a^* = \frac{1}{a \sin \gamma}, \quad \text{etc.}$$

Equation (4.24) could also be obtained simply by applying the cosine formula to the triangle OP^*A^*. This equation establishes the case for the three-dimensional monoclinic and hexagonal lattices by an argument similar to that for the derivation of equations (4.10) and (4.11).

4.5.2.2   A confirmation of the validity of the reciprocal lattice construction for the most general case can be based on a proof by C. W. Bunn [11]. Thus in Fig. 4.9(c) $OP = d$ and $OP^* = 1/d$. If P^*T is drawn perpendicular to OX, we have by similar triangles

$$\frac{OT}{OP^*} = \frac{OP}{a/h} \quad \text{so that} \quad OT = OP OP^* \frac{h}{a} = \frac{h}{a}$$

which is a reciprocal distance.

The conditions that

$$OX^* \text{ is perpendicular to } OY, OZ$$
$$OY^* \text{ is perpendicular to } OZ, OX$$
$$OZ^* \text{ is perpendicular to } OX, OY$$

make OX perpendicular to OY^*, OZ^*, as in the diagram. Hence, for a given value of h all values of P^* lie on a plane through T at a fixed perpendicular distance $= OT = h/a$ from the Y^*OZ^* plane, and thus the successive values of $h = 0, 1, 2, \ldots$ give a series of parallel planes. A similar construction provides a series of planes parallel to Z^*OX^* and another parallel to X^*OY^*. These planes intersect in points such as P^* each corresponding to a unique combination of h, k, l and forming the reciprocal lattice. It is worth noting that the intercept of the plane containing P^* and T with OX^* (not OX) is a point T^* and $OT^* = ha^*$, but a^* is not simply equal to $1/a$ in this general case.

The reciprocal lattice is still a three-dimensional concept but it has many useful properties, e.g.

1. It gives expressions for interplanar distances which are more easily calculable.
2. It is of considerable value in connection with the interpretation of diffraction patterns. (Later chapters.)
3. In the crystal lattice a zone $[uvw]$ will lie along the intersections of all planes which obey the relation

$$hu + kv + lw = 0 \quad \text{(equation (4.20))}$$

The perpendiculars to these planes radiate from it like radii to a circle. These directions therefore all define an array of points in one plane in the reciprocal lattice. Hence a zone $[uvw]$ is represented in the reciprocal lattice by this

130

plane—just as (*hkl*) now represents a single point. (In Fig. 4.9(c) the zone [100] is in the direction OX and the corresponding reciprocal plane is Y^*OZ^*.)

It is now useful to consider Fig. 4.8(b)—the reciprocal lattice for a hexagonal crystal based on the space lattice indicated in Fig. 4.8(a). The a^* and b^* axes are at right angles to b and a respectively so the angle between them is $60°$ (and not $120°$). Reciprocal indexing follows as indicated and the position of certain zone axes is indicated. Corresponding diagrams should be drawn for projections of the monoclinic, orthorhombic, tetragonal and cubic systems.

4.5.3 The Stereographic and Gnomic Projections

4.5.3.1   The location of perpendiculars to planes in a zone leads naturally to the methods of projection mentioned—in particular the stereographic projection which has several applications of metallurgical interest. Thus we consider a single zone [*uvw*] and its set of plane normals or their *poles*—where the normals intersect a sphere—corresponding to different values of (*hkl*) but all necessarily lying in a plane perpendicular to the zone axis. According to Fig. 4.9 these reciprocal directions (OP^*) are longer for higher values of (*hkl*) but, *except where there is a common factor*, the directions are separate and unique. Hence we only need to draw them as radii to a circle (like a wheel) and their positions round the axis uniquely define their directions *and indices*. The crystal can be regarded as defined by any number of these zones set up at various angles to each other about the centre of a sphere of which they are *great circles*. This situation is indicated in Fig. 4.10(a). In this (perspective) diagram rectangular co-ordinate axes are chosen and one zone is shown tilted at an angle to OY but containing OX. One normal to a plane in this zone is ON lying in the YOZ plane and another is OM. The two methods of representing the distribution of planes and zones arise as follows:

(a) Stereographic projection: N (the pole) is joined to P and the point N_P is chosen to represent it on a circle in the XY plane, i.e. an 'equatorial plane'. Points such as M would intersect this plane at points as M_P (x, y) in Fig. 4.10(b) and lie on a curve (see below). Hence all points on the sphere can be projected on to this circle.

(b) Gnomic projection: $X'OY'$ is a plane tangent to the sphere. N is projected to N_G and M to M_G. Since the circle containing M and N lies in a plane through OX the points for this zone must all lie on a straight line N_GM_G parallel to OX'. Hence zones become straight lines which intersect in points corresponding to the appropriate values of h, k and l. This projection is useful for planes whose normals make a small or moderate angle with OZ but an obvious disadvantage is the extension to infinity as these angles approach $90°$.

4.5.3.2 The stereographic projection does not have this disadvantage and there are several relationships which can be derived in connection with Figs. 4.10(a) and (b) and are further illustrated in Fig. 4.11. If these results are considered carefully *they will greatly facilitate the subsequent use of the method.*

The co-ordinates of M are $(x_1y_1z_1)$ say and so by the equation to a sphere

$$x_1{}^2 + y_1{}^2 + z_1{}^2 = r^2 \tag{4.25}$$

The zone axis makes an angle ρ with OZ and the plane normal OM makes an

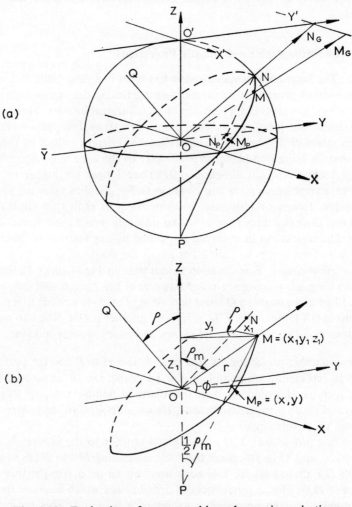

Fig. 4.10 Derivation of stereographic and gnomic projections.
(a) the location of a zone with axis OQ in relation to the plane or projection (XOY)
(b) angles, etc., for locating points of projection (M_P)

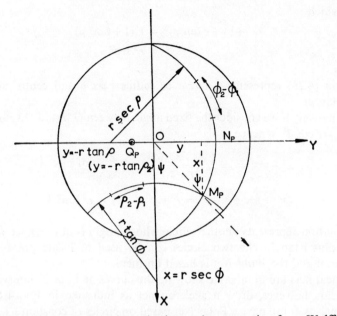

Fig. 4.11   The stereographic projection and construction for a Wulff net.

angle ϕ with OX. The process of projection involves scaling down the coordinate $x_1 y_1$ to xy in the ratio of $r+z_1$ to r. This can be seen by reference to similar triangles. Thus:

$$\frac{x}{x_1} = \frac{y}{y_1} = \frac{\sqrt{(x^2+y^2)}}{\sqrt{(x_1^2+y_1^2)}} = \frac{r}{r+z_1} \qquad (4.26)$$

The same kind of relationship applies to the corresponding squares and we may also make use of a well-known addition or subtraction rule

$$\left(\frac{a}{b} = \frac{c}{d} = \frac{ma \pm nc}{mb \pm nd}, \quad \text{etc.}\right)$$

$$\frac{x^2}{x_1^2} = \frac{y^2}{y_1^2} = \left(\frac{r}{r+z_1}\right)^2 = \frac{x^2+y^2+r^2}{(x_1^2+y_1^2+z_1^2)+r^2+2rz_1}$$

$$= \frac{x^2+y^2+r^2}{2r^2+2rz_1} \qquad (4.27)$$

Reference to Fig. 4.10(b) shows that $z_1 = y_1 \tan \rho$. Hence from (4.26)

$$\frac{r}{r+z_1} = \frac{y}{y_1} = \frac{y \tan \rho}{y_1 \tan \rho} = \frac{y \tan \rho}{z_1} = \frac{r-y \tan \rho}{r}$$

and from (4.27)

$$x^2+y^2+r^2 = 2r^2 \left(\frac{r}{r+z_1}\right) = 2r^2 \left(\frac{r-y \tan \rho}{r}\right)$$

$$= 2r^2 - 2ry \tan \rho$$

From which

$$x^2 + (y + r \tan \rho)^2 = r^2(1 + \tan^2 \rho)$$
$$= r^2 \sec^2 \rho \qquad (4.28)$$

Equation (4.28) represents a *circle* of radius $r \sec \rho$ and centre at ($x=0$, $y = -r \tan \rho$).

Alternatively if we consider the fixed angle between OM and OX, viz. ϕ, we have $x_1 = r \cos \phi$. Whence

$$x^2 + y^2 + r^2 = 2r^2\left(\frac{r}{r+z_1}\right) = 2r^2 \frac{x}{x_1} = 2rx \sec \phi$$

so $\qquad (x - r \sec \phi)^2 + y^2 = r^2(\sec^2 \phi - 1) = r^2 \tan^2 \phi \qquad (4.29)$

This equation represents another circle whose origin is at ($x = r \sec \phi$, $y=0$), with radius $r \tan \phi$. The two circles can be used to locate points on the stereogram and the *Wulff Net* is based on them.

Practical nets are obtainable showing the curves at 1° or 2° intervals. The reader can, however, draw a skeleton net as indicated in Fig. 4.11. It is sometimes useful to have a Polar Net based on circles of constant ρ (ρ or ρ_m). These have a radius equal to $OM_P = r \tan \frac{1}{2}\rho$. (The angle subtended at P is half that at the centre.) Also from Fig. 4.10 let ρ_m = angle between OZ and OM, then

$$x_1 = r \cos \phi$$
$$y_1 = z_1 \cot \rho = r \cos \rho_m \cot \rho$$

but also $\qquad y_1 = r \sin \phi \cos \rho$

so that $\qquad \cos \rho_m = \sin \phi \sin \rho$

Also in the polar projection, M_P could be located by the angle $M_P OX$ or ψ (say) where

$$\tan \psi = \frac{y}{x} = \frac{y'}{x'} = \tan \phi \cos \rho$$

4.5.3.3 The Wulff net has many convenient properties and uses and is more generally useful than the polar net.

1. M_P lies on a great circle 90 (stereographic) degrees away from Q_P the projection of Q where OQ is the zone axis.
2. The angle between two planes is obtained by measuring the angle $\phi_2 - \phi_1$ along this circle—see Fig. 4.11.
3. If the crystal is rotated about OX (constant ϕ) then M_P moves along the corresponding circle through the angular distance $\rho_2 - \rho_1$.
4. It follows that the great circle projections can be used for rotations about

an axis anywhere in the diagram and are constant ρ (zone axis) curves. The constant ϕ curves can be used for rotations about any axis in the plane of the diagram (ρ_m—plane—varies along these curves).

5. Just as a great circle can be used to represent a zone when its radii are the normals to the planes in the zone, so a great circle can equally well be used to represent a plane and the radii in it are then zone axes. In either case equation (4.20) is obeyed.

6. Such a diagram can be used to show the poles of the principal planes of a crystal and the circles corresponding to principal zones. It is convenient in the cubic case to use the plane (001) as shown in Fig. 4.12(a) for a standard projection. Other planes and zones can be found by interpolation using calculations of angles or the zone law (4.20).

7. In these diagrams the projections of certain spherical triangles can be picked out and these are found to be repeated by the operation of the symmetry elements. In Fig. 4.12(a) for example the corner of the standard triangle could be the projections or poles of (001), (111) and (110). The rest of the projection could then be derived by repeating this triangle by rotating or reflecting it according to the symmetry elements. It is for example easy to see that the four-fold axis of symmetry at the centre causes the pattern to repeat after a 90° rotation, whilst certain planes normal to the diagram reflect one half into the other. It is useful to confirm that a (111) pole is the same number of degrees (54·74°) measured on the stereographic scale from each of the cube axes and to check this and other angles from (4.21). In Fig. 4.12(b) the six-fold axis is normal to, and mirror planes or two-fold axes in, the plane of the projection.

8. It follows that a stereogram could also be used to represent point group symmetry. The symmetry elements placed in suitable positions repeat a point or pole accordingly. Thus a three-, four- or six-fold axis is conveniently placed at the centre. Mirror planes may reflect one half of a diagram into the other or lie in the plane of the diagram. Figs. 4.12(a) and (b) can be studied from this point of view. The most general point (*hkl*) is represented by different non-zero indices and it is very illuminating to find all the locations corresponding to the multiplicity of the set {*hkl*}, or {*hkil*} in Fig. 4.12(b).

9. The transition from two to three dimensions and back can be considerably assisted by studying a terrestrial globe and maps based on circular projections. Some of these are equal area projections but the stereographic projection is not. This will be appreciated by reference to Fig. 4.12 where the triangles referred to correspond to the same angular separations but have obviously different areas on the same projection.

10. There are a number of uses in connection with diffraction or the microscopical study of metals which will be referred to in later chapters (e.g. the use of 'pole figures', Chapter 7).

4.5.3.4 In view of its importance the significance of equation (4.20) and its relation to planes and zone axes is summarised in Table 4.2. This also brings out the way in which the stereographic projection connects with crystal space on the one hand and reciprocal space on the other.

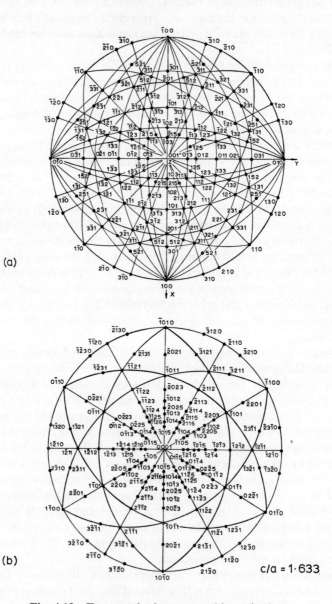

(a)

(b)

c/a = 1·633

Fig. 4.12 Two standard stereographic projections.

(a) cubic crystal on (001)
(b) hexagonal crystal on (0001)

TABLE 4.2   *Summary of Relationships for the Equation $hu + kv + lw = 0$*

Symbols or relationship	Crystal space	Stereographic projection	Reciprocal space
(hkl)	Plane	$\leftrightarrow$ Great circle *or* Point = projection of its pole	$\leftrightarrow$ Lattice point—or, more exactly, the vector joining the origin to this point
$[uvw]$	Zone axis	$\leftrightarrow$ A pole projected *or* Great circle	$\leftrightarrow$ Plane
$hu + kv + lw = 0$			
(a) $[uvw]$ constant	Equation defines planes in the zone	$\leftrightarrow$ Planes represented by points on the great circle for the zone	$\leftrightarrow$ Equation defines lattice points which lie in the reciprocal plane
(b) (hkl) constant	Equation defines zone axes contained in the plane	$\leftrightarrow$ Points representing these zones lie on the great circle for the *plane*	$\leftrightarrow$ Equation defines the planes which contain the reciprocal vector

References

1. PHILLIPS, F. C., *An Introduction to Crystallography*, 3rd edition. Longmans, London 1963.
2. TAYLOR, A., *X-ray Metallography*. Wiley, London and New York, 1961.
3. BARRETT, C. S. and MASSALSKI, T. B., *Structure of Metals*, 3rd edition. McGraw-Hill, Toronto, New York and London, 1953.
4. DE JONG, W. F., *General Crystallography. A Brief Compendium*. W. H. Freeman and Co., San Francisco, 1959.
5. HENRY, N. F. M. and LONSDALE, K. (eds.), *International Tables for X-Ray Crystallography*, vol. I, *Symmetry Groups*. Kynoch Press, Birmingham, England, for the International Union of Crystallography, 1952.
6. PEARSON, W. B., *A Handbook of Lattice Spacings and Structures of Metals and Alloys*. Pergamon Press, London, 1958 and 1967.
7. SMITHELLS, C. (ed.), *Metals Reference Book*. Butterworth, London, 1967.
8. A.S.T.M., E. 157–61T, 1961.
9. KASPER, J. S. and LONSDALE, K. (eds.), *International Tables for X-Ray Crystallography*, vol. II, *Mathematical Tables*. Kynoch Press, Birmingham, England, for the International Union of Crystallography, 1959.
10. ANDREWS, K. W., DYSON, D. J. and KEOWN, S. R., *Interpretation of Electron Diffraction Patterns*, 2nd edition. Hilger, London, 1971.
11. BUNN, C. W., *Chemical Crystallography*, 2nd edition, appendix 4. Clarendon Press, Oxford, 1961.

Chapter 5

FUNDAMENTAL BASIS OF DIFFRACTION

5.1 GENERAL PRINCIPLES—LAÜE CONDITIONS, BRAGG'S LAW, ETC.

5.1.1 Introduction

In the previous chapter the external and internal symmetries of crystals were seen to be inter-connected, the former being a consequence of the latter. The discovery of X-ray diffraction in 1912 made possible for the first time the study of the internal arrangements of atoms and the full determination of crystal structure. Many metallurgically important structures have since been determined and although metallurgists engaged in structural research may still find it necessary to use X-ray diffraction methods for this fundamental purpose, it is more usual for the structure to be taken for granted. For this reason the main metallurgical applications of diffraction methods come into the second and third stages indicated in Chapter 1, and the present chapter falls into place after the general description of crystallographic principles and elementary facts about structure and its geometry.

In addition to the use of X-rays it is also possible to investigate or interpret the structure of materials with the aid of electrons or neutrons which, although they are particles, also behave as waves and so give similar diffraction effects to X-rays. For this reason it is desirable to consider the diffraction of waves by crystals without at first specifying what kind of waves are involved. It is then possible to recognise the difference between the three kinds of waves and the extent to which they can provide the same or different information.

It should be pointed out here that electron diffraction techniques were first applied in the metallurgical field to thin films on oblique surfaces. In recent years many new metallurgical uses have been developed for transmission *through* thin films. This is due to the wide accessibility and use of instruments which are primarily designed as electron microscopes. Although it has been necessary to separate the two applications here (as in Chapters 6 (this Vol.) and (2) of Vol. 2), advantage should be taken whenever possible of the dual function of such instruments. Neutron diffraction is not generally available and requires access to a nuclear reactor. Its potentialities for the fundamental study of metallurgical materials will however be briefly indicated.

(General Refs. [2] and [3] of Chapter 4; [1], [2], [3] (and others) of this

138

chapter for X-ray diffraction, [4] of this chapter, [41] and [42] of Chapter 6 for electron, and [5], [6] of this chapter for neutron diffraction.)

5.1.2 The Laüe Conditions and Bragg's Law

5.1.2.1 In Fig. 5.1(a) a row of equally spaced atoms is indicated by the small circles with a separation distance of *a*. A plane wave is supposed to approach this row of atoms so that the crests and troughs of the wave are parallel to the line of atoms. Such a wave in two dimensions would have the section shown in the top right-hand corner and is similar to a surface wave on water. In fact it is possible to demonstrate diffraction by such waves (using a row of

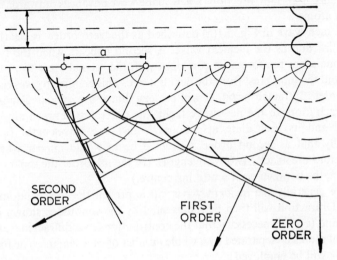

(a)

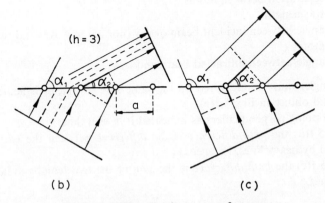

(b)   (c)

Fig. 5.1 Diffraction by a row of atoms.

(a) showing how the scattered wavelets reinforce each other in certain directions

(b) and (c) constructions for determining the relationships for diffraction obstacles with small equally spaced gaps). The wave itself might be represented by a sine or cosine function. When it reaches the row of atoms each atom becomes the centre of a new spherical wave or 'wavelet' which progresses outwards in all directions—these waves are shown on one side only to avoid confusion. These small spherical wavelets undergo destructive interference which is complete where the trough of one coincides with the crest of another. They do, however, reinforce each other where crest and crest exactly coincide. As the waves spread out their radii rapidly increase and such crests become progressively less curved so that they could soon be replaced by a new plane wave represented by a tangent to the spherical wavelets. Hence in certain directions a new plane wave advances outwards from the row of atoms.

One such wave in Fig. 5.1(a) described as the 'zero order' is clearly in the same direction as the incident wave. A second appears when the first crest from one atom reinforces or combines with the second crest from the next atom and the third crest from the next but one. This condition gives rise to the 'first order' diffracted beam. It will be seen that if a perpendicular is drawn from each atom to the new wave front, the difference between any perpendicular and its immediate neighbours is just one wavelength. It follows naturally that a 'second order' diffracted beam would appear when these separations are exactly two wavelengths. (In the nth order the separation is n wavelengths between each scattering centre.)

More generally the incident wave will approach the row of atoms at an angle. Diffraction will then be represented by the conditions shown in Figs. 5.1(b) and (c). It is necessary that the contributions to a diffracted wave from adjacent atoms be separated by a whole number of wavelengths. The following symbols will be employed:

a = separation between atoms
λ = wavelength
α_1 = angle between incident beam or ray (normal to wave) and the row of atoms
α_2 = angle between diffracted beam and row

These angles must be measured on the same side so that α_1 is acute in Fig. 5.1(b) and obtuse in Fig. 5.1(c).

By dropping perpendiculars as indicated it is seen that:
In Fig. 5.1(b) the *path difference* is the *difference* between the two lengths indicated by heavy lines, i.e. $a \cos \alpha_2 - a \cos \alpha_1$.
In Fig. 5.1(c) the *path difference* is the *sum* of the two lengths indicated by heavy lines, i.e.

$$a \cos \alpha_2 + a \cos (180° - \alpha_1) = a \cos \alpha_2 - a \cos \alpha_1$$

Hence in either case:

$$a(\cos \alpha_2 - \cos \alpha_1) = h\lambda \tag{5.1}$$

where h is a whole number. The choice of this symbol is appropriate for reasons which will appear.

It is to be noted that:

1. Equation (5.1) would be true for beams *in any directions* which make the angles α_1 or α_2 with the row of atoms (a 'one-dimensional lattice'). Although the incident beam will have a fixed direction, the diffracted beam will be represented by a *cone* of semi-angle α_2. The diagrams in Fig. 5.1 represent a section in one plane containing the incident direction. (Fig. 5.2 further illustrates this point.)

2. If the wave separation or path difference for a pair of adjacent atoms is h, then the separation for alternate atom pairs separated by $2a$ is $2h$ as in Fig. 5.1(c). On the other hand, suppose the distance a between a pair of atoms is divided into h equal parts, then by simple geometry an atom placed at any of these points would scatter *in phase*, i.e. a whole number of wavelengths from the atoms at the principal (lattice) points. Such points correspond to path differences of $1, 2, 3, \ldots, (h-1)$ wavelengths. This condition is illustrated in Fig. 5.1(b) for $h=3$ where the directions of normals to the wave fronts for these points are represented by broken lines.

3. The *scattering* has so far been assumed to radiate from the atoms represented by points. These might, for example, be the nuclei. In practice the scattering for the three different types of wave involves interaction with the nuclei and/or with the surrounding electrons in different ways. This affects some of the information that may be obtained (as will be seen). The assumption of point atoms does not, however, affect the derivation of many of the basic relationships such as (5.1) and others derived below.

5.1.2.2 The diffraction cones from the row of atoms would appear as in Fig. 5.2(a). If a photographic film or plate is placed parallel to the row then the intersections of the cones with this plate will give a series of lines which are hyperbolae (Fig. 5.2(b)). If now atoms are considered on a two-dimensional lattice (not necessarily rectangular), it is easy to appreciate that one such row at right-angles to the first would give a second series of cones. In reality the net is made up of many rows but their separation is so small on the atomic scale that the beams of diffracted rays are still 'narrow' on the experimental scale. For rows of atoms in this second direction the condition for diffraction analogous to (5.1) is

$$b(\cos \beta_2 - \cos \beta_1) = k\lambda \qquad (5.2)$$

For diffraction to occur from the net it is necessary that both (5.1) and (5.2) be satisfied *simultaneously*. Thus diffracted rays will be restricted only to those directions along the *intersections of cones*. These are straight lines which, on recording the pattern in a plane parallel to both rows, would intersect this plane in points corresponding to the intersections of two sets of hyperbolae, as in Fig. 5.2(c).

141

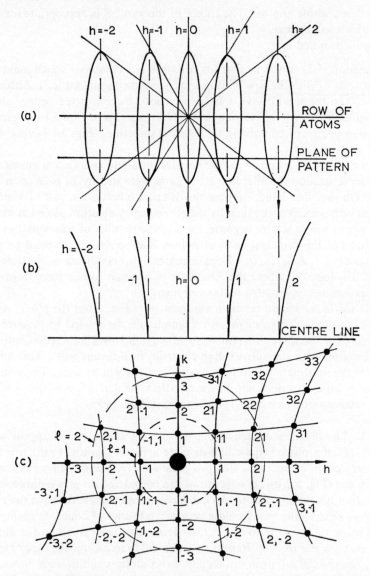

Fig. 5.2 Diffraction principle extended from one to two (and three) dimensions.

(a) diffraction cones
(b) (half) diffraction pattern from row
(c) diffraction pattern from two (and three) dimensional lattice

If a third set of rows of atoms is now added at an angle to the plane of the other two a three-dimensional array is obtained. The cones in this case would intersect the plane of the diffraction pattern in ellipses, or circles if the

142

direction is exactly perpendicular (broken circles in Fig. 5.2(c)). In general there are three intersections between the three sets of curves and no diffraction pattern is obtained unless the angles of incidence ($\alpha_1, \beta_1, \gamma_1$) and/or λ are altered so that *three* 'Laüe conditions' (after Max von Laüe) are simultaneously fulfilled, viz.

$$a(\cos \alpha_2 - \cos \alpha_1) = h\lambda \quad \text{equation (5.1)}$$
$$b(\cos \beta_2 - \cos \beta_1) = k\lambda \quad \text{equation (5.2)}$$
$$c(\cos \gamma_2 - \cos \gamma_1) = l\lambda \quad \quad \quad \quad \text{(5.3)}$$

The triple condition for diffraction in three-dimensions corresponds to the *simultaneous intersection* of the three sets of diffraction cones which would exist for the separate rows of atoms alone.

5.1.2.3   The significance of equations (5.1), (5.2) and (5.3) can be more fully appreciated by reference to Fig. 5.3 which leads naturally to a further fundamental condition. For this purpose a single wavelength is assumed. This wave is, however, a disturbance in three spatial dimensions. It is easy to envisage a two-dimensional 'plane' wave on the surface of a fluid. The wave motion is the up and down movement of particles on the surface, and the wave advances in straight wave fronts as at the top of Fig. 5.1(a). In the three-dimensional plane wave the wave front is a plane surface at all points of which the disturbance has the same magnitude. Thus, the parallel planes containing successive wave crests are separated by one wavelength. The direction of motion is perpendicular to these planes. In Fig. 5.3(a) such a plane wave approaches a crystal lattice with atoms at the lattice points and gives a diffracted wave in the direction indicated. The three Laüe conditions must be satisfied.

Certain conclusions follow:

1. By analogy with Fig. 5.1(c) this means that a ray path through the corner A is h wavelengths different from a ray path through the corner O. The difference at B is k wavelengths, and that at C, l wavelengths. By the same reasoning used previously in connection with Fig. 5.1(b) (for $h=3$) it follows that atoms placed at points A', a/h from O along OA, B', b/k from O along OB, and C', c/l from O along OC, would give path lengths exactly one wavelength different from the path through O.

2. By simple geometry the path difference from O is exactly one wavelength for *any point* on the *plane* passing through these three points as in Fig. 5.3(b). Since h, k, l are whole numbers all such planes are parallel to (or indistinguishable from) the crystallographic planes with the same indices as in Fig. 4.7. (Hence the particular choice of symbols for the equations (5.1)–(5.3).)

3. By accepting (without proof) an optical principle which states that the ray or beam must follow the shortest possible path (more strictly a 'stationary'

path) it is also easy to establish that the paths from a wave front on the incident beam to a wave front on the diffracted beam must:

    (a) be contained in planes perpendicular to the plane $A'B'C'$ or (hkl),
    (b) follow paths which make equal angles with the plane $A'B'C'$ or (hkl).

4. This last condition is identical with that required for *reflection* in a mirror

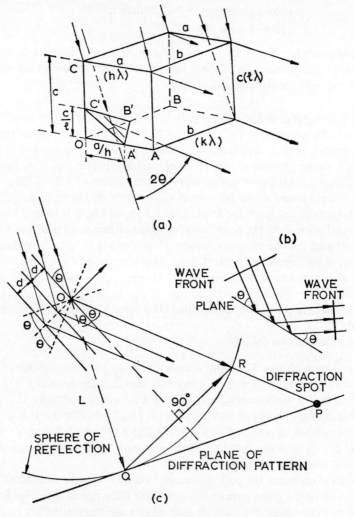

Fig. 5.3 Diffraction by space lattice.
  (a) simultaneous fulfilment of three Laüe conditions (path differences of $h\lambda$, $k\lambda$, $l\lambda$)
  (b) equal paths
  (c) construction for Bragg law derived from (a) showing reciprocal relations

and is represented by Fig. 5.3(c). Here the plane through O is indicated together with other equally spaced planes separated by the appropriate interplanar spacing d as defined in the previous chapter. The path difference between any pair of planes is obtained by dropping perpendiculars from a point such as O onto the ray paths for the next plane. The extra distance along the path through O in Fig. 5.3(c) is thus equal to $2d \sin \theta$ and is one wavelength, θ being the angle between the direction of the wave and the plane. The same result would apply if the difference was two or more wavelengths. In general

$$n\lambda = 2d \sin \theta \qquad (5.4)$$

where n is an integer. This is the *Bragg Equation* or *Law* and thus contains the three Laüe conditions. From equations (4.10)–(4.14) (or the general triclinic equation) d is a function of squares or products of h, k, l. Hence if the indices have a common factor this can be divided into d. Consequently there is no distinction in equation (5.4) between the nth order diffraction from an (hkl) plane or the first order from (nh, nk, nl). The latter has a value of $d = 1/n$ of that for the former.

It is therefore usual to employ the planar indices without a common factor to describe the planes for which the diffraction conditions are satisfied. Indices without brackets and with common factors if necessary are then used to describe the individual diffracted beams or *reflections*. The latter term is justified by the mirror-like arrangement of the ray path in Fig. 5.3(c).

In much applied diffraction work values of d_{hkl} derived from reflections hkl are generally employed even if a common factor is involved and so n is generally dropped. The equation used is:

$$\lambda = 2d \sin \theta \qquad (5.5)$$

or

$$d = \frac{\lambda}{2} \operatorname{cosec} \theta \qquad (5.6)$$

5.1.2.4   An alternative and significant way of writing equation (5.5) is

$$\frac{1}{d} = \frac{2}{\lambda} \sin \theta \qquad (5.7)$$

This is an expression using the reciprocal spacing as defined in the previous chapter. The relationship is brought out in Fig. 5.3(c) by considering how the individual diffracted ray from the planes would pass along the direction OR and would form a diffraction spot on a flat plate or film at P. We have $OQ = L$ perpendicular to the plane of the pattern containing P. Make $OR = OQ$. Then

$$\left.\begin{array}{l} QR = 2L \sin \theta \\ QP = L \tan 2\theta \end{array}\right\} \qquad (5.8)$$

Let $L = M(1/\lambda)$ where M is a constant (magnification). Then

$$QR = M(2/\lambda) \sin \theta = M(1/d)$$

145

Hence a circle of radius $1/\lambda$ (or $L=M/\lambda$), centre O would cut off a chord QR of length $1/d=d^*$ (or M/d). But QR is perpendicular to the diffracting planes and therefore is not only proportional to the reciprocal lattice vector but is also parallel to it, so that it represents it in both *magnitude and direction*.

Furthermore if θ is small, as it may be in electron diffraction work, then R and P are very close together. Thus from (5.8) $2L \sin \theta \simeq 2L\theta \simeq L \tan 2\theta$. The spots on a diffraction pattern are then defined by distances QP representing the reciprocal lattice vectors for the various planes giving the diffraction spots. The *diffraction pattern therefore corresponds closely to (a plane of) the reciprocal lattice*. The correspondence decreases as d^* and θ increase, so that it is closer with shorter wavelengths than with longer. The connection is already apparent in Fig. 5.2(c) where the network of spots is almost exactly a reciprocal lattice parallelogram (shown as a rectangular net) in the centre but becomes more distorted away from it. With short wavelengths used in 100 kV electron diffraction (see section 5.3) the distortion is small for a considerable number of rows in each direction. In general the diffraction pattern is more simply related to the reciprocal lattice rather than to the crystal lattice and is usually a systematic distortion of it, following a regular pattern which can be solved according to the geometrical arrangement used to record the pattern.

In simple cases it is not necessary to use the principle directly but its use is always implicit if not explicit.

Examples of its use are:

1. Methods for indexing some patterns depend on (5.7). Thus:

$$\sin^2 \theta = \left(\frac{\lambda^2}{4}\right)\left(\frac{1}{d^2}\right) \tag{5.9}$$

where $1/d^2$ is defined in terms of reciprocal lattice cell edges, etc., as in equations (4.10)–(4.14). Values of $\sin^2 \theta$ are therefore directly resolvable into the components of reciprocal lattice vectors—a method referred to in the section on powder methods (section 6.4).

2. The circle of radius equal to or proportional to $1/\lambda$ in Fig. 5.3(c) becomes a sphere in three dimensions. This is called the *sphere of reflection* or the *Ewald sphere* after its originator P. P. Ewald. Diffraction occurs when the sphere passes through the point (Q) corresponding to the undiffracted direct beam and the reciprocal lattice point (R) corresponding to the diffraction indices *hkl*.

3. Consider the triangle OQR in Fig. 5.3(c). Let $\mathbf{k}_0$ be a vector parallel to OQ, and $\mathbf{k}$ be a vector parallel to OR, so that each is numerically $=1/\lambda$, i.e.

$$|\mathbf{k}_0| = |\mathbf{k}| = \frac{1}{\lambda}$$

or in some work $2\pi/\lambda$, but the present use is preferable for our purpose. Then in the triangle OQR the vectors indicated become

$$\mathbf{OQ}+\mathbf{QR} = \mathbf{OR} \quad \text{(as in any triangle)}$$

Therefore

$$M\mathbf{k_0} + M\mathbf{d}^* = M\mathbf{k} \quad (M \text{ constant})$$

or
$$\mathbf{k_0} + \mathbf{d}^* = \mathbf{k}$$

or
$$\mathbf{k_0} - \mathbf{k} = \mathbf{d}^* = \mathbf{g} \tag{5.10}$$

$\mathbf{g}$ being *an alternative symbol* sometimes used for a reciprocal vector (e.g. in some electron diffraction work). The vector equation (5.10) is the simplest expression of Bragg's Law since

$$|\mathbf{k_0} - \mathbf{k}| = \frac{2 \sin \theta}{\lambda}$$

The vectors $\mathbf{k_0}$ or $\mathbf{k}$ are of considerable interest since they correspond to the waves in magnitude $(1/\lambda)$ and direction and exist in reciprocal space.

5.2 FACTORS AFFECTING THE INTENSITY

5.2.1 General

It is now necessary to consider the factors which determine the intensity of diffracted beams or their recorded patterns. There are certain factors which do not require detailed consideration but others do because they either concern or reveal differences between (a) structures, (b) source of wave used and (c) experimental arrangements. The distinction between X-rays, electrons and neutrons, as waves, follows logically, in regard to the nature of the waves, and in differences in their interactions with matter.

5.2.2 A Preliminary Note on Waves and Their Combination (*Required for later sections*)

5.2.2.1 It is useful to consider the equation to a wave like that in Fig. 4.1(a). A simple harmonic plane wave can be represented either by a sine or cosine function, e.g.

$$\psi = A \sin 2\pi \left(\frac{x}{\lambda} - \frac{vt}{\lambda} \right) \tag{5.11}$$

where ψ = height of disturbance at any point
$\quad x$ = distance in direction of motion
$\quad \lambda$ = wavelength
$\quad v$ = wave velocity ($=c$ for light or X-rays)
$\quad t$ = time
$\quad A$ = amplitude.

The intensity $I \propto A^2$. Also $v = \nu\lambda$ where $\nu =$ frequency, and a 'wave number' k is sometimes used as above, i.e. $k = 1/\lambda$ and so

$$\psi = A \sin 2\pi(kx - \nu t) \tag{5.12}$$

Such a wave as (5.11) (or (5.12)) moves with time as indicated in Fig. 5.4(a).

147

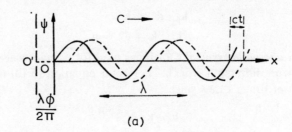

(a)

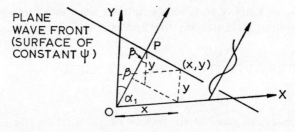

(b)

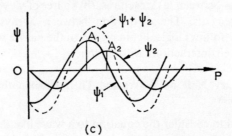

(c)

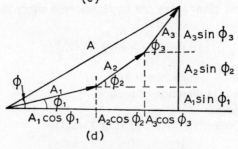

(d)

Fig. 5.4 Addition of waves.

(a) wave motion parallel to Ox
(b) wave motion in direction of OP
(c) addition of waves
(d) addition of amplitudes

If the origin is not at O but at O' then a *phase angle* ϕ is needed, where the distance $OO'=(\lambda\phi/2\pi)$ ($t=0$ in (5.11)).

For a direction of propagation OP in the plane of two axes OX and OY, see Fig. 5.3(b), the plane wave front is represented by the straight line on which all points (xy) give rise to a single value of ψ at a constant t. In Fig. 5.3(b) this condition, i.e. the equation to the plane itself, can be written

$$x \cos \alpha + y \cos \beta = \text{constant}$$

at this time. In three dimensions we have to add $z \cos \gamma$ or

$$lx+my+nz = \text{constant} \quad (\psi, t \text{ constant})$$

where l, m, n are 'direction cosines' for the wave normal (as used for another purpose in Chapter 4). These relationships may readily be proved. Hence the three-dimensional equivalent of (5.11) is the phase angle:

$$\psi = A \sin \left[2\pi \left(\frac{lx+my+nz}{\lambda} - \frac{vt}{\lambda} \right) - \phi \right] \tag{5.13}$$

There need be no confusion about the symbols l, m, n as the use is clear from the context. Also the x, y, z axes may be the crystal axes or some other suitable co-ordinate system.

5.2.2.2 Consider now the effect of two waves travelling in the same direction with the *same wavelength*—like waves scattered at different points in a lattice as in Fig. 5.3(a). The ψ values at any point can be added to give the combined effect—the 'principle of superposition'. Thus in Fig. 5.3(c)

$$\psi = \psi_1 + \psi_2$$

Let the variable part of $(5.13) = X(x, t) = $ a function of $x(y, z)$ and t, and is the same for both waves.

Then

$$\begin{aligned}
\psi = \psi_1 + \psi_2 &= A_1 \sin (X - \phi_1) + A_2 \sin (X - \phi_2) \\
&= A_1 \cos \phi_1 \sin X - A_1 \sin \phi_1 \cos X \\
&\quad + A_2 \cos \phi_2 \sin X - A_2 \sin \phi_2 \cos X \\
&= [A_1 \cos \phi_1 + A_2 \cos \phi_2] \sin X \\
&\quad - [A_1 \sin \phi_1 + A_2 \sin \phi_2] \cos X
\end{aligned}$$

But the resultant wave will have its specific amplitude A and phase ϕ, which are therefore given by:

$$A \cos \phi = A_1 \cos \phi_1 + A_2 \cos \phi_2 \tag{5.14a}$$

$$A \sin \phi = A_1 \sin \phi_1 + A_2 \sin \phi_2 \tag{5.14b}$$

true for the addition *of any number of waves*. Squaring and adding (5.14a) and (5.14b) $(\cos^2 \phi + \sin^2 \phi = 1)$

$$A^2 = \left[\sum_n A_r \cos \phi_r \right]^2 + \left[\sum_n A_r \sin \phi_r \right]^2 \tag{5.15}$$

where $r = 1, 2, 3, \ldots, n$.

5.2.2.3 Considerable abbreviation is possible by using Demoivre's theorem which states that

$$\cos x + i \sin x = \exp ix \quad \text{where} \quad i = \sqrt{-1}$$

If now (5.14b) is multiplied by i and added to (5.14a)

$$A(\cos \phi + i \sin \phi) = A_1(\cos \phi_1 + i \sin \phi_1) + A_2(\cos \phi_2 + i \sin \phi_2) + \cdots$$

i.e.
$$A \exp i\phi = \sum_n A_r \exp i\phi_r \tag{5.16}$$

The intensity of the combined wave $I \propto A^2$ and this does not indicate the value of ϕ directly. ϕ is not determinable unless the separate A_r and ϕ_r values are known. In Fig. 5.4(d) the quantities from (5.14) or (5.15), viz. $A_1 \cos \phi_1$, $A_2 \cos \phi_2$, etc., are plotted horizontally and the sine terms vertically. If the first sum $= C$ and the second S then by Pythagoras' theorem

$$A^2 = C^2 + S^2 = (C + iS)(C - iS)$$

But $C + iS$ is given by equation (5.16) whilst $C - iS$ gives

$$\sum_n A_r \exp -i\phi_r = A \exp -i\phi \tag{5.17}$$

Hence A^2 has two factors $[A \exp i\phi]$, $[A \exp (-i\phi)]$ and the second is called the 'complex conjugate' of the first. Also the numerical value of A, $|A| = \sqrt{(C^2 + S^2)}$. Alternatively we can regard A as a complex quantity or vector so that

$$A = C + iS, \qquad A^* = C - iS, \qquad A^2 \quad \text{or} \quad |A|^2 = A.A^*$$

It follows that both amplitude and phase can be represented by equations such as (5.16) and (5.17) providing the real intensity is obtained at the end by multiplying the complex amplitude by its complex conjugate.

Another useful device employing the Demoivre relation is sometimes used and should be distinguished from the above application since occasionally the two appear together. Equations such as (5.11) to (5.13) could equally well have been represented by the cosine formula since this is merely the sine wave with a phase difference of 90° (or $\pi/2$). It is possible to combine waves in many ways but in particular by adding $\cos X$ and $i \sin X$ the expression $\exp iX$ (where X is the function of x, y, z, t) is obtained. This can be used for calculations and then the sine or cosine parts separated at the end. This use applies to the variable part of the wave function and the other to the amplitude. The combined use could involve expressions such as

$$A \exp i(X \pm \phi) = A \exp (\pm i\phi) \exp iX$$

5.2.2.4 One other point about waves is required later. It involves the addition of waves of *different wavelengths*, in particular of closely similar wave-

150

lengths. For this purpose the form (5.12) happens to be most convenient. For simplicity let $A_1 = A_2 = 1$, then

$$\psi = \psi_1 + \psi_2 = \sin 2\pi(k_1 x - \nu_1 t) + \sin 2\pi(k_2 x - \nu_2 t)$$

$$= 2 \cos 2\pi \left[\frac{(k_1 - k_2)x}{2} - \frac{(\nu_1 - \nu_2)t}{2} \right] \sin 2\pi \left[\frac{(k_1 + k_2)x}{2} - \frac{(\nu_1 + \nu_2)t}{2} \right] \quad (5.18)$$

This expression sufficiently illustrates an essential point for the second term is simply a wave of mean frequency and mean wave number. The first term can then be regarded as a variable amplitude for this wave—the variation with x and t occurs as though it was another wave of wave number $\Delta k/2$ and frequency $\Delta \nu/2$ where

$$\Delta k = k_1 - k_2 = \text{difference in wave number}$$

$$\Delta \nu = \nu_1 - \nu_2 = \text{difference in frequency}$$

Since for any wave $\lambda \nu = v = \text{velocity of wave}$, the velocity of this amplitude-wave is $\Delta \nu / \Delta k$ which, in the limit, becomes $d\nu/dk$. The wavelength is $1/\Delta k = \lambda_2 \lambda_1/(\lambda_2 - \lambda_1)$, which is longer than either wavelength λ_1 or λ_2 and $\rightarrow \infty$ as the difference $\rightarrow 0$. The same general conclusion is true when several such waves are added together forming a 'wave packet' or group. The wavelengths might, for example, follow a Gaussian distribution. This general case is considered by Coulson [7]. (See also Pain [8].) The condition is reflected in the finite width of spectral lines and the impossibility of defining a strictly unique wavelength or frequency.

5.2.2.5 These results now enable a basic distinction to be made between X-rays (or any other electromagnetic radiation) and waves associated with particles such as electrons or neutrons.

(a) For X-rays the velocity of the wave is that of light and is constant for all wavelengths *in vacuo* but not in a refracting (i.e. dispersing) medium when it does vary with λ or ν. We note, however, that the photons associated with these waves considered as particles have zero 'rest mass' (see below: $m_0 = 0$).
(b) Particles such as electrons and neutrons have associated wavelengths and frequencies which both vary with the velocity of the particles under all conditions so that a constant product $\lambda \nu$ or ν/k is not obtained. The essential relationships may be demonstrated (not proved) as follows:

Let the particles have a velocity u. (The symbols v or V have sometimes been used but both have other applications. We must also distinguish v from ν, e.g. remember to multiply two Greek letters $\lambda \nu$ to give velocity v.) Since the energy of a group of waves is greatest where its amplitude is greatest it is natural to *identify the particle velocity with the group velocity*, since in either case the energy is then transmitted at the same speed, viz.

$$\frac{d\nu}{dk} = u \quad (5.19)$$

151

The particles should also obey the relativistic relations

$$E = mc^2 \tag{5.20}$$

and

$$m = \frac{m_0}{\sqrt{(1-u^2/c^2)}} \tag{5.21}$$

(variation of mass m with velocity u).

According to the quantum theory, however,

$$E = h\nu \tag{5.22}$$

Therefore from (5.22) and (5.20)

$$h \, d\nu = c^2 \, dm$$

To eliminate dm we consider the momentum $p = mu$ since (like (5.21)) this involves m and u as variables. Therefore

$$dp = u \, dm + m \, du$$

Hence

$$\frac{dp}{h \, d\nu} = \frac{u}{c^2} + \frac{m}{c^2} \frac{du}{dm} = \frac{u}{c^2} + \frac{m}{c^2} \left(\frac{c^2 - u^2}{mu} \right) = \frac{1}{u}$$

in which du/dm has been obtained from (5.21). Therefore $h \, d\nu/dp = u$ and from (5.19) $dk = dp/h$, so that

$$kh = p = mu$$

or

$$\lambda = \frac{h}{mu} \tag{5.23}$$

which is the fundamental connection between wavelength and momentum used in wave mechanics. The equation (5.23) receives experimental confirmation in electron diffraction where the wavelengths found by experiment depend on the electron velocities determined by the voltages used to accelerate them (section 5.3.2).

Another basic relationship follows:

Combine (5.22) and (5.23) and then (5.20):

$$v = \lambda\nu = \left(\frac{h}{mu}\right)\left(\frac{E}{h}\right) = \left(\frac{h}{mu}\right)\left(\frac{mc^2}{h}\right) = \frac{c^2}{u}$$

or

$$uv = c^2 \tag{5.24}$$

This relationship indicates that since $u < c$, $v > c$, i.e. the wave velocity for these particles is greater than that of light whilst the group velocity is less. Equations (5.23), (5.24) enable all the various quantities to be related to each other which define the wave–particle properties.

5.2.3 Summary of Factors

5.2.3.1 Further *essential differences* between the waves also emerge in considering the way in which the *intensity* of a diffracted beam is affected by some of the various factors now considered:

1. There may be more than one of the possible sets of planes {*hkl*} contributing to the individual diffraction spot or line (photographed or recorded). The number of planes contributing gives the *planar factor p* which may equal the multiplicity factor (previous chapter) or a fraction of it.

2. There is a reduction in intensity due to absorption of the waves or particles as they pass through the material. This is due to a variety of causes. Hence the *absorption factor (A)*—section 5.2.4.

3. The intensity actually recorded may be affected by the *geometry* of the experimental arrangement, e.g. with polycrystalline samples the *probability* that the grains will be in a position to contribute to diffraction.

4. Effects due to the absence or presence of polarisation.

   (a) In X-rays the incident beam is generally *unpolarised*. This leads to a term $\frac{1}{2}(1 + \cos^2 2\theta)$ which is known (unfortunately) as the *polarisation factor* and which derives initially from the scattering by an electron as in 5(i) below.

   (b) Electron waves can also be polarised (Chapter 14 of Ref. [4]) but this fact was difficult to establish experimentally and it is open to some doubt how it would affect diffracted intensities. In any case, with the usual wavelengths the term $\frac{1}{2}(1 + \cos^2 2\theta)$ approximates to $\frac{1}{2}$ since θ is small (0–3°, say).

   (c) For neutron diffraction polarisation occurs in connection with magnetic scattering.

5. There is an intrinsic contribution from either or both of:

   (i) a single electron

   (ii) the nucleus, or each of the nuclei present if more than one kind.

   In the neutron case only:

   (iii) the magnetic moment of an atom with unpaired spins—as in ferromagnetic materials.

   Thus:

   (a) For X-rays only (i) is relevant.

   (b) For electrons (i) and (ii).

   (c) For neutrons (ii) and (iii).

6. For practical purposes the nucleus may be regarded as occupying a point. The electrons surrounding the nucleus contribute to the relatively large size of the atom. Hence particularly in regard to X-rays: if we regard a wave path (from the incident wave front to a diffracted wave front) through the atom centre as having zero phase angle ($\phi = 0$), then contributions to the intensity arise from paths which are longer or shorter (on either side of the central path) giving corresponding phase differences. The amplitude contributions to

153

the net intensity must therefore be added by the methods of section 5.2.2. This gives the *atomic scattering factor f* (different for different atoms) for the atomic electrons to which the nuclear contribution can be added for electron diffraction—section 5.2.5. In effect this treatment replaces the diffraction intensity spread over the whole atom by a single beam of resultant amplitude considered as passing through the atom centre.

7. The same process is used to add up the intensity (f) and phase contributions from atoms at different points in the unit cell—in effect replacing them by a single wave of resultant amplitude considered to pass through the origin of the unit cell. If the atomic factors are all the same (elements) then f can be taken outside the summation. Otherwise the fs are different for different atoms (compounds, solid solutions). This gives the *structure factor F* which is considered in section 5.2.6. Note $I \propto F^2$ or more exactly $|F|^2 = FF^*$.

8. Since a crystal contains many unit cells the contributions from the separate unit cells can be added to give a resultant amplitude factor which indicates *crystal size or shape* if the number of cells is small enough in all directions or some directions only (e.g. plates or needles)—developed in section 5.2.7. For many purposes this factor is ignored but it comes into crystal size determination and is important in some applications of electron diffraction as in Chapter 6.

9. There is a temperature factor given by $\exp[-(B \sin^2 \theta)/\lambda^2]$ where B is a function of the temperature derived from specific heat theory and is characteristic of each kind of atom. The factor applies to diffraction by the three types of wave and represents a decrease in intensity with increasing temperature.

5.2.3.2 All these factors may be summarised schematically as follows:

$$
\begin{array}{cc}
(1) & (2) \\
\end{array}
$$
$$I \propto [\text{planar factor } p] \times [\text{absorption factor } A]$$
$$
\begin{array}{cc}
(3) & (4) \\
\end{array}
$$
$$\times [\text{geometrical factor}] \times [\text{polarisation factor}]$$
$$(5), (6), (7)$$
$$\times [\text{separate and/or combined effects giving } |F|^2]$$
$$
\begin{array}{cc}
(8) & (9) \\
\end{array}
$$
$$\times [\text{crystal size or shape}] \times [\text{temperature}] \qquad (5.25)$$

These factors have thus been divided:

(i) Those for which the essential points are simple and have been made above (Nos. 1 and 4).

(ii) One for which consideration is deferred until required (No. 3).

(iii) One which requires no further consideration in connection with the simpler metallurgical applications (No. 9).

(iv) The remainder requiring more detailed consideration as in the follow-

ing sections. Some of them are more immediately important in connection
with X-ray diffraction.

5.2.4 Absorption

Both incident and diffracted beams are reduced in intensity as they pass
through the crystalline material being studied. This fact is of interest not only
in connection with diffraction but also in fluorescence analysis and radiography (Chapter 8). The mechanism of absorption is different for the
different radiations (or waves) used in diffraction but the essential principles
are the same. The fractional decrease in intensity is given by $-(dI/I$ as the
radiations (or waves) pass through a thickness dx of material. We define the
linear absorption coefficient μ where $-(dI/I) = \mu \, dx$. From which

$$I = I_0 \exp(-\mu x) \tag{5.26}$$

for the intensity I at a distance x. The actual value of μ depends on the density.
If for example the same absorbing atoms occupied double the volume they
would have half the absorption coefficient (for example consider a beam 1 cm^2
cross-section and a path of 1 cm and halve the mass or number of atoms in
this volume). If ρ is the density, μ/ρ is a constant of the material and is called
the mass absorption coefficient. Equation (5.26) is then replaced by

$$I = I_0 \exp -(\mu/\rho)\rho x \tag{5.27}$$

The absorption of a mixture or compound is obtained from the formula:

$$\frac{\mu}{\rho} = \frac{w_1}{100} \left(\frac{\mu}{\rho}\right)_1 + \frac{w_2}{100} \left(\frac{\mu}{\rho}\right)_2 + \frac{w_3}{100} \left(\frac{\mu}{\rho}\right)_3 + \cdots \tag{5.28}$$

where w_1, w_2, are weight percentages of the various elements (or weight
fractions if the 100 is omitted from each denominator).

This information is sufficient to calculate the absorption of any specimen
being studied providing:

(a) The geometry of the incident and diffracted rays is known relative to the
specimen surface and shape.

(b) The necessary values of μ/ρ have been obtained from tables available in
the literature—in particular the *International Tables*, Volume III [11], although other tabulations are available [3].

In regard to (a) specific examples are:

(i) Transmission through a parallel sided crystal or polycrystalline mass
(the simplest using equation (5.27)).

(ii) Diffraction from similar specimens in which the diffracted rays leave
the same surface as the entrant rays (e.g. Chapter V of Ref. [9]).

(iii) Cylindrical (powder specimens)—especially Ref. [9], Appendix, Table
II [10, 11]; and for neutrons Ref. [5], pp. 95–101.

In regard to (b) reference to the tables gives actual values but the variation of the coefficient with wavelength requires further consideration in connection with the nature of X-rays in particular as in section 5.3.

5.2.5 Development of the Atomic Scattering Factor

5.2.5.1 Here we consider the scattering contribution of a single atom and how it may be built up from the respective contributions given the factor (5) of equation (5.25). The expression for scattering of X-rays by a single electron contains a number of constants and for an unpolarised incident beam the term $\frac{1}{2}(1+\cos^2 2\theta)$ already separated as the polarisation factor. It is thus sufficient to consider the atomic scattering factor as represented by

$$f = \frac{\text{(amplitude of wave scattered by all the electrons in an atom)}}{\text{(amplitude scattered by one electron)}}$$

The most elementary way in which this might be derived would be to imagine the electrons to be stationary and evenly distributed in Bohr orbits and then to add the amplitudes for different ray paths using equation (5.16), i.e. Fig. 5.4(d). In practice this is too artificial and the electrons are represented by an 'electron cloud' surrounding the nucleus. The electron density is supposed to be equal at equal distances from the centre, i.e. spherically symmetrical—this is not always the case—and is given by $\rho(r)$/unit volume so that a shell of surface area $4\pi r^2$ and thickness dr contains $4\pi r^2\rho(r)\ dr$ electrons. Multiplying by the appropriate phase factor (as in (5.16)) for any point in this shell and integrating successively all over the shell and then over all values of r we obtain the factor for the whole atom, thus:

In Fig. 5.5 part of Fig. 5.3(a) is taken and a spherical atom shown at the origin. The shell at radius r is indicated. As shown previously a ray path through θ is separated by one wavelength from a path through any point on the plane through P and other points on the axes such that

$$OP = d = \frac{\lambda}{2\sin\theta} \quad \text{(for given } h, k, l\text{)}$$

The corresponding phase angle in equation (5.16) would be exactly 2π. By a simple further construction of similar triangles (left to the reader) resembling that used in connection with Fig. 5.1(b) a parallel plane through P' would correspond to a path difference proportional to OP' and phase angle

$$\phi = 2\pi\frac{OP'}{OP}$$

This plane would cut the spherical shell (of constant $\rho(r)$) in a circle, the area of which is shaded. The volume of the ring for constant $\rho(r)$ which lies in this

156

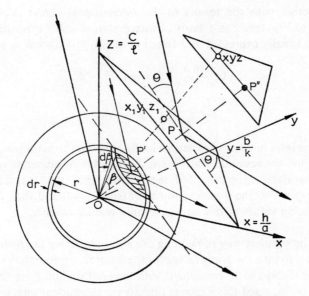

Fig. 5.5   Derivation of atomic and structure factors.

plane is proportional to its circumference, i.e. $2\pi r \sin \beta$ and its width $= r \, d\beta$. The phase angle for this value of β is

$$2\pi \frac{OP'}{OP} = \frac{2\pi r \cos \beta}{d} = \frac{4\pi r \sin \theta}{\lambda} \cos \beta = Kr \cos \beta$$

where
$$K = \frac{4\pi \sin \theta}{\lambda}$$

The amplitude contribution is thus

$$2\pi r^2 \sin \beta \, d\beta \exp (iKr \cos \beta) \, \rho(r) \, dr$$

Integration replaces the summation of (5.16). The first integration is with respect to β which varies from 0 to π.

$$f = 2\pi \int_0^\infty \int_0^\pi \rho(r) r^2 \sin \beta \exp (iKr \cos \beta) \, d\beta \, dr$$

$$= 2\pi \int_0^\infty \rho(r) r^2 \left[\frac{-\exp (iKr \cos \beta)}{iKr} \right]_0^\pi dr$$

$$= 2\pi \int_0^\infty \rho(r) r^2 \left\{ \frac{-1}{iKr} \left[\cos Kr(-1) - \cos Kr + i \sin Kr(-1) - i \sin Kr \right] \right\} dr$$

i.e.
$$f = 4\pi \int_0^\infty \rho(r) r^2 \frac{\sin Kr}{Kr} \, dr \qquad (5.29)$$

The calculation of $\rho(r)$ involves atomic theory and some indication of this is

157

given together with the results in the *International Tables*, Vol. III [11]. Equation (5.29) stands as a basis of interpretation and experiment. In particular f is usually expressed as a function of $(\sin \theta)/\lambda$ since $K = (4\pi \sin \theta)/\lambda$.

Since
$$\underset{x \to 0}{\text{Lt}} \frac{\sin x}{x} = 1 \text{ when } \theta \to 0$$

$$f = 4\pi \int_0^\infty \rho(r) r^2 \, \mathrm{d}r \tag{5.30}$$

which is the total number of electrons in the atom and so is the *atomic number* Z. For an ion the resultant f at $\theta = 0$ is $Z \pm$ number of electrons removed or added. The standard curves and tables (Ref. [11], pp. 121, 132 *et seq.*) thus correspond to the atomic number at $(\sin \theta)/\lambda = 0$ and then f decreases with increasing values of this parameter in a regular manner.

5.2.5.2 The simplest way of defining the atom scattering factor for electron diffraction is to take the X-ray factor for the electrons surrounding the nucleus as given by (5.29) and to combine it with the contribution from the nucleus. As would be expected this depends directly on the nuclear charge Z. Stated without proof the contributions are

(i) $-\dfrac{e^2}{2mu^2} \dfrac{1}{\sin^2 \theta} f$

(ii) $+\dfrac{e^2}{2mu^2} \dfrac{1}{\sin^2 \theta} Z$

The '$-$' and '$+$' signs are required by the difference in sign of the charges on the electrons and nucleus respectively. u is the velocity of the electrons and appears because λ is replaced by h/mu according to (5.23). The net result is

$$f(\theta) = \frac{e^2}{2mu^2} \frac{1}{\sin^2 \theta} (Z - f) \tag{5.31}$$

(where f is the X-ray factor and $f(\theta)$ is the factor for electrons). Typical curves for $|f(\theta)|^2$ to which the intensity is proportional are given by Pinsker [4].

5.2.5.3 In ordinary neutron diffraction the nucleus is involved and a quantity defined as the *scattering cross-section*.

$$\sigma = \frac{\text{outgoing current of scattered neutrons}}{\text{incident intensity of neutrons}}$$

$$= 4\pi b^2 \quad \text{where } b \text{ is a complex quantity}$$

($b = \alpha + i\beta$) with dimensions of length. The variation of σ with b is important since b is different for different isotopes of the same element. These will be randomly distributed through the crystal and so not be separately distinguishable, but their average effect is calculated. Secondly, the values of σ

158

which correspond to f (X-rays) or $f(\theta)$ (electrons) do not vary systematically with atomic number (in fact a systematic component and an irregular component add together). Hence adjacent elements may have sufficiently different atomic scattering factors to enable their distinction by their effect on the structure factor, and light elements, especially hydrogen, can be distinguished in the presence of much heavier atoms.

In addition magnetic atoms make the contribution previously noted, viz.:
(a) A term representing scattering by randomly oriented paramagnetic ions (or atoms).
(b) If these ions take up a regular orientation or pattern of orientations this produces a regular effect on the intensities of the diffracted beams according to the structure factor. Hence the use of neutron diffraction for the study of magnetic materials.

5.2.6. The Structure Factor

5.2.6.1   In Fig. 5.5 the phase angle was zero for the ray path through O. The atomic factor f for X-rays could thus be taken as the amplitude of a single ray through O. The finite sized atom with its scattering electrons spread over the whole volume has been replaced by a point. Consequently any other atoms in the unit cell may be considered to have an amplitude f corresponding to ray paths through their centres. Thus consider an atom centre at the point (x, y, z). We ignore its finite size and assign it an atomic factor f for a ray path through this point. We then find its phase relative to O, exactly as for the former case. Thus the point (x, y, z) lies on a plane through P'' for which plane all points would have the same phase given by $2\pi(OP''/OP)$. If the point (x, y, z) is joined by a straight line to O all points on the line have proportional co-ordinates and so (x_1, y_1, z_1) in the plane through P is given by:

$$\frac{x}{x_1} = \frac{y}{y_1} = \frac{z}{z_1} = \frac{OP''}{OP} \quad \text{(similar triangles)}$$

but

$$\frac{hx_1}{a} + \frac{ky_1}{b} + \frac{lz_1}{c} = 1 \quad \text{(equation (4.8))}$$

Multiply each term by the appropriate fraction from the previous line and by 2π. Thus

$$\left.\begin{array}{l} \phi = 2\pi\left(\dfrac{hx}{a} + \dfrac{ky}{b} + \dfrac{lz}{c}\right) = 2\pi\dfrac{OP''}{OP} \\[3mm] \phi = 2\pi(hx + ky + lz) \end{array}\right\} \quad (5.32)$$

or

if fractional co-ordinates are used. It follows that the amplitudes and phases for the separate atoms can be added according to equations (5.15) or (5.16) (Fig. 5.4(d)). That is

$$F = \sum f \exp\left[2\pi i(hx + ky + lz)\right] \quad (5.33)$$

and

$$I \propto FF^* = |F|^2$$

where in F^* there is a $-$ sign in front of $2\pi i(hx+ky+lz)$. F is the *Structure Factor*. Thus in a precisely similar manner to the determination of f, the contents of the unit cell have in effect been replaced by an effective scattering mass at O for which the diffracted amplitude and phase are given by F and the intensity multiplied by $|F|^2$.

Either the exponential form (5.33) or the trigonometrical equivalent can be used. For example

$$F = \sum f \cos 2\pi(hx+ky+lz) + i \sum f \sin 2\pi(hx+ky+lz) \qquad (5.34)$$

We note here that if any structure has a centre of symmetry (see Chapter 4 and Fig. 4.1) then for every atom at a point (x, y, z), there is another equal atom at $(\bar{x}, \bar{y}, \bar{z})$. Therefore every sine term is cancelled out by a term of opposite sign $[\sin(-x) = -\sin x]$. Hence in this case only the cosine terms remain.

5.2.6.2 The following examples illustrate essential applications of the structure factor to some of the simple metallurgical structures already described in Chapter 4.

1. The body-centred cubic *structure* is a special case of a structure with two equal atoms per unit cell at (x, y, z) and $(\frac{1}{2}+x, \frac{1}{2}+y, \frac{1}{2}+z)$. Generally we take the origin at the first so $(x, y, z) = (000)$. The general result applies to any body-centred lattice and to any number of atoms. Thus

$$\begin{aligned}
F &= \sum f \exp\{2\pi i(hx+ky+lz)\} \\
&\quad + f \exp[2\pi i\{(\tfrac{1}{2}+x)h+(\tfrac{1}{2}+y)k+(\tfrac{1}{2}+z)l\}] \\
&= \sum \exp\{2\pi i(hx+ky+lz)\}[f+f \exp\{2\pi i(h+k+l)/2\}]
\end{aligned}$$

The term in square brackets is common to all terms in $\sum$ and for the simple two-atom case:

$$\begin{aligned}
F &= f[1+\exp\{2\pi i(h+k+l)/2\}] \qquad (5.35) \\
&= f[1+\cos \pi(h+k+l)]
\end{aligned}$$

h, k, l being integers, the sine term is 0, and the

$$\begin{aligned}
\text{cosine term} &= +1 \quad \text{if} \quad h+k+l \quad \text{even} \\
&= -1 \quad \text{if} \quad h+k+l \quad \text{odd.}
\end{aligned}$$

Hence for *all body-centred structures* only those diffracted rays occur for which

$$\underline{h+k+l = \text{even number}}$$

f stands for the X-ray atom factor but could equally well be the electron or neutron factor. For the cubic case the interplanar spacings depend on $h^2+k^2+l^2=N$, say—see equations (4.14) or (4.15). For even $h+k+l$ this sum must also be even since either all three are even or two are odd. (The squares of even numbers are even and of odd numbers odd.) $N = 2, 4, 6, 8$.

2. If in the body-centred cubic structure one atom is different from the other the two positions are no longer identical as explained in section 4.3.2.2 and the CsCl structure is obtained—general formula AB. f is now different for the two atoms. Therefore when $h+k+l$ is even $F=f_A+f_B$, when $h+k+l$ is odd $F=f_A-f_B$. Since $f_A \neq f_B$ the 'odd reflections' no longer vanish. In a *random solid solution* f is the weighted mean of the factors for the separate atoms which are indistinguishable. In the *ordered solid solution AB*, the ordered state is detected by the appearance of the weaker odd reflections. This effect is very useful in the study of the order/disorder transformation.

(N.B. The sum N of three squares cannot equal any number given by $4^m(8n+7)$ where m and n are the integers 0, 1, 2, 3, etc. Hence in *any cubic structure* there are no diffracted rays and no planes for which

$$h^2+k^2+l^2 = 7, 15, 23, \ldots \qquad n = 1, 2, 3, \ldots; m = 0$$
$$28, 60, 92, \ldots \qquad n = 1, 2, 3, \ldots; m = 1$$
$$112, 240, 368, \ldots \qquad n = 1, 2, 3, \ldots; m = 2 \text{ etc.})$$

3. The face-centred cubic structure (section 4.3.2.1) common to several metals contains the smallest number of atoms that can be accommodated on a lattice centred on three faces. For identical atoms there is a single value of f:

$$F = f[e^0 + \exp\{2\pi i(\tfrac{1}{2}h+\tfrac{1}{2}k)\} + \exp\{2\pi i(\tfrac{1}{2}k+\tfrac{1}{2}l)\} + \exp\{2\pi i(\tfrac{1}{2}l+\tfrac{1}{2}h)\}]$$

the co-ordinates being $(0, 0, 0)$, $(\tfrac{1}{2}, \tfrac{1}{2}, 0)$, $(0, \tfrac{1}{2}, \tfrac{1}{2})$, $(\tfrac{1}{2}, 0, \tfrac{1}{2})$. Again the expanded form shows sine terms zero. The cosine terms can be added to give a product but this is not necessary in the simple four-atom formula. Thus

$$F = f[1 + \cos \pi(h+k) + \cos \pi(k+1) + \cos \pi(l+h)] \qquad (5.36)$$

Again cosines of even multiples of π are $+1$, and cosines of odd multiples of π are -1.

h, k, l can be all odd, in which case the sum of any pair is even, i.e. each cosine $= +1$. h, k, l can be all even, i.e. each cosine $= +1$. For these cases $F=4f$. If, however, any one integer is odd it will make two of the cosines negative. If two are odd, e.g. k and l, then $k+l$ will be even but $h+k$ and $l+h$ will be odd. In both these cases $F=0$. Therefore in *any face-centred lattice* those reflections occur for which h, k, l *are all odd or all even*. In the *cubic case* those interplanar spacings occur for which:
With even h, k, l

$$N = \sum h^2 = 4p \quad \text{where } p \text{ is an integer}$$

With odd h, k, l

$$N = \sum h^2 = 8q+3 \quad \text{where } q \text{ is an integer}$$

161

(the square of an even number is a multiple of 4 and of an odd number one more than a multiple of 8, e.g. $(2r+1)^2 = 4r(r+1)+1)$. That is, $N = 3, 4, 8, 11, 12, \ldots$.)

4. The ordered structures that are sometimes formed from face-centred cubic solid solutions and previously described are of the two kinds AB_3 or AB.

(i) AB_3. The A atoms occupy the corners (000) and the B atoms the three face-centres.

$$F = f_A + f_B[\cos \pi(h+k) + \cos \pi(k+l) + \cos \pi(l+h)]$$

For h, k, l $\qquad$ all even $\Big\}F = f_A + 3f_B$
$\qquad\qquad\qquad$ all odd

For one or two odd $\qquad F = f_A - f_B$

Hence the additional diffraction intensities will be weak.

(ii) AB. Here the A atoms occupy say (000) and $(\frac{1}{2}\frac{1}{2}0)$ and the B atoms $(0\frac{1}{2}\frac{1}{2})$ and $(\frac{1}{2}0\frac{1}{2})$. We noted that this provides alternate layers of A and B atoms and the structure no longer needs cubic symmetry—it is tetragonal with c normal to the layers and although we could now use a smaller unit cell (a face-centred tetragonal cell is not necessary) it is useful to keep the four co-ordinate positions so

$$F = f_A[1 + \cos \pi(h+k)] + f_B[\cos \pi(k+l) + \cos \pi(l+h)]$$

For all odd or all even $\qquad F = 2(f_A + f_B)$

For h, k both odd or both even but l with opposite sign

$$F = 2(f_A - f_B)$$

For h, k with opposite signs $\qquad F = 0$

Hence in addition to the departure from cubic symmetry, the appearance of some but not all of the 'forbidden reflections' in also a proof of ordering of this kind.

5. The close-packed hexagonal structure (see section 4.3.2.1) has atoms at positions (000) and $(\frac{1}{3}, \frac{2}{3}, \frac{1}{2})$

$$\begin{aligned} F &= f[1 + \exp\{2\pi i(\tfrac{1}{3}h + \tfrac{2}{3}k + \tfrac{1}{2}l)\}] \\ &= f[1 + \cos 2\pi[\tfrac{1}{3}(h+2k) + \tfrac{1}{2}l] + i \sin 2\pi[\tfrac{1}{3}(h+2k) + \tfrac{1}{2}l]] \end{aligned} \qquad (5.37)$$

The sine term vanishes if $\frac{1}{3}(h+2k) + \frac{1}{2} =$ integer or half integer ($\sin \pi = \sin 2\pi = 0$), i.e. $h + 2k =$ multiple of 3. Then if

$$l \text{ is even} \quad F = 2f$$

$$l \text{ is odd} \quad F = 0$$

Thus in the close-packed hexagonal structure there are 'missing reflections' for $h+2k=3n$ and $l=2m+1$ where m and n are integers.

5.2.6.3 In some alloys ordering may lead to the appearance of additional reflections at positions which are between two successive integral values of h (k or l) such as $\frac{1}{2}$ or $\frac{1}{3}$ of the separation. These indices can be made integral by taking a larger unit cell containing the corresponding number of multiples. The effect is simply that the ordering leads to the larger repeat pattern in order to accommodate the different atoms.

An example is the Fe_3Al type *superlattice*. The original body-centred cube is sufficient for the random solid solution but the ordered structure is obtained in units of $2 \times 2 \times 2 = 8$ of these cells, in which alternative small cube centres are Al atoms and the remainder iron. The cube corners are Fe. In the Heusler type of alloy such as Cu_2MnAl, Cu_2MnSn, Fe_2TiAl, the second and third atoms occupy the alternative small cube centres. The co-ordinates referred to the larger cell are now combinations of $\frac{1}{4}$s and/or $\frac{3}{4}$s. As an exercise the reader may work out the structure factor for such a unit of general formula A_2BC.

5.2.6.4 In the case of neutron diffraction involving magnetic atoms it may also be necessary to have multiples of the basic unit cell since the atoms with opposing (or oriented) spins act like atoms of different scattering factors. In this case the *intensities* of diffracted beams from the ordinary nuclear diffraction and from the magnetic scattering must be added, i.e.

$$F^2 = F_{\text{nuc}}^2 + q^2 F_{\text{mag}}^2$$

using the enlarged unit cell if necessary where q is a quantity which measures the 'magnetic interaction'. Here a symbol p is used for the magnetic scattering factor for an atom with corresponding nuclear factor b. The result is thus equivalent to a general atomic factor of $\sqrt{(b^2 + p^2q^2)}$. If the magnetic moments are parallel to the (hkl) plane, $q^2 = 1$, but if they are perpendicular, $q^2 = 0$. If they are random the mean value is $q^2 = \frac{2}{3}$. (Other possibilities are described by Bacon.) These observations indicate how the intensities and the existence of superlattice diffractions may provide information about the magnetic structure.

5.2.6.5 In the full determination of crystal structure (which will not be described here) a number of steps are taken. The diffraction pattern itself may reveal the crystal system or this may need support from other physical methods. Some procedures for indexing patterns will be given in the following chapter. The lattice type can be distinguished by noting the presence or absence of certain reflections as has already been indicated in section 5.2.6.2 in connection with the body-centred and face-centred lattices—see also equations (5.35) and (5.36). From (5.36) it is seen that if a lattice is centred on

163

one face only the condition is that the two indices referring to axes in that plane must be both odd or even. Hence:

Lattice type	Symbol	Present reflections	Systematic absences
Primitive	P	all	none
Face-centred	F	h, k, l all odd or all even	h, k, l mixed odd and even
Base-centred	A	$k+l$ even	$k+l$ odd
	B	$l+h$ even	$l+h$ odd
	C	$h+k$ even	$h+k$ odd
Body-centred	I	$h+k+l$ even	$h+k+l$ odd

These relationships were deduced from the structure factor equation, but can also be inferred in a general way by reference to Figs. 5.1(b) and 5.3(a). We have noted that the path differences are constant for atoms anywhere on a plane (hkl) for a reflection of the same indices. In a primitive lattice the atoms at various positions have been shown to give diffraction equivalent to a single 'atom' at the origin with resultant amplitude and phase given by F. From Fig. 5.1(b) it is seen that if an identical atom is put halfway between two others for which the path difference is one wavelength, then the path difference for this atom will be exactly a half wavelength and the diffraction intensity becomes zero.

The same argument applies to a plane halfway between O and $A'B'C'$ in Fig. 5.3(a). In a simple centred lattice, with atoms at the lattice points reflections are present when the planes hkl contain the face or body-centred positions, since then $hx+ky+lz=0, 1, 2, 3, \ldots$ (see equation (4.9) of Chapter 4) but absent when $(hx+ky+lz)=\frac{1}{2}, \frac{3}{2}, \frac{5}{2}, \ldots$, since then there are identical atoms on an interleaving set of planes at half-integral values. The diffraction from these planes is exactly out of phase $(\lambda/2)$ from the diffraction by atoms at the lattice corners. For example in a body-centred lattice 100 is absent because there are planes through cell faces and centres, but 200 is present because there is now one wavelength path difference between planes in these positions. In the face-centred lattice 100 is again absent, so is 110 (present in body-centred) but 111 is present (absent in body-centred). It is easy to check that 111 planes contain the face-centred atoms on integral planes. The same principles apply to systematic absences due to the presence of symmetry elements. It has been noted above that the presence of a centre of symmetry eliminates the sine term from the expanded form of F. In the case of the other symmetry elements in Fig. 4.1 it is easy to check that the reflections will not vanish for any choice of indices. Thus point group symmetry elements do not give systematic absences, but glide planes and screw axes do. Examples were given in Fig. 4.4.

(a) A glide plane (parallel to) 001 has a translation $b/2$ parallel to Oy. Every atom at x, y, z, is matched by one at $x, \frac{1}{2}+y, \bar{z}$. The structure factor has corresponding terms in pairs. The z terms become unity if $l=0$. Hence for $hk0$ reflections

$$F = \exp\left[2\pi i(hx+ky)\right] + \exp\left[2\pi i(hx+ky+\tfrac{1}{2}k)\right]$$

(omitting f)

$$= \exp\left[2\pi i(hx+ky)\right]\left[1+\exp\left(2\pi ik/2\right)\right]$$
$$= 0 \quad \text{when } k \text{ is odd}$$
or
$$= 2\exp\left[(2\pi i)(hx+ky)\right] \quad \text{when } k \text{ is even}$$

(b) A glide plane (001) with translations $\frac{1}{2}a$, $\frac{1}{2}b$ similarly requires $h+k$ even.
(c), (d) For the screw axis 4_1 illustrated every atom at x_1, y_1, z_1 is matched by identical atoms at the positions shown. The screw translation only affects the z co-ordinate and so planes perpendicular to z contain equal numbers of atoms at $\frac{1}{4}c$ separations. $(00l)$ reflections therefore occur when $l=4n$. For other values of l in $(00l)$ the four terms in F go in pairs with opposite signs and so F vanishes.
(e) The screw axis 2_1 shown similarly requires $l=2n$ for $(00l)$ reflections. These principles are thus of great value in helping to determine the space group. A complete table of such conditions is given by Barrett (Ref. [3] of Chapter 4, Table VII, p. 164) and in other standard works including the *International Tables*, Ref. [5] of Chapter 4.

5.2.7 Effect of Crystal Size and Shape

5.2.7.1 For the present purpose we consider the diffraction from a regular crystal containing

N_1 unit cells parallel to OX
N_2 unit cells parallel to OY
N_3 unit cells parallel to OZ
Total $N = N_1 N_2 N_3$

Each unit cell is represented by a wave of amplitude (and phase) F passing through its origin. Just as we used equation (5.15) or (5.16) to sum up the contribution of the electrons in an atom, and then the contribution of the atoms in a unit cell, we use the same procedure to sum up the amplitude and phase contribution of each unit cell. The corner of any unit cell is represented by co-ordinates whose lengths are whole numbers of cell edges—identical with the indices of the zones defined in Chapter 4, i.e. uvw. In actual lengths the distances are $x=ua$, $y=vb$, $z=wc$ and the phase difference from (000) is thus

$$\phi_{uvw} = 2\pi\left(\frac{hx}{a}+\frac{ky}{b}+\frac{lz}{c}\right) = 2\pi(hu+kv+lw)$$

and if fractional co-ordinates are used the result is the same since then $x = u$, $y = v$, $z = w$. If the six indices are all exactly integers the phase angle is exactly an integral multiple of 2π, i.e. $\equiv 2\pi$, but we ignore this point and assume that uvw alone are integers. If now we use equation (5.15) to add up the contributions of all the unit cells we have

$$\text{amplitude} \propto \sum F \exp \left[2\pi i(hu + kv + lw)\right]$$

Actually
$$u \text{ can have all values } 0\text{–}N_1$$
$$v \text{ can have all values } 0\text{–}N_2$$
$$w \text{ can have all values } 0\text{–}N_3$$

and we can take out F the common factor. The sum is really a triple sum, i.e. all values of v occur with each value of w, and all values of w with each combination of u and v.

$$\sum_{u=0}^{N_1} \cdot \sum_{v=0}^{N_2} \cdot \sum_{w=0}^{N_3} \exp \left[2\pi i(hu + kv + lw)\right]$$

$$= \sum_{u=0}^{N_1} \cdot \sum_{v=0}^{N_2} \cdot \sum_{w=0}^{N_3} \exp (2\pi ihu) \exp (2\pi ikv) \exp (2\pi ilw)$$

$$= \left\{ \sum_{0}^{N_1} \exp (2\pi ihu) \right\} \left\{ \sum_{0}^{N_2} \exp (2\pi ikv) \right\} \left\{ \sum_{0}^{N_3} \exp (2\pi ilw) \right\}$$

The sum in each bracket is simply a geometrical progression ($u = 0, 1, 2, \ldots$) which on addition becomes:

$$\frac{1 - \exp (2\pi iN_1 h)}{1 - \exp (2\pi ih)}$$

and correspondingly for the second and third terms. The intensity I varies as the product of these factors with their complex conjugates. That is

$$I \propto \frac{[1 - \exp (2\pi iN_1 h)][1 - \exp (-2\pi iN_1 h)]}{[1 - \exp (2\pi ih)][1 - \exp (-2\pi ih)]}$$

$$= \frac{1 - \cos 2\pi N_1 h}{1 - \cos 2\pi h} = \frac{\sin^2 \pi N_1 h}{\sin^2 \pi h}$$

with similar terms in k and l. Thus

$$I \propto \frac{\sin^2 \pi N_1 h}{\sin^2 \pi h} \frac{\sin^2 \pi N_2 k}{\sin^2 \pi k} \frac{\sin^2 \pi N_3 l}{\sin^2 \pi l} \qquad (5.38)$$

Alternatively, the sums (of such geometrical progressions) can be shown (by using Demoivre's relationship) to equal

$$\frac{\sin \pi N_1 h}{\sin \pi h} \times (\text{exponential phase factor})$$

and on multiplying by the complex conjugate the second factor gives unity.

166

When $h(k, l)$ is exactly integral, however, this expression becomes $0/0$ and so is indeterminate. Its value may be found exactly by considering that h is a variable, e.g. let $h + \Delta h$ be the non-integral value, h being retained as integer and Δh the difference. Now:

$$\frac{\sin \pi N_1 (h + \Delta h)}{\sin \pi (h + \Delta h)} = \pm \frac{\sin \pi N_1 \Delta h}{\sin \pi \Delta h}$$

$$= \pm \frac{\pi N_1 \Delta h + \text{terms in } \Delta h^3 \text{ and higher powers}}{\pi \Delta h + \text{terms in } \Delta h^3 \text{ and higher powers}}$$

on dividing numerator and denominator by Δh and taking the limit when $\Delta h \to 0$ we obtain

$$\operatorname*{Lt}_{\Delta h \to 0} \frac{\sin \pi N_1 (h + \Delta h)}{\sin \pi (h + \Delta h)} = \operatorname*{Lt}_{\Delta h \to 0} \frac{\sin \pi N_1 \Delta h}{\sin \pi \Delta h} = N_1$$

Hence for the intensity at the exact integral values

$$I \propto N^2 = N_1{}^2 N_2{}^2 N_3{}^2 \tag{5.39}$$

5.2.7.2   The idea of *non-integral* values can now be considered further. It is of great interest as a useful extension of the *reciprocal lattice* concept. Thus a diffracted beam with indices h, k, l could be represented by a point of the reciprocal lattice but it is now seen that the function (5.38) has a finite size extending on either side of the integral values of h, k, l. Consider the term $\sin \pi N_1 (h + \Delta h)$, ignoring the question of sign because it is squared in (5.38). The magnitude increases from 0, as Δh increases, to 1 and then decreases until the *next integral* multiple of π is reached. This occurs at $N_1 h + 1 = N_1 (h + 1/N_1)$, i.e. $\sin \pi N_1 (h + \Delta h) = 0$ again when $\Delta h = 1/N_1$, and subsequently when $\Delta h = 2/N_1, 3/N_1$, etc., n/N_1 (n integer).

In between each zero minimum there must be a maximum when the sine in the numerator $= 1$ (± 1). The denominator changes much more slowly. Now, when

$$\sin \pi N_1 (h + \Delta h) = \pm 1 = \sin \pm \tfrac{1}{2} \pi = \sin (2m\pi \pm \tfrac{1}{2}\pi)$$

$$N_1 (h + \Delta h) = N_1 h + n + \tfrac{1}{2} = N_1 \left(h + \frac{2n+1}{2N_1} \right)$$

i.e. $$\Delta h = \frac{2n+1}{2N_1}$$

Here,

$$\frac{\sin \pi N_1 (h + \Delta h)}{\sin \pi (h + \Delta h)} = \frac{\sin \pi N_1 \Delta h}{\sin \pi \Delta h} = \frac{1}{\sin \pi \Delta h} \simeq \frac{1}{\pi \Delta h} = \frac{2N_1}{\pi(2n+1)}$$

for small values of Δh. This height is $2/[\pi(2n+1)]$ of the height of the peak at $h(\Delta h = 0)$. Hence the secondary amplitudes get progressively smaller by the fractions $\simeq 2/(3\pi), 2/(5\pi), 2/(7\pi)$ ($n = 1, 2, 3, \ldots$) of the principal amplitude

167

height but the approximation used gets *less accurate* as n increases—or N_1 gets smaller. The amplitude ratio of $2/(3\pi)$ means that the intensity ratio is $4/(9\pi^2)=0\cdot045\simeq1/22$ which can be ignored for most practical purposes. We note that for $n=0$ the result is a point in the main peak since there is no minimum between this and the centre. The main peak may thus be said to have a *width* corresponding to $\Delta h= \pm 1/N_1$, i.e. $2/N_1$. The bulk of the diffraction intensity is therefore spread between these limits which become wider, the fewer the number of unit cells in this direction. Hence this *width in reciprocal space is a direct measure of crystallite size*. Furthermore a crystal that has a shape will have the broader spread of diffraction intensity in the direction in which the crystal is smallest. Hence:

	Crystal shape	Reciprocal effect
1.	Sphere larger	Sphere smaller
	↓	↓
	smaller	larger
2.	Cube larger	Octahedron smaller
	↓	↓
	smaller	larger
3.	Platelets, e.g. N_1 small, N_2, N_3 large	Rods, parallel to Ox^*
4.	Needles, e.g. N_1 large, N_2, N_3 small	Plates, parallel to Oy^* and Oz^*

An example of the use of these principles is given by Geisler and Hill [12]. These principles underlie the determination of crystal size and shape by diffraction methods such as line broadening in X-ray powder photographs. This diffraction broadening becomes appreciable for crystal sizes below 10^{-5} cm.

5.3 SUMMARY OF THE CHARACTERISTICS OF THE THREE SOURCES OF WAVES

There remain certain further essential details of the nature of X-ray and the other waves used which will now be described. In section 5.3.4 a comparative table is set out summarising the essential differences.

5.3.1 X-rays

The effective use of X-rays depends upon the possibilities presented by the X-ray spectrum available from a given metal which forms the target or anti-

168

cathode in an X-ray tube. In this tube electrons are accelerated by a high voltage and on impact with the atoms in the target displace electrons from the inner shells of the atoms. The vacancies left are then filled by electrons from adjacent shells, which lose energy according to the fundamental relation

$$\Delta E = h\nu \qquad (5.40)$$

where ΔE is the difference between the energies of the electron in the two states and ν is the frequency of the radiation emitted (h is Planck's constant). In many common materials there are three *characteristic* wavelengths (or frequencies) described by $K\alpha_1$, $K\alpha_2$, $K\beta$, since the inner shell of atoms is the K shell. The α_1 and α_2 wavelengths are close together and are often used in this way. It is possible to eliminate the β wavelength if desired by using the absorption property described below. The wavelength decrease with atomic number according to Moseley's law. L and M wavelengths may also be obtained and may come within the useful diffraction range when the target material is a heavier element such as tungsten, or be of value in X-ray (fluorescent) spectrographic analysis (Chapter 8).

In addition to the characteristic wavelengths a band of 'white radiation' is emitted—the wavelengths in which are spread over a range which cuts off rather abruptly at the upper energy or frequency limit, i.e. shorter wavelength side. If the X-ray tube voltage is low enough, only this white radiation is emitted but the characteristic wavelengths appear as soon as the voltage is raised high enough. The whole spectrum is shown for a typical metal in Fig. 5.6(a). The white radiation arises from the energy given up by an electron which strikes an atom in the target. The energy of such an electron is equal to eV where e is the electronic charge and V the applied voltage. If all the energy is given up in one collision then an X-ray quantum of maximum energy is obtained. Therefore

$$eV = h\nu_{\max} = \frac{hc}{\lambda_{\min}} \qquad (5.41)$$

In general the kinetic energy is not all given up at once and several collisions may occur. Hence quanta of energy are obtained in which $h\nu$ varies over a wide range of frequency, ν lower than $\nu_{\max}$ in (5.41). (Incidentally equation (5.41) is the basis of one of the most reliable direct methods for determining Planck's constant.)

In addition to showing the wavelengths of the X-rays emitted from a typical target material, i.e. copper, Fig. 5.6(a) also shows how the absorption for another element, i.e. nickel, varies with wavelength. This property is responsible for the typical rise in μ (see equation (5.26)) with increasing wavelength; in fact in such a region

$$\frac{\mu}{\rho} = \text{const. } Z^4 \lambda^3$$

169

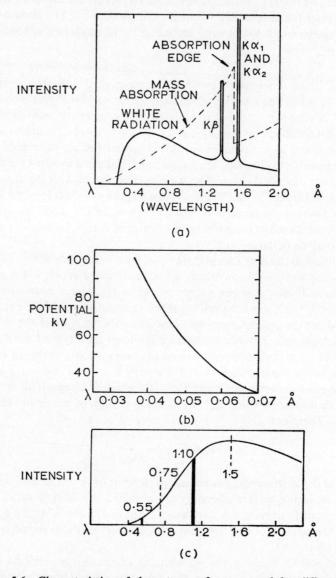

Fig. 5.6 Characteristics of three types of waves used for diffraction.

(a) X-rays; illustrated for copper radiation with mass absorption for nickel

(b) electrons; variation of λ with voltage

(c) neutrons; spectrum showing selection of λ

where Z is the atomic number, but at each step or *absorption edge*, the constant itself is replaced by a new constant. The reason for this step is that the energy of the X-rays near the edge on the short wavelength (higher energy) side is sufficient to displace the electrons from the K shell of nickel. Hence the $K\beta$ copper wavelength has sufficient energy for this purpose and is highly absorbed. The $K\alpha$ wavelengths have insufficient energy and are very much less absorbed. Hence a nickel screen can be used to eliminate the β wavelength when it is unwanted as in powder photography where 'monochromatic' X-rays are desired as far as possible. Other methods of obtaining better monochromatisation are indicated in the next chapter. In general the element which provides a suitable β filter has atomic number one lower than that of the target metal emitting the X-rays. An oxide can be used if necessary since absorption is a property of the atoms. Typical filter materials:

Source of X-rays	Cr	Mn	Fe	Co	Ni	Cu	Zn	Mo	Ag
β Filter	V	Cr	Mn	Fe	Co	Ni	Cu	Zr (O_2)	Pd (or Rh)

A table of wavelengths, and further details of the absorption properties, is given by Taylor [2]. Full wavelength tables are available in the standard publications such as Ref. [11]. We note here that typical X-ray characteristic wavelengths are between 0·5 and 2·5 Å.

5.3.2 Electrons

The fact that although electrons are particles they also behave as waves has already been pointed out and the essential relationship for this property is equation (5.23) which connects the wavelength with the particle velocity (section 5.2.2.3). The kinetic energy of the particles accelerated in a high voltage is given by $\frac{1}{2}mu^2 = eV$ (cf. equation (5.41)) so that from (5.23)

$$\lambda = \frac{h}{mu} = \frac{h}{\sqrt{(2meV)}} \quad \text{Å} \tag{5.42a}$$

with $e = 4\cdot803 \times 10^{-10}$ e.s.u., $m(=m_0) = 9\cdot109 \times 10^{-28}$ g, $h = 6\cdot626 \times 10^{-27}$ erg s, and with V in volts (1 V = 1/300 e.s.u.) and λ in Å

$$\lambda = \frac{12\cdot236}{V} \text{ Å} \approx \sqrt{\left(\frac{150}{V}\right)} \text{ Å}$$

This variation is shown in Fig. 5.6(b). It is, however, less accurate above a few thousand volts. For example λ would require a $2\frac{1}{2}$ per cent correction at 50 kV. This is because mass m in (5.42a) varies with velocity according to the

171

relativity equation (5.21). The exact relationship required is (see Ref. [4] and Ref. [40b] of Chapter 6)

$$\lambda = \frac{h}{\sqrt{[2m_0 eV(1 + eV/2m_0 c^2)]}} \; \text{Å} \tag{5.42b}$$

$$\lambda = \frac{12\cdot236}{\sqrt{[V(1 + 0\cdot9788 \times 10^{-6} \, V)]}} \; \text{Å}$$

Some values of λ used in electron microscopy and diffraction are:

kV	40	60	80	100	200	500	1000
λ(Å)	0·0602	0·0487	0·0418	0·0370	0·0251	0·0142	0·0087

5.3.3 Neutrons

The relationships for neutrons are similar except that the mass m_0 is 1850 times that of the electron, i.e. $1\cdot67 \times 10^{-24}$ g. In the electron case the velocity is obtained by accelerating in an electrical field. In the neutron case the neutrons are liberated during fission and pass through a graphite (or some other) 'moderator' which slows them down by elastic collisions, between neutrons and atoms, until they finally achieve kinetic energies and speeds which correspond to the temperature of the graphite. The neutrons then have mean kinetic energies corresponding to the temperature according to the relations

$$\tfrac{1}{2}mu^2 = \tfrac{3}{2}kT \tag{5.43}$$

k being Boltzmann's constant $= 1\cdot3803 \times 10^{-16}$ erg deg^{-1}, $T =$ absolute temperature. Using this equation and (5.23)

$$\lambda = \frac{h}{\sqrt{(3mkT)}} \tag{5.44}$$

Thus, at 0°C, $\lambda = 1\cdot55$ Å; at 100°C, $\lambda = 1\cdot33$ Å.

These wavelengths are similar to the X-ray wavelengths and are of the right order for diffraction experiments.

The velocity u in equation (5.43) is, however, the root mean square velocity. If the neutrons are allowed to pass out of the atomic pile through a collimator tube they will contain a wide band of wavelengths containing the one given by (5.44). The distribution (for which an equation is given by Bacon [5]) is shown in Fig. 5.6(c). In order to obtain an effectively monochromatic beam of neutron waves a single crystal of a known structure is set at a suitable angle to the collimated beam. Only those neutron waves are diffracted for which the wavelengths are near to a value of λ given by the Bragg law, i.e. $\lambda = 2d \sin \theta$ where θ is the angle between incident beam and reflecting plane. Selection of λ

TABLE 5.1 *Summary of Diffraction by X-rays, electrons, neutrons*

	Electromagnetic radiation	Particles obeying wave/quantum relationships		Reference to equation or sections, etc.
	X-rays	Electrons	Neutrons	
General or common properties				
Wave velocity	c = velocity of light	u		5.2.2.4
Particle velocity	c = velocity of light	v		
Relationship of λ		$uv = c^2$		Equations (5.20)–(5.24)
Variation of mass m	None ($m = m_0 = 0$)	$m = m_0/\sqrt{(1-(u^2/c^2))}$		
Energy	$E = hv = hc/\lambda$	$E = hv = mc^2$		
Wavelength	$\lambda = c/v$	$\lambda = h/(mv)$		
Specific properties				
Rest mass	$m_0 = 0$	$m_0 = 9.11 \times 10^{-28}$ g	$m_0 = 1.67 \times 10^{-24}$ g	
Approx. range of λ	(a) White radiation $\Big\}$ 0·3–2·5 Å (b) Characteristic	0·03–0·07 Å	0·4–2·0 Å	
Determined by	(a) $eV \geqslant hv\ (= hc/\lambda)$ (b) $\Delta E = hv\ (K\alpha_1, K\alpha_2, K\beta, \text{etc.})$	Accelerating voltage $\lambda = \sqrt{(150/V)}$	Choice of λ from wide band due to varying velocities of thermal neutrons. (Mean velocity determined by (5.43))	5.3 and Fig. 5.6
Fixed wavelength selection	(i) Use of characteristic wave-lengths with or without β screen (ii) Use of crystal monochromator	Stabilisation of voltage V	Use of crystal monochromator	
Diffraction Effect of:				
(a) Nucleus	Nil	Combined scattering factor $f(\theta) = \dfrac{e^2}{2mu^2}\dfrac{1}{\sin^2\theta}(Z-f)$ (equation (5.31))	Scattering cross-section of nucleus	$\Big\}$ 5.2.3.1 and 5.2.5
(b) Electrons surrounding nucleus	Atomic scattering factor = $f \propto \sin\theta/\lambda$	Nil	(No direct effect of extra-nuclear electrons)	
(c) Magnetic moment of atom	Nil		(i) Effect of randomly oriented para-magnetic ions (ii) Effect of regular arrangements Ordered structures where atoms are close or adjacent in periodic table. Magnetic structure of metals and alloys —spins and type of ordering, etc.	$\Big\}$ 5.2.5.3 and 5.2.6.4
Some metallurgical applications (See Table 1.2, p. 18)	Crystal structure, ordering, texture, particle size, identity of phases, etc.	Identity of phases, orientation relationships (particle size), etc.		

is thus by choice of the angle θ. For this purpose a calcium fluoride crystal is typical—the plane being (111). Since the same value of θ would correspond to a reflection of half the wavelength, $\frac{1}{2}\lambda$, by a plane of spacing $\frac{1}{2}d$, i.e. (222) in this case, it is better not to chose the peak wavelength which is about 1·5 since $\frac{1}{2}\lambda$ is then 0·75 and has appreciable intensity. A somewhat lower $\lambda = 1·10$ gives $\frac{1}{2}\lambda = 0·55$ which has a much lower intensity. Hence the reflection contains a much smaller proportion of the unwanted wavelength. The width of this band about the selected value of λ depends on the extent to which the collimated beam is not exactly parallel (divergence) and this gives a variation in λ of $\pm 0·025$ Å about the mean (in a typical arrangement). Crystals are also used for producing monochromatic X-ray beams.

5.3.4 Notes on Table 5.1

5.3.4.1 Table 5.1 provides a summary of the essential similarities and differences between the three types of wave as already seen in Fig. 5.6 and described in the previous paragraphs. The table also brings in other information which was necessarily developed earlier in the chapter and thus indicates the diffraction behaviour. Finally there is a brief indication of some metallurgical applications for which the specific waves are particularly useful. In the case of X-rays and electrons these will be developed later. There will, however, be *no further consideration of neutron diffraction here* beyond this stage—in view of its more specialised nature and relatively small availability.

5.3.4.2 One further point emerges from the table and from Fig. 5.6. The electron wavelengths are generally much smaller than the X-ray or neutron wavelengths. From the Bragg equation this means that very much smaller values of θ occur for the same values of d and the diffraction pattern is much more concentrated round the original incident beam direction and in practice the pattern is only recorded for relatively small angles of θ. In the other cases values of θ up to 90° may occur. Since $\sin\theta \leqslant 1$, $d \geqslant \lambda/2$, hence the diffraction pattern is not recorded for values of d smaller than this limit (highest indices). At the other extreme, if $\lambda > 2d$, this would require $\sin\theta > 1$ and so if the largest value of d (e.g. for a (100) plane in some structure) is less than $\lambda/2$ no diffraction will occur at all, i.e. if the wavelength is progressively increased the θ values increase to the limit and the diffraction pattern vanishes as θ passes 90°.

References

1. HENRY, N. F. M., LIPSON, H. and WOOSTER, W. A., *The Interpretation of X-Ray Diffraction Photographs*, 2nd edition, Macmillan, London, 1960.
2. TAYLOR, A., X-ray diffraction, Chapter V, p. 253 of *The Physical Examination of Metals*, eds. B. Chalmers and A. G. Quarrell. Arnold, London, 1960.

3. CULLITY, B. D., *Elements of X-Ray Diffraction*, Addison-Wesley, Reading, Massachusetts, 1956.
4. PINSKER, Z. G., *Electron Diffraction*, trans. J. A. Spink and E. Feigl, Butterworth, London, 1953. (See also Ref. [41] of Chapter 6.)
5. BACON, G. E., Neutron diffraction, Chapter VII, p. 385 of *The Physical Examination of Metals*, ed. B. Chalmers and A. G. Quarrell, Arnold, London, 1960.
6. BACON, G. E., Neutron diffraction, 2nd edition. Clarendon Press, Oxford, 1962.
7. COULSON, C. A., *Waves*, 5th edition. Oliver and Boyd, Edinburgh and London, 1949.
8. PAIN, H. J., *The Physics of Vibrations and Waves*. Wiley, London and New York, 1968.
9. PEISER, H. S., ROOKSBY, H. P. and WILSON, A. J. C., *X-Ray Diffraction by Polycrystalline Materials*. Revised reprint. Chapman and Hall, London, 1959.
10. COMPTON, A. H. and ALLISON, S. K., *X-Rays in Theory and Experiment*, 2nd edition. Macmillan, London, 1935.
11. MACGILLAVRY, C. H. and RIECK, G. D., *International Tables for X-Ray Crystallography* (see Ref. [5], of Chapter 4) vol. III, *Physical and Chemical Tables*, Kynoch Press, Birmingham, 1962.
12. GEISLER, A. H. and HILL, J. K., *Acta Crystall.*, 1948, **1**, 238.

Chapter 6

PRACTICAL DIFFRACTION METHODS (X-ray and Electron)

6.1 SUMMARY

Since the Laüe conditions, or the Bragg Law, are only fulfilled for a given (*hkl*) plane in a specific orientation, diffraction will not occur unless the wavelength and orientation are correctly arranged or adjusted. The methods described in this chapter therefore arise as follows:

1. A single crystal in a *monochromatic* X-ray beam is rotated or oscillated so that from time to time planes will move into a position for diffraction to occur, i.e. λ is constant and the angle between incident beam *and* plane (*hkl*) varied to include θ. This is the method of section 6.2.

2. A single crystal in a fixed position but irradiated by a beam of X-rays containing a band of wavelengths (the white radiation). Here planes of a given *hkl* and *d* spacing have a fixed orientation between plane and beam direction. Diffraction occurs for the wavelength which satisfies the Bragg equation with this angle as θ, i.e. θ specific for each plane and appropriate (different) values of λ selected from the spectrum. This is the Laüe method of section 6.3.

3. For a random polycrystalline or powder sample a fixed wavelength can be used. Diffraction can be made more frequent by rotation or oscillation. Otherwise a particular diffracted beam occurs because some of the numerous randomly oriented grains happen to be in the correct position to satisfy the diffraction condition for the corresponding plane. In effect λ is constant and the angle of incidence again varied to include θ. This is the basis of the powder methods (section 6.4).

4. The electron diffraction procedure can either be similar to the X-ray powder method (3) or similar to the single crystal method (1) with oscillation and fixed wavelength, according to the conditions of section 6.5.

The next section indicates certain relationships which are of more general interest besides being useful in the interpretation of a particular type of pattern.

6.2 X-RAY: ROTATING CRYSTAL METHOD (WITH SOME GENERAL RELATIONSHIPS)

6.2.1 Geometrical and Other Principles of the Experimental Arrangement

6.2.1.1 The essentials of this arrangement are shown in Fig. 6.1(a). Charac-

176

teristic X-rays from which the β component has usually been removed are collimated by passing through a small hole in a horizontal cylindrical rod of brass or other suitable metal. A fine pencil of practically parallel X-rays thus

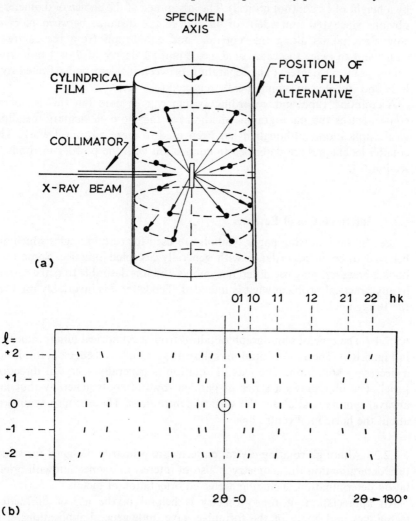

Fig. 6.1   Rotating crystal method.
  (a) general agreement
  (b) diffraction pattern: schematic for $[uvw]=[001]$ and a simple ortho-
      rhombic lattice

encounters the specimen at the centre of the camera. The specimen is usually a single crystal but it can be polycrystalline as for example a fibre, wire or rod specimen which may have a preferred orientation. Provision is made for rotation of the specimen about a vertical axis.

6.2.1.2   The diffraction pattern may be recorded on a flat film, but a cylindrical film is more general. The film holder usually has a diameter of 5·73 cm so that the cylindrical film when flattened out is a rectangle exactly 18·0 cm long by a height of 13 cm (not critical). The advantage of the choice of diameter is obvious since 180 mm≡360° of 4θ and so the distance between pairs of equivalent points along the 'equator' and equidistant from the centre is equivalent to twice the value of θ, i.e. 2 mm≡1 degree of θ or 1 mm from centre point=1 degree. (The separation between incident and reflected rays is 2θ and so between symmetrically reflected rays 4θ.)

A counting tube and recording system can be used but this is not as convenient as the photographic method for the more elementary metallurgical applications, although it has been used in more advanced work. The counter techniques are described in connection with the powder methods in section 6.4.

6.2.2   Interpretation of Patterns

(*Note*: In the following pages symbols are used for certain angles which are believed to be in accordance with generally accepted practice. Some textbooks, however, may use different symbols and it is desirable to make certain in any practical problem what is intended. The letter θ is invariably used for the Bragg angle.)

6.2.2.1   The type of photograph obtained from a cylindrical film is indicated in Fig. 6.1(b). The axis of rotation is shown as a simple cell edge but this is not a necessary condition. The axis of rotation is generally a crystal direction [*uvw*]. The film shows a series of parallel rows of spots generally (but not always) symmetrical about the vertical centre lines. There is also symmetry about the horizontal centre line.

6.2.2.2   Although rotating crystal patterns are primarily designed for structure determination the geometry is also of interest in connection with orientation determination, including preferred orientation of metals.

An appreciation of this geometry is helpful in the use of diffraction techniques and some of the formulae have quite general applications. The essential details are shown in Fig. 6.2.

In Fig. 6.2(a), incident X-rays in the direction X–C–O meet the crystal at C and either a cylindrical or flat film at O. A sphere of radius $r=XC=CO$ is centred at C. As the crystal rotates about a vertical axis CV, diffraction occurs for a plane (*hkl*) and the diffracted ray travels in the direction CPF, when angle $OCP=2\theta$ for this plane. If the radius of the sphere is proportional to $1/\lambda$, OP is proportional to $1/d$ or d^* on the same scale. This is an Ewald sphere as described in section 5.2.2.

178

Let $\qquad CP = CO = r = M(1/\lambda)$

therefore $\qquad M = r\lambda$

and $\qquad OP = Md^* = r\lambda d^*$

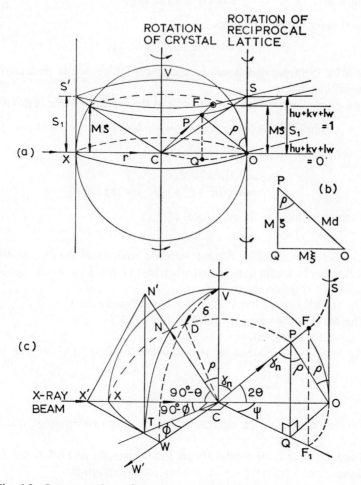

Fig. 6.2 Interpretation of rotating crystal patterns (and some other diffraction relationships).

The diffraction spot is recorded at F on the cylindrical film of radius r which just encloses the sphere, or at the corresponding point on a flat film.

If P and F lie in the equatorial plane through XCO, then the values of h, k, l are those which satisfy equation (5.6), viz.

$$hu + kv + lw = 0$$

The diffraction spots are located at points on the cylindrical film in the appropriate positions. In this case $2\theta = 2\phi = \psi$ in Fig. 6.2(c) and the distance from O on the films are

(a) cylindrical film $\qquad\qquad r\psi = 2r\theta$

(b) flat film $\qquad\qquad\qquad r\tan\psi = r\tan 2\theta$ $\Bigg\}$ $\qquad\qquad$ (6.1)

The next layer of spots requires

$$hu + kv + lw = 1 \quad (\text{or} -1)$$

These arise from parallel layers of reciprocal lattice points separated by a reciprocal distance ξ.

For a given direction $[uvw]$ in the crystal the repeat distance is given by

$$I_{uvw} = \sqrt{[a^2u^2 + b^2v^2 + c^2w^2 + 2bcvw\cos\alpha + 2cawu\cos\beta + 2abuv\cos\gamma]} \quad (6.2)$$

which for the rectangular case reduces to

$$I_{uvw} = \sqrt{[a^2u^2 + b^2v^2 + c^2w^2]}$$
$$= a\sqrt{(u^2 + v^2 + w^2)} \quad \text{for the cube}$$

The corresponding ζ is the reciprocal of I_{uvw}.

6.2.2.3 Before proceeding further with the analysis of the diffraction geometry it might be useful to note that when $[uvw] = [001]$, i.e. a cell edge is used as the axis of rotation then:

(a) The central row of spots are given by $(hk0)$ reflections.

(b) The next row corresponds to spots for which $l=1$ (or -1).

(c) I_{uvw} = the c axis itself.

If ζ, or $(\lambda\zeta)$, is obtained by measurement or from a chart (see below) then

$$I = \frac{1}{\zeta} = \frac{\lambda}{(\lambda\zeta)} \qquad\qquad (6.3)$$

so that cell edges can be determined from inter-row spacings ($l = \pm 1, 2, 3, \ldots$).

(d) Since the three Laüe conditions are simultaneously true it follows that for rotations about a cell axis, e.g. c, we may use the equation:

$$l\lambda = c(\cos\gamma_2 - \cos\gamma_1) \quad \text{with} \quad \gamma_1 = 90°$$

so $\qquad\qquad\qquad\qquad\qquad c = \dfrac{l\lambda}{\cos\gamma_2} \qquad\qquad (6.4)$

γ_2 is then the semi-angle of the cone SCS', i.e. $\angle FCV$ in Fig. 6.2(a).

(e) For the nth row for any axis $[uvw]$, not a cell edge,

$$nI_{uvw} = \frac{n\lambda}{\cos\gamma_n} = n\frac{\lambda}{(\lambda\zeta)} \qquad\qquad (6.3a)$$

180

Here, γ_n is not a Laüe angle but simply the cone semi-angle referred to, i.e.

$$FCV \ (\cos \gamma_n = M\zeta/r = \lambda\zeta \ \text{and} \ \cot \gamma_n = S_n/r).$$

6.2.2.4 In order to define the position of the point P (on the Ewald sphere) a perpendicular is taken from P to the equatorial plane which it meets at Q. Then $OP = Md^* = r\lambda d^*$ (as defined earlier) and

$$PQ = M\zeta = r(\lambda\zeta)$$
$$OQ = M\xi = r(\lambda\xi)$$

and
$$OP^2 = PQ^2 + OQ^2$$

so
$$(d^*)^2 = (\zeta)^2 + (\xi)^2 \tag{6.5}$$

It is clear that if the co-ordinates ζ and ξ are obtained from the photographs values of d^* follow and indexing and other interpretations can be effected or completed. These co-ordinates actually relate to the positions of spots on the films as determined by co-ordinates x and y measured from O. Bernal [1] devised charts for the direct reading of ζ and ξ from films. These co-ordinates must be divided by λ to give true reciprocal distances in terms of Å^{-1}. Hence they are really $(\lambda\zeta)$ or $(\lambda\xi)$ charts if we accept the convention here that d^* is in terms of Å^{-1} and is not itself to be divided by λ. Transparent charts can be obtained from the Institute of Physics and details for their construction are given in the *International Tables*. The treatment of the relationships below necessarily includes $(\lambda\zeta)$ and $(\lambda\xi)$ where, in the other cases, λ is omitted.

6.2.2.5 *Cylindrical Film Geometry*: The y co-ordinate of F is equal to the (constant) layer line separation S_1 which is related to $M\zeta = r(\lambda\zeta)$, by the ratio

$$\frac{y^2}{r^2} = \frac{S_1^2}{r^2} = \frac{\{r(\lambda\zeta)\}^2}{r^2 - \{r(\lambda\zeta)\}^2} = \frac{(\lambda\zeta)^2}{1 - (\lambda\zeta)^2}$$

Therefore
$$(\lambda\zeta)^2 = \frac{y^2}{r^2 + y^2} \tag{6.6}$$

The x co-ordinate of F is given by

$$x = r\psi \quad \text{where } \psi \text{ is the angle } OCF'$$

This is related to lengths in the triangle OCQ by the cosine formula:

$$OQ^2 = OC^2 + CQ^2 - 2OCOQ \cos \psi$$

and
$$OQ = r(\lambda\xi), \ OC = r, \ CQ^2 = r^2 - PQ^2 = r^2\{1 - (\lambda\zeta)^2\}$$

Therefore
$$(\lambda\xi)^2 = 1 + \{1 - (\lambda\zeta)^2\} - 2\{1 - (\lambda\zeta)^2\}^{1/2} \cos \psi \tag{6.7}$$

(removing r^2 from each term).

Substituting for $(\lambda\zeta)$ from (6.6), and $\psi = x/r$,

$$(\lambda\xi)^2 = 1 + \frac{r^2}{r^2 + y^2} - 2\frac{r}{\sqrt{(r^2 + y^2)}} \cos \frac{x}{r} \tag{6.8}$$

181

Hence $(\lambda\zeta)$ can be calculated from y or y/r and $(\lambda\xi)$ from both x and y. Alternatively, a standard chart can be constructed by selecting fixed values of $(\lambda\zeta)$ and so y, to give the horizontal lines. Points along these lines for constant $(\lambda\xi)$ can be found by substituting in (6.8) and using that equation to find x/r.

6.2.2.6 Flat Film Geometry.

Only a film perpendicular to CO is considered. The layer lines are now the intersections of the cones of semi-angle γ_n with a plane and so are hyperbolae. The minimum separation for a layer line is still as S_1 and from (6.6)

$$(\lambda\zeta) = \frac{S_1}{(r^2 + S_1{}^2)^{1/2}}$$

The x co-ordinate of any point is $r \tan \psi$ (see (6.1)). The y co-ordinate corresponds to the distance FF' in Fig. 6.2(c) but for the plane rather than the cylinder. With such a plane, similar triangles CQP, $CF'F$ would give

$$\frac{FF'}{PQ} = \frac{CF'}{CQ}$$

since

$$y = FF', \qquad y^2 = \frac{r^2(\lambda\zeta)^2}{r^2\{1-(\lambda\zeta)^2\}} \frac{r^2}{\cos^2 \psi}$$

$$\frac{y^2}{r^2} \frac{\{1-(\lambda\zeta)^2\}}{(\lambda\zeta)^2} = \frac{1}{\cos^2 \psi} = 1+\tan^2 \psi = 1+\left(\frac{x}{r}\right)^2$$

From which:

$$(\lambda\zeta)^2 = \frac{y^2}{x^2+y^2+r^2} \tag{6.9}$$

and

$$y^2\left\{\frac{1-(\lambda\zeta)^2}{(\lambda\zeta)^2}\right\} - x^2 = r^2 \tag{6.10}$$

Also

$$\cos \psi = \frac{r}{(x^2+r^2)^{1/2}}$$

Equation (6.9) enables $(\lambda\zeta)$ to be calculated from the co-ordinates of a point whilst (6.10) could be used to give the hyperbolae of constant $(\lambda\zeta)$. From (6.7), which applies to any film system, and (6.9)

$$(\lambda\xi)^2 = 1+\frac{r^2+x^2}{r^2+x^2+y^2} - 2\left(\frac{r^2+x^2}{r^2+x^2+y^2}\right)^{1/2}\left(\frac{r^2}{x^2+r^2}\right)^{1/2}$$

i.e.

$$(\lambda\xi)^2 = 1+\frac{r^2+x^2}{r^2+x^2+y^2} - \frac{2r}{\sqrt{(r^2+x^2+y^2)}} \tag{6.11}$$

It follows that the second co-ordinate $(\lambda\xi)$ can be found by measuring x and y on the film. As for the cylindrical case a chart can be completed from lines of constant $(\lambda \zeta)$ (equation (6.10)) by locating points for constant $(\lambda \xi)$ along the curves. The required equation derives directly again from (6.7) (with $x = r \tan \psi$):

$$\frac{x^2}{r^2} + 1 = \frac{1}{\cos^2 \psi} = \frac{4\{1 - (\lambda \zeta)^2\}}{[2 - \{(\lambda \xi)^2 + (\lambda \zeta)^2\}]^2} \tag{6.12}$$

This gives the x co-ordinates of points of intersection and so $(\lambda \xi)$ curves can be contracted from these intersections.

6.2.2.7 Constant θ Curves.

The values of ζ and ξ can be used to give d^* by (6.5) and to index the patterns of crystals where the symmetry has been established and indexing is now required. It follows from the Bragg Law that values of d can be found from θ (and λ) alone, so that the values of θ corresponding to x, y values are also important.

For the cylindrical film:

Again from (6.7) $\cos \psi = \cos \dfrac{x}{r} = \dfrac{2 - \{(\lambda \xi)^2 + (\lambda \zeta)^2\}}{2\{1 - (\lambda \zeta)^2\}^{\frac{1}{2}}}$

$$= \frac{1}{2}\left\{1 + \left(\frac{y}{r}\right)^2\right\}^{1/2}[2 - (\lambda d^*)^2]$$

(from (6.5) and (6.6)). But $\lambda d^* = 2 \sin \theta$, therefore

$$\tfrac{1}{2}[2 - (\lambda d^*)^2] = 1 - 2 \sin^2 \theta = \cos 2\theta$$

Therefore $\cos^2 \left(\dfrac{x}{r}\right) \sec^2 2\theta - \left(\dfrac{y}{r}\right)^2 = 1 \tag{6.13}$

which is the equation to the closed curve on a cylindrical film for a fixed value of θ, or θ in terms of x and y (by re-arranging (6.13))

$$\cos 2\theta = \frac{r}{\sqrt{(x^2 + y^2)}} \cos \frac{x}{r} \tag{6.14}$$

For the flat film:

Circles formed by the intersection of cones of semi-angle 2θ are given by

$$R = r \tan 2\theta \tag{6.15}$$

with $R = \sqrt{(x^2 + y^2)}$. So

$$\left. \begin{array}{l} \tan 2\theta = \sqrt{(x^2 + y^2)}/r \\ \cos 2\theta = r/\sqrt{(x^2 + y^2 + r^2)} \end{array} \right\} \tag{6.16}$$

or

These results enable θ to be calculated directly, or charts constructed. A chart (from The Institute of Physics) provides curves of constant θ according to (6.13) and other angles considered below.

6.2.2.8 *The Angles* δ, ρ *and* ϕ: These angles are useful in orientation problems especially where a stereographic projection may be needed. They can be indicated in terms of other quantities (x, y etc.) or use can be made of inter-relations including θ. Thus:

(i) δ is the angle VCD between the diametral planes through XCO containing respectively the vertical CV and the plane normal CN. From Fig. 6.2(c) it will be appreciated that on a *flat film* δ could be measured directly as it would be given by

$$\tan \delta = x/y \tag{6.17}$$

(ii) ρ is the angle between the normal CN and the vertical and so between d^*, i.e. OP and OS. Hence from Fig. 6.2(b)

$$\tan \rho = \xi/\zeta \tag{6.18}$$

(iii) ϕ is the angle in the equatorial plane (or stereographic projection) between two great circles through V, viz. the one normal to XCO and the one containing N.

In the case when $\rho = 90$, $\phi = 90 - \theta$ and $2\theta = \psi$ (as in section 6.2.2.2).

(iv) By constructing a tangent plane at T perpendicular to CT and taking its intersection with CN and CX as N' and X' as in Fig. 6.2(c), a relationship follows between the angles subtended at C, viz.

$$\cos N'CX' = \cos (90 - \theta) = \frac{CN'^2 + CX'^2 - N'X'^2}{2CN'CX'}$$

$$\text{Numerator} = (N'T^2 + CT^2) + (X'T^2 + CT^2) - N'X'^2 = 2CT^2$$

Therefore

$$\cos (90 - \theta) = \frac{2CT^2}{2CN'CX'} = \frac{r}{CN'} \frac{r}{CX'} = \cos (90 - \rho) \cos (90 - \phi)$$

i.e. $$\cos (90 - \theta) = \cos (90 - \rho) \cos (90 - \phi)$$

or $$\sin \theta = \sin \rho \sin \phi \tag{6.19}$$

(v) The cosine relation for angles subtended at the centre of a sphere, C, is quite general for a spherical triangle such as NTX with one angle 90°. A similar triangle is VDN and angle $DCN = \theta$, $VCN = \rho$, $VCD = \delta$. Therefore

$$\cos \rho = \cos \delta \cos \theta \tag{6.20}$$

(vi) Since the plane PCQ is perpendicular to the equatorial plane in Fig. 6.2(c) the same relation applies viz.

$$\cos 2\theta = \cos \psi \cos (90° - \gamma_n) = \cos \psi \sin \gamma_n$$

This leads directly, by the appropriate substitutions, to equations (6.14) or (6.16).

(vii) Equation (6.20) enables ρ to be found from θ and δ. Equation (6.19) enables ϕ to be found from θ and ρ. By eliminating either ρ or θ from these two equations (or otherwise by suitable construction, etc.)

$$\tan \delta = \tan \rho \cos \phi \qquad (6.21a)$$

$$\tan \theta = \tan \phi \sin \delta \qquad (6.21b)$$

In particular if θ and δ are measured ϕ is obtained and so with (6.20) the location of the pole of (hkl) on a stereographic projection can be completed for a *flat film* without any difficulty. On a cylindrical film $\tan \rho$ can be immediately obtained from (6.18), i.e. ξ/ζ, which enables $\sin \phi$ to be obtained from (6.19).

Alternatively:

$$2r \sin \theta = r(\lambda d^*) = OP = OQ \operatorname{cosec} \rho = PQ \sec \rho$$

or $\qquad 2 \sin \theta = (\lambda\xi) \operatorname{cosec} \rho = (\lambda\zeta) \sec \rho \qquad (6.22)$

There are thus several relationships to chose from according to the angles or quantities obtained from the diffraction pattern and the nature of the problem.
(viii) A chart is available (from the same source as the ζ/ξ chart) giving curves of constant θ, constant ρ and constant ϕ for cylindrical films. This enables values of these angles to be read off directly to a reasonable accuracy for most purposes. (See also section 6.3.2.1.)

6.2.3 Oscillating Crystal Method

When patterns are complicated and have a large number of reflections from a complete rotation it may be useful to obtain a pattern by oscillating through 5, 10 or 15° say. In Fig. 6.2(a) the full pattern is obtained by rotating through 360° equivalent to rotating the reciprocal lattice about OS so that all points within a certain region cut the sphere. Alternatively, one may imagine the reciprocal lattice to be kept fixed and the sphere itself to rotate round OS (tracing out a torus). In the oscillating crystal pattern therefore only those reflections occur for which the reciprocal lattice points cut the sphere during the oscillation.

This procedure not only reduces the number of spots on one film, it also reduces the number of cases where two (or more) spots are superimposed (see 5.2.3.1). This occurs in the cubic case for different possible permutations—p (planar factor)—of the same indices, e.g. $(hk0)$ and $(kh0)$. It also occurs when the sum of squares $(h^2 + k^2 + l^2)$ happens to be the same, e.g. (411), (330). In other systems accidental coincidences may arise due to the particular values of the axial ratio(s) for the crystal under examination. In such cases the planar factor p for a composite spot (from (hkl) planes from different sets) is $p = p_1 + p_2$ but the intensities would be completely independent and should be added. There is clearly considerable advantage in being able to separate these composite spots when they do occur. This application is of less direct interest in metallurgical work but could be required in the study of complex alloy phases or preferred orientation of complex structures.

Some applications of diffraction to metallurgical problems are described

185

in the next chapter, and these include uses of the rotating crystal arrangement requiring the geometry described in this section.

6.3 X-RAY: THE LAÜE METHOD

6.3.1 Experimental Basis

From section 6.1, this is a single crystal method in which the crystal is not moved. It is therefore necessary to provide for the satisfaction of the Bragg law by varying λ, since for any plane (hkl) d is fixed and there is a fixed angle θ between it and the beam direction. This condition is met by employing the band of wavelengths in the white radiation usually from a tungsten or molybdenum tube. In a typical arrangement a 'pinhole' system or collimator (0·5 mm diameter is typical) is used to focus the X-rays—a variety of collimator sizes can be kept for dealing with different materials.

In the usual film arrangements either:

(a) the film is placed at a suitable distance away from the specimen and perpendicular to the incident beam direction on the opposite side to the collimator as illustrated for a single diffraction in Fig. 6.3(a) ($45° > \theta \geqslant 0$, or $2\theta < 90°$). This is the *transmission method*. Or,

(b) the film has a central hole through which the collimator passes so that it records diffraction spots which have been reflected through more than 90°, i.e. $\theta > 45°$ and $2\theta > 90°$. This is the *back reflection method* illustrated in Fig. 6.3(b). The construction lines shown in Fig. 6.3 will be used to explain the patterns.

6.3.2 Interpretation of Patterns

6.3.2.1 Fig. 6.3 shows the position of certain directions and points in a plane containing the X-ray beams—incident and diffracted—and the diffraction spot P. It is seen that this plane contains a line G–O–P in (a) or D–X–P in (b) on which certain points lie and have the following significance and distances from O or X.

(i) P the diffraction spot is either,
  (a) $r \tan 2\theta$ from 0 or
  (b) $r \tan (180° - 2\theta) = -r \tan 2\theta$ from X $\left.\right\}$ (cf. equations (6.1), (6.15))
(ii) CNG is normal to the plane (hkl) giving the diffraction spot.

  G is therefore a point on a gnomic projection on a plane normal to XCO in (a) (XG in (b)) $OG = r \tan (90° - \theta) = r \cot \theta$
(iii) A sphere of radius r centre O gives the circle indicated: N is the pole of (hkl).

Therefore N_{s_1} in either diagram is a stereographic projection of this pole.

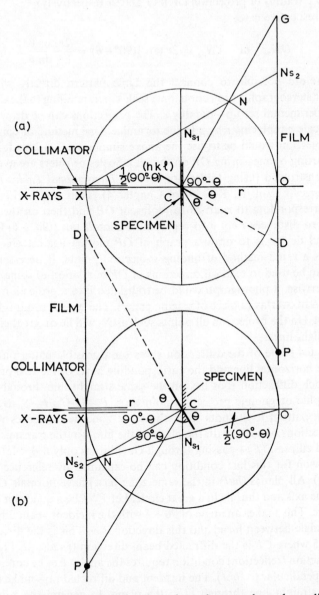

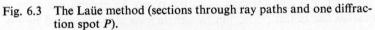

Fig. 6.3   The Laüe method (sections through ray paths and one diffraction spot *P*).

(a) transmission case

(b) back-reflection case

187

Alternatively N_{s_2} can be used since this merely increases the size of the projection by 2 (radius of projection circles r and $2r$ respectively).

In the respective cases

$$ON_{s_2} \quad \text{or} \quad XN_{s_2} = 2r \tan \tfrac{1}{2}(90° - \theta) = \frac{2r \cos \theta}{1 + \sin \theta}$$

It is therefore possible to connect the Laüe pattern directly with either gnomic or stereographic projection, on a scale corresponding to the dimension of the experimental set-up—notably r. The projections can be drawn on any suitable scale p replacing r by p in the formulae. One method of constructing either projection would be to use the above simple relationships for the distances starting by measuring OP on the film. Scales or rulers are available or can be constructed (using tables provided in *International Tables for X-ray Crystallography*, Vol. II, Ref. [9] of Chapter 4). Thus one could mark off points corresponding to $r \tan 2\theta$ to represent OP and then on the opposite side of zero distances $r \cot \theta$ to represent OG or $2r \tan \tfrac{1}{2}(90° - \theta)$ for ON_{s_2}. One useful device is to draw a graph of OP plotted against OG or ON_s. This gives a rapid method of finding projection points. If necessary a protractor can be used to read off directions of OP in relation to some reference line. Otherwise, a photograph could be translated into a projection by using a transparent overlay (a piece of tracing paper). The type of result is indicated in Fig. 6.4. On the projection all points such as N_2 will lie on great circles and G on straight lines.

In Fig. 6.4 some of the diffraction spots on a possible pattern are shown below the horizontal line and the corresponding parts of the two projections above. Each diffraction spot lies on the same straight line through O as its stereographic or gnomic projection point (e.g. P_a–S_a–G_a, P_b–S_b–G_b). In the diffraction pattern itself spots from the same zone lie on curves which are the intersections of cones with the plane of the film—in the transmission case illustrated ellipses, (E)—passing through O. (Ellipses only if $\phi < 45°$.)

The reason for this last condition can be explained by reference firstly to Fig. 6.2(c). All planes (hkl) in the same zone $[uvw]$ have normals CN at 90° to the zone axis and thus lie in a great circle. Let VNT be a quadrant of such a zone circle. This makes an angle of $90 - \phi$ with the incident beam direction so that the angle between $[uvw]$ and this direction is ϕ. This is the direction XC in Fig. 6.5 where CP is the diffracted beam direction (as also in Fig. 6.2(c)). The diffraction (reflection) condition requires the angles θ to be contained in a plane perpendicular to (hkl). The incident and diffracted rays make the angle θ with a straight line through C in the plane. In general the zone axis is *some other line* in this plane, but by simple geometry (suggested by Fig. 6.5) this direction also makes equal angles ϕ with the two X-ray beam directions. Since ϕ is the fixed angle between XC and $[uvw]$, all the diffracted rays must lie on a cone of semi-angle ϕ and axis $[uvw]$. This is also confirmed by noting that if ϕ is constant in equation (6.19) and ρ is varied from 0 to 90° (as in

188

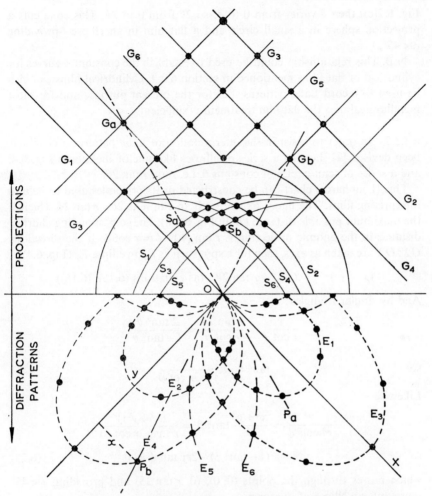

Fig. 6.4 Essential relation between transmission Laüe pattern and projections (illustrated for a (hypothetical) case with rect-angular (showing simple relation between some diffraction spots and corresponding points on the stereographic (s) or gnomic (G) projection)).

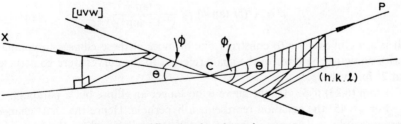

Fig. 6.5 Diffraction by planes in a zone.

189

Fig. 6.2(c)) then θ varies from 0 to ϕ, so 2θ from 0 to 2ϕ. This cone cuts a projection sphere in a small circle and a flat film in an ellipse—providing $\phi < 45°$.

N.B. This relationship could be used to establish the constant ϕ curves for cylindrical or flat films mentioned in section 6.2.2. Cylindrical films can also be used to record Laüe patterns, but for the present purpose and for most metallurgical uses the flat film treatment is sufficient.

6.3.2.2   In order to assist in the interpretation of Laüe patterns a chart has been devised [2] showing a series of ellipses like one of the sets in Fig. 6.4, and a series of semi-circles of constant θ, i.e. radii $r \tan 2\theta$.

This (Leonhardt) chart can be constructed or its principles used to help in interpreting films. The major axis of such an ellipse is $2b = r \tan 2\phi$ which is the maximum possible distance OP in Fig. 6.3(a). The corresponding shortest distance in the gnomic projection is $r \tan (90-\phi) = r \cot \phi$. If the directions OX, OY are taken as axes, then for a spot such as P_b on ellipse E_4 (Fig. 6.4):

$$(x^2+y^2)^{1/2} = OP_b = r \tan 2\theta \quad \text{(as equations (6.15), (6.16))}$$

And by similar triangles:

$$\frac{y}{r \cot \phi} = \frac{r \tan 2\theta}{r \cot \theta} = \frac{2 \tan^2 \theta}{1-\tan^2 \theta}$$

so
$$\tan^2 \theta = \frac{y}{y+2r \cot \phi}$$

Likewise

$$\frac{y^2}{r^2 \cot^2 \phi} = \frac{x^2+y^2}{r^2} \tan^2 \theta = \frac{(x^2+y^2)y}{r^2(y+2r \cot \phi)}$$

so
$$x^2+y^2(1-\tan^2 \phi)-2ry \tan \phi = 0 \tag{6.23}$$

which passes through the points $(0, 0)$, $(0, r \tan 2\phi)$ and providing $\phi < 45°$ represents an ellipse of the form

$$\frac{x^2}{a^2}+\frac{(y-b)^2}{b^2} = 1 \quad \text{or} \quad \frac{x^2}{a^2}+\frac{y^2}{b^2}-\frac{2y}{b} = 0 \tag{6.24}$$

where $b = \frac{1}{2}r \tan 2\phi$ and

$$a = \sqrt{(br \tan \phi)} = \frac{r \tan \phi}{\sqrt{(1-\tan^2 \phi)}}$$

It is a useful exercise to construct one or more of these ellipses to compare with an actual pattern. In the chart referred to the curves were constructed at 2° intervals.

From (6.23) if $\phi = 45°$ the curve is no longer an ellipse but a parabola.

For $\phi > 45°$ the equation represents a hyperbola. Hence the chart changes progressively as ϕ increases from 0 to 90° through 45° where the curves become open. In the hyperbolic region there are two branches. The one of interest for transmission passes through (0, 0) and the other through (0, $r \tan 2\phi$) (y negative). The relationship is of interest in Fig. 6.6(a) where it is clear that a (vertical) plane normal to XCO would cut the lower part of the same (double) cone below CO hence giving the two branches. The lower branch does not appear on the transmission film but would obviously do so if the X-ray beam direction reversed. Hence it appears as the curve for *back-reflection* Laüe patterns as in section 6.3.2.4.

6.3.2.3   In the construction of Fig. 6.4 a hypothetical and somewhat simplified pattern was assumed in order to establish the principles. This diagram may be used to illustrate the following further points:

1. The diagram assumes the X-ray beam perpendicular to the pattern at O is along a four-fold axis of a tetragonal (or cubic) crystal. There are therefore two major sets of ellipses with axes along OX and OY (E_1, E_3, E_5 and E_2, E_4, E_6). These lead, as they should, to the sets of zone circles in the stereographic projection (S_1, etc.) and the rectangular zone lines (G_1, etc.) in the gnomic projection.

2. Spots nearer the centre correspond to zone circles nearer to the plane of projection or gnomic zones further from O.

Intersections of ellipses, etc., which may be occupied by diffraction spots, necessarily coincide with intersections of the zone circles or lines. It is therefore possible to read off from a Wülff net (section 4.5.3) the angles between zones in a series such as S_1, S_3, S_5. If the poles of these zones are plotted at 90° along the diameter intersecting them at their mid-points, and likewise the poles of the zones S_2, S_4, S_6, then several inter-zonal angles can be determined direct from the projection. In the case of a known structure, especially a cubic metal, it would then be easy to index the zones from these angles (which in the cubic case are independent of axial ratios—equation (4.21)). The point of intersection of two zones can be found by the procedure of equation (4.19) applied to the zone case. In a pattern such as Fig. 6.4, OX, OY are also zones for which the poles of planes in the stereographic projection lie along diameters of the projection circle.

3. If in Fig. 6.4, the stereographic projection is omitted and the gnomic projection extended to the centre, it will be apparent that the rectangular network also corresponds to a plane in a reciprocal lattice on a suitable scale. The reciprocal lattice vectors do in fact lie in the same directions as the G points from C as in Fig. 6.4.

Since in this example the X-ray beam is in the direction OZ normal to the plane of the diagram, any Ewald sphere would be tangential to this plane at O and not cut it at any other point so that diffraction spots for $l=0$ cannot occur in this pattern. Since λ varies the sphere will have radii varying between M/λ_{max} and M/λ_{min} (where M is a constant as in section 6.2.2.1). Spheres anywhere between these limits will cut points in the layer for $l=1$ also normal to

191

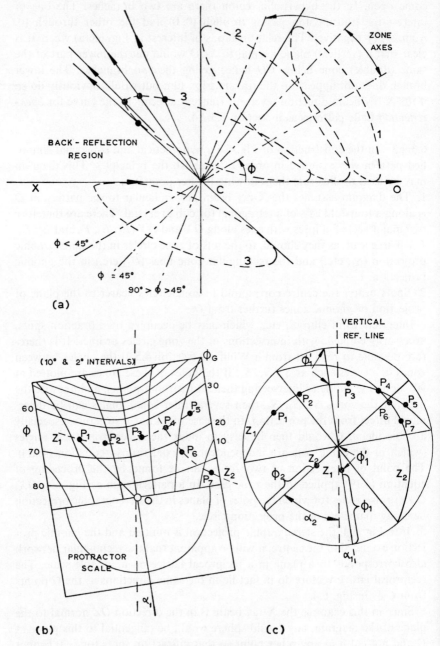

Fig. 6.6 Laüe patterns.
(a) cones formed by diffracted rays from zones
(b) essentials of Greninger chart
(c) corresponding stereographic projection

OZ. These points, for varying hk, will give the corners of the network in the gnomic projection, since a simple network could only arise from one reciprocal lattice plane normal to OZ. Points in the plane for $l=2$ would, by the geometry of the construction, fall in the middle of a square corresponding to half indices $\frac{1}{2}h$, $\frac{1}{2}k$. Clearly such points could occur in the diffraction pattern ($l=2$, $l=3$, etc.) but the main network is usually delineated by the lowest index (>0) in the series. Hence:

Fig. 6.4 is incomplete since points could also occur inside the squares or along the lines. In the latter case they are on the ellipses (or hyperbolae) but not at intersections. In the former case they appear inside the prominent ellipses.

It is worth noting that if the X-ray beam direction is along a cube edge [001] then the normal of a reflecting plane (hkl) is [hkl] and this makes an angle of $(90-\theta)$ with [001], i.e. XC (as in Fig. 6.2(c)). By equation (4.21)

$$\cos(90-\theta) = \sin\theta = \frac{0\times h+0\times k+1\times l}{\sqrt{(h^2+k^2+l^2)}\sqrt{(0^2+0^2+1^2)}}$$

$$= \frac{l}{\sqrt{(h^2+k^2+l^2)}}$$

If l is small, i.e. 1, 2 or 3, then the *smallest* values of θ require the *largest* values of the denominator, i.e. h and k. This is consistent with the construction in Figs. 6.3 and 6.4 and the point made at the beginning of (2) above. Further details and interpretation of Laüe photographs are given by Henry, Lipson and Wooster (Ref. [1] of Chapter 5).

4. In this case therefore either projection can be used and each provides useful information which may assist in indexing or orientation determination. It is found, however, that when a prominent zone axis does not lie along OZ the interpretation becomes more difficult. Hence it has been suggested that the best way to use transmission Laüe patterns for orientation determinations is to build up a collection of standard patterns. The possibility arises of determining the orientation of phases in a metal foil by tilting with respect to the X-ray beam until a recognisable zone appears (a fluorescent screen could be used).

6.3.2.4 The transmission method is satisfactory for films or slices of metal or other material which are thin enough to allow the passage of X-rays, or for single crystals. It can thus be used to determine the orientation of a single grain in a foil. Since, however, many metals are not conveniently reduced to this condition it is desirable to have the back-reflection Laüe method available and in particular adapted for orientation determination.

The paths of diffracted rays giving back-reflection spots lie on cones as in Fig. 6.6(a). These cones must have $\phi>45°$, so $2\phi>90°$. The diagram also shows the cones for $\phi<45°$. These are the cones which give equation (6.23) or (6.24). Thus a vertical plane perpendicular to CO will record the diffraction spot patterns in ellipses ($\phi<45°$), a parabola ($\phi=45°$) or hyperbolae ($90°>\phi>45°$)

on the transmission side as explained in section 6.3.2.2. All the cones must contain the line CO on this side. The back-reflection spots therefore lie on the intersections of cones such as No. 3 (Fig. 6.6(a)) with a vertical plane between X and C. The curves are hyperbolae given basically by (6.23) (with $\phi > 45°$) but are the alternative branch (not the one passing through $(0, 0)$) and are regarded as being transposed from the lower cone on the transmission side to the upper cone on the back-reflection side. This involves a change in the sign of y. The curve thus passes through $(0, -r \tan 2\phi)$, $(-r \tan 2\phi$ is positive).

The equation may be re-written:

$$x^2 - y^2(\tan^2 \phi - 1) + 2ry \tan \phi = 0 \qquad (6.23(a))$$

The zone axes in Fig. 6.6(a) need not be in the same plane through XCO. If we take a vertical plane through CO as reference then the zone axis lies in a plane making an angle α with it. Charts for the location of zones by the angles α and ϕ have been devised by Greninger [3], [4], [5]. The general principles are illustrated in Fig. 6.6(b) where curves of constant ϕ are indicated at $10°$ and $2°$ intervals. These are given by equation (6.23(a)). The chart is rotated until a row of spots, e.g. $P_1, P_2, \ldots$, lie along a zone line. This gives the two angles α and ϕ. The angle on the chart is $90°$ minus the ϕ of equation (6.23) which has been kept here (for consistency with the low angle equation and Fig. 6.2(c)). Greninger's angle is therefore given here by ϕ_G. The corresponding zone Z_1 may be plotted on a stereographic projection as in Fig. 6.6(c). It may be possible to find other rows such as Z_2 (at angle α_2) and with intersections such as P_4. The chart also contains lines which correspond to equal angles along the zones, i.e. along the great circles of the projection. It follows that interplanar angles can be read off directly. These angles correspond to different values of $90 - \rho$, i.e. the angle NCT in Fig. 6.2(c), except that CV is now parallel to the x axis and CW to the y axis. The two sets of curves in the chart thus represent lines of constant ϕ (or ϕ') and constant ρ respectively. The observed angles must correspond to those possible for the crystalline phase giving the diffraction pattern and might in certain cases assist in identification. Use is naturally made of the zone laws, viz. $hu + kv + lw = 0$ (equation (4.20)) and the multiplication rule (4.19). The orientation follows when the plane of the projection is either indexed itself or its distance from a standard zone read off. A direction in the plane completes the description. Further details of the use of such charts are given by Barrett and Massalski (Ref. [3] of Chapter 4).

Additional Notes:

1. The charts are obtainable from the Institute of Physics (London).

2. The equations stated for Greninger charts given in Vol. II of the *International Tables* (Ref. [9] of Chapter 4) actually refer to the intersections of cones containing plane normals and not diffracted rays.

3. The zone lines given by (6.23(a)) can be also defined as follows. If $OP_1 = R$ is the line joining O to a typical diffraction spot, the angle $P_1O\bar{X}$ is the angle δ of Fig. 6.2(c) then $R^2 = x^2 + y^2$ and $y = R \sin \delta$.

Therefore from equation (6.23(a))

$$x^2 + y^2 = R^2 = R^2 \sin^2 \delta \tan^2 \phi - 2Rr \sin \delta \tan \phi$$

So
$$R = \frac{2r \sin \delta \tan \phi}{1 - \sin^2 \delta \tan^2 \phi} \tag{6.25}$$

Greninger gives a table of R values and $(90 - \delta)$ for which the symbol α is again used.

4. The lines of constant ρ (or $90 - \rho$) are not so easy to represent by an equation but points on these curves which are also on a given zone line must necessarily obey the relationship

$$\sin \theta = \sin \rho \sin \phi = \cos (90 - \rho) \cos \phi'$$

for the angles on the chart (equation (6.19)) with $R = r \tan 2\theta$.

6.3.3 Connection with Crystallographic Symmetry

It was noted, in connection with Fig. 6.4, that the transmission Laüe pattern, from a cubic or tetragonal crystal, will exhibit four-fold symmetry if the incident X-ray beam is itself directed along the four-fold axis. This illustrates a quite general principle which may be useful in certain cases to give an indication of the identity of a phase, its crystal system and some other information about its symmetry. One could take diffraction patterns along the apparent symmetry axes of a crystal or tilt a crystal or specimen in various ways until it showed evidence of simple symmetry. 2-, 3-, 4- or 6-fold symmetry would show accordingly whilst mirror planes contained in these axes would give reflection symmetry about the corresponding lines in the plane of the pattern.

The patterns cannot give any indication of whether the axes were simple or screw axes, or the planes other than simple mirror planes. They therefore only reveal point group symmetry as in Table 4.1. They do not, however, distinguish all the 32 crystal classes, because where a centre of symmetry is absent the pattern appears to add one. Thus, from Table 4.1, all patterns from triclinic crystals have a centre and 1, $\bar{1}$ cannot be distinguished. The same applies to monoclinic crystals all of which would appear to have the symmetry $2/m$ (confirmed by adding a centre of symmetry to 2 or to m). The orthorhombic crystals similarly all give patterns corresponding to the presence of three mirror planes at 90° and (necessarily) two-fold axes along their intersections—adding a centre of symmetry to the first two classes.

The other crystal systems are each divided into two groups of crystal classes. Within each of these groups the same simple principle applies, e.g. tetragonal 4, $\bar{4}$, $4/m$ for one such group. In the trigonal case 3, $\bar{3}$ go together, in the hexagonal 6, $\bar{6}$ and $6/m$ and in the cubic 23 and $m3$.

It follows that altogether there are eleven different groups which can be distinguished from each other by Laüe photographs.

6.3.4 A Combined Laüe and Oscillation Method

If a Laüe photograph is taken with unfiltered radiation containing a band of 'white' wavelengths and also characteristic wavelengths it is possible to superimpose an oscillation pattern from the same specimen recording the two exposures on the same film. The specimen is kept stationary for a suitable exposure time and then further exposed whilst being oscillated in the same camera. It then records spots which correspond to points on layer lines, or since the axis of oscillation may not be any specific zone direction to points on powder diffraction rings (see next section). The constant factor is the angle ρ between the normal to a plane and the axis of oscillation. A Laüe spot may therefore be joined to a characteristic spot by a streak which lies along a line of constant ρ as on the charts referred to earlier (section 6.2.2.8). The characteristic spot will lie on a recognisable ring or curve of constant θ. This can usually be assigned indices for a set $\{hkl\}$. The angles ρ and ϕ can be read off from a chart or calculated for the connected Laüe spot. Consequently a corresponding point can be plotted on a stereographic projection. Three points can be used to determine the orientation of the crystal (in the fixed position). This method has been used to determine the orientation of diamonds in wire-drawing dies. Further details of the method are given by Henry, Lipson and Wooster (Ref. [1] of Chapter 5).

6.4 THE X-RAY POWDER METHODS

6.4.1 General Introduction

The powder methods require particular consideration because they provide the means of employing X-ray diffraction to solid polycrystalline materials or powders. In many metallurgical applications the crystal structure is known and taken for granted. Extensive compilations are available containing such information about metallurgical phases and systems (Refs. [6, 7] of Chapter 4 and Ref. [9] of this chapter). New alloy phases do, however, continue to be found and the powder patterns may be used to determine the structure. This is admittedly easiest for the simpler structures in systems of higher symmetry (especially cubic, tetragonal and hexagonal or rhombohedral). It may be essential when only very small quantities of a material are available in powder form or where it is very difficult to obtain single crystals large enough for handling.

The basic geometry follows quite readily from the principles already described. The experimental methods which will then be outlined classify in three ways:

1. According to the way the pattern is recorded:
   (a) Photographic methods.
   (b) Counter diffractometry.

196

2. According to the X-ray beam geometry:
   (a) Using parallel or collimated X-rays.
   (b) Using focusing methods.
3. According to the part of the total pattern recorded, viz.
   (a) Virtually the whole range of θ values as in cylindrical powder methods.
   (b) Transmission method—low values of θ.
   (c) Back-reflection method—high values of θ.
   (d) 'Glancing angle' or 'side-reflection' method—intermediate values of θ.

These basic methods will be described first assuming photographic techniques and collimated incident X-rays. The photographic focusing methods can then be conveniently considered together. Another separate section then outlines the use of counter diffractometry arrangements.

The originators of the powder methods were Debye and Scherrer [6] and Hull [7]. Two standard texts (Ref. [9] of Chapter 5 and Ref. [8] of this chapter) are devoted specifically to these methods which are also substantially covered by others (Refs. [2, 3] of Chapter 4 and Ref. [3] of Chapter 5).

6.4.2 Basic Geometry

The powder methods employ the characteristic K wavelengths from which the $K\beta$ has usually been removed by a 'β screen', i.e. a material with an absorption edge between the $K\alpha$ and $K\beta$ of the radiation (see previous chapter, section 5.3.1). Common anticathode metals in the X-ray tube include Cr, Fe, Co, Ni and Cu. The β screen is usually the element preceding the source metal in the Periodic Table, i.e. one atomic number less. One or other of the available types of monochromator may be used. These have wider applications in connection with focusing arrangements.

The essential feature of powder diffraction is that the X-ray beam irradiates large numbers of small crystals some of which will be in a position to diffract according to the Bragg law, i.e. planes with specific indices hkl happen to make the correct angle, θ, with the incident beam. The diffracted ray will make an angle 2θ with incident beam direction. All such rays can lie anywhere on a cone of semi-angle 2θ. Fig. 6.2(c) again provides a useful starting point. The crystals that can give diffracted rays for a specific value of θ have normals CN all making the fixed angle $90 - \theta$ with XC and give diffracted rays such as CPF all making the angle 2θ with CO. The construction for powder photographs is hence shown in Fig. 6.7(a). It is simply derived from Fig. 6.2(c) by rotating the plane containing CN and CF about the line XCO, through 360°. The distribution of diffracted intensity around the cone of semi-angle 2θ will be continuous if the number of grains with suitable oriented planes is large enough. Otherwise discontinuous or 'spotty' rings would result (e.g. on a flat film in the transmission position). In many powder cameras the specimen is therefore oscillated or rotated in order to bring as many planes as possible into the diffracting position.

197

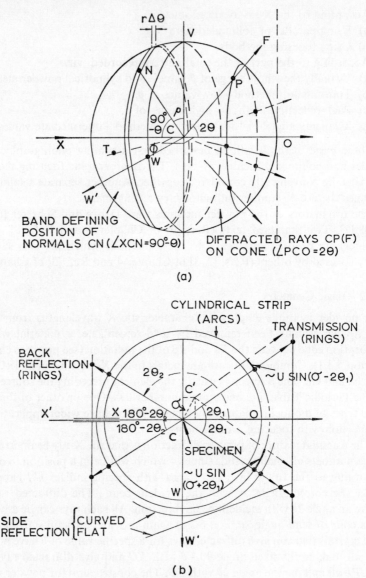

BAND DEFINING
POSITION OF
NORMALS CN (∠XCN=90°-θ)

DIFFRACTED RAYS CP(F)
ON CONE (∠PCO=2θ)

(a)

CYLINDRICAL STRIP
(ARCS)

TRANSMISSION
(RINGS)

BACK
REFLECTION
(RINGS)

~U SIN(σ-2θ₁)

X 180°-2θ₁
180°-2θ₂

2θ₁
2θ₁

O

SPECIMEN

~U SIN
(σ+2θ₁)

SIDE   {CURVED
REFLECTION {FLAT

(b)

Fig. 6.7.
(a) geometry of powder methods
(b) experimental arrangements (direct unfocused beam)

The construction of Fig. 6.7(a) refers particularly to a collimated (by a small cylindrical tube with round bore or slits) 'parallel' beam of incident X-rays. Even with a small circular hole a beam is never strictly parallel: there is always a finite angular spread. This is equivalent to saying that the direction

XC in Fig. 6.7 (or Fig. 6.2(c)) can lie anywhere in a small cone. It is preferable to allow for this by assuming that XC is a straight line, but diffraction can occur for a small angular spread of the direction CN represented here by $\Delta\theta$. This detail is required later to complete the formula for the intensity of a powder diffraction line. The next step from Fig. 6.7(a) is to recognise the relation of this geometry to the experimental arrangement summarised in section 6.4.1 and Fig. 6.7(b). Thus:

(a) The film may lie on a cylinder at a constant distance from the centre C where the specimen is itself rod-like or cylindrical. This film intersects the diffraction cones of semi-angle 2θ in curves which become straight lines at $\theta = 45°$ and close in to small rings near O on X.
(b) A flat film in the transmission position would record circles.
(c) A flat film in the back-reflection position likewise records rings.
(d) A curved film in the side-reflection or glancing angle methods records cone–cylinder intersections, and a flat film, cone–plane intersections, i.e. simple hyperbolae.

A fuller description of these different experimental arrangements and essential details of the interpretation of patterns are now provided. (The significance of the point C' and rays represented by broken lines is referred to in section 6.4.7.3.)

6.4.3 Cylindrical Powder Cameras

6.4.3.1 Horizontal and vertical arrangements: In these arrangements the camera is sometimes fixed with the cylinder axis in a horizontal position. In this case the specimen must be a rigid rod or wire. Cameras with the axis vertical can use specimens in which powder is stuck (by Canada balsam or some other adhesive) onto a hair held straight by a small lead weight. Alternatively, in either case, the powder can be glued to a small glass fibre or packed into a capillary tube made of a suitable glass containing elements of low atomic number and low absorption (Lindemann glass). Metallic rod or wire specimens can be used directly.

6.4.3.2 Types of film arrangement: Determination of θ: It is necessary to be able to determine the angle θ for each line from its position on the film. Since the film must originally have been flat it must be fitted round the cylinder as in Fig. 6.7(b) and there will be a hole or a gap at X where the X-ray beam collimator must pass into the cylinder and opposite where the direct beam leaves. In the smaller diameter cameras, 9 cm is a standard diameter, a single strip is used. Cameras of 19 cm diameter are sometimes used in metallurgical research and these take two lengths of film since a single length would be difficult to handle and process. Powder photographs can obviously also be obtained in the cylindrical rotating crystal camera. In particular the useful diameter 5·73 cm (referred to in section 6.2.1) makes 1 mm exactly equal to 2 degrees of the total diffraction cone angle (4θ). A cylindrical powder

199

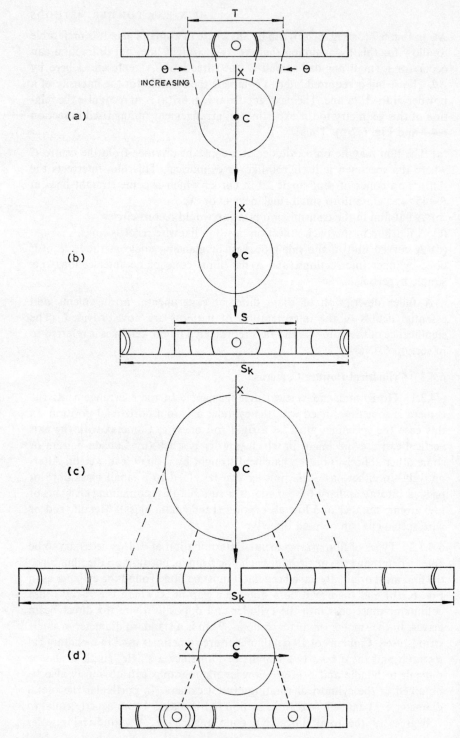

camera is available with exactly twice this diameter so that 1 mm is then equivalent to 1°. This arrangement of dimensions is useful for approximate measurements of θ but, for accurate determinations, the films require to be calibrated and the line positions accurately measured. Provision for this is made according to the choice of the film arrangement and must allow for film shrinkage after processing and drying.

Fig. 6.8 illustrates the film arrangements in most general use (especially the first three) and these are:

(a) Arrangement due to van Arkel [10]. In this case the X-ray beam collimator passes through a hole in the film which closes round onto a metal block at the opposite side. The main undiffracted beam passes out of the camera through a hole in this block, which, together with the similar block containing the collimator, holds the top and bottom flanged discs accurately in position so providing the cylindrical frame round which the film can be accurately held by metal strips or other means. In this arrangement the distance T is measured between the lines on either side of the centre corresponding to the same diffraction cone. The required value of θ is given by

$$\theta = \frac{\pi}{2} - \frac{T}{4R} \quad \text{radians} \tag{6.26}$$

where R is the effective radius of the camera. Relative errors in θ get smaller as θ increases and this would imply that small errors in the assumption of constant effective R or in its determination, would not be critical. In any case the same error in θ produces progressively smaller errors in interplanar spacing (or lattice parameter) as θ approaches $\frac{1}{2}\pi$. This point is developed in a later paragraph.

(b) The Bradley–Jay Camera [11]. This has the reverse arrangement from (a). A film calibrating device is necessary and this consists of a small curved metal plate with sharp knife edges which cast sharp shadows at the high-angle ends of the film. When the film is processed and dried the sharp cut-off shadow at each end of the exposed portion corresponds to a fixed angle θ_k. This angle can be found.

(i) By direct measurement on an accurately graduated turntable.
(ii) By direct measurement of the distance between the edges across the plate (on a travelling microscope).
(iii) By calibration with a substance of accurately known lattice parameters. Quartz was originally used for this purpose, but the direct methods are more generally preferred.

Fig. 6.8 Cylindrical powder cameras.
(a) Van Arkel
(b) Bradley–Jay
(c) Bradley–Bragg (19 cm diameter)
(d) Straumanis–Ieviņš

In this case it is not possible to calculate angles from the camera radius since varying film shrinkage means that the true R differs from film to film, and this would produce an increasing error in θ as $\theta \rightarrow \frac{1}{2}\pi$. Instead the individual distance between the knife-edge shadows, i.e. S_k, can be found for each film, and then θ taken as proportional to S according to

$$\frac{\theta}{\theta_k} = \frac{S}{S_k} \quad \text{or} \quad \theta = S\left(\frac{\theta_k}{S_k}\right) \tag{6.27}$$

Thus θ_k is a camera constant (fixed—but to be checked periodically) and S_k is a constant for each film. Here θ, and θ_k, can be in degrees. This method could be—and is—applied to the van Arkel arrangement, the angles calculated from (6.27) being then subtracted from $\frac{1}{2}\pi = 90°$ to give θ.

(c) The large 19 cm diameter camera described by Bradley and Bragg [12] requires the two lengths of film and calibration of two knife-edges. Calculation is assisted by converting the low angle knife-edge constant to a length of arc and using equation (6.27) as for the Bradley–Jay arrangement.

(d) The arrangement due to Straumanis and Ieviņš [13], shown here, has the film gap half-way between the collimator and beam exit on one side, so that there is a complete length of film from $2\theta = 0$ to $2\theta = 180$ on the other side. In this way the pair of lines on either side of the low angle and high angle holes respectively can be used to locate the exact positions for 0 and 180°, and the θ values for these and the intervening lines then accurately determined in proportion to their lengths from zero. Cold worked metals, or poorly crystalline materials in particular, and others with complex structures giving many lines will tend to show weak and diffuse lines towards high values of 2θ and so may be lost in the background. It may therefore happen that the location of the 180° position is difficult or impossible. Wilson [14] has pointed out that this difficulty can sometimes be met by placing the gap nearer to the low angle position so that sharper lines with 2θ appreciably less than 90° appear on the higher angle side of the gap. Thus, the gap could be at $2\theta \simeq 45/50°$, and some of the sharper lines lie on either side of it.

6.4.3.3 Calculation of d and Lattice Parameters from θ: When θ has been found the interplanar spacing is calculated from the Bragg equation (equations (5.5), (5.6))

$$\lambda = 2d \sin \theta \quad \text{or} \quad d = \frac{1}{2}\lambda \operatorname{cosec} \theta$$

This may provide all the information desired. Parrish et al. [15] have constructed charts for different radiations which give d to a reasonably high degree of accuracy direct from θ or 2θ.

Alternatively the lattice parameters of the unit cell may be found from the appropriate formula, e.g. equations (4.10)–(4.15) in Chapter 4. For purposes of calculation direct from θ these equations can be combined with the Bragg equation. For cubic crystals

$$a = \frac{1}{2}\lambda\sqrt{N} \operatorname{cosec} \theta \quad \text{where} \quad N = h^2 + k^2 + l^2 \tag{6.28}$$

Tables of $\frac{1}{2}\lambda\sqrt{N}$ have been provided by Parrish and Irwin [16]. Values of $\sqrt{N}$ are also given in Vol. II of the *International Tables* (Ref. [9] of Chapter 4) and in some text books.

In the case of hexagonal crystals, $\sqrt{N}$ in (6.28) is replaced by the corresponding function including the axial ratio c/a:

$$a = \frac{\lambda}{2}\sqrt{\left/\left\{\frac{4}{3}(h^2+hk+k^2)+\frac{l^2}{(c/a)^2}\right\}\right.}\,\text{cosec }\theta \qquad (6.29)$$

(The expression follows easily from equation (4.11).) An alternative form gives c as a function of the same variables (indices, axial ratio and θ). Tables of the function under the square root sign for each case have been described by Massalski and King [17]. For calculations of this kind the initial choice of c/a may involve some error. This can be dealt with in the methods for obtaining accuracy to be considered separately in section 6.4.7.

A corresponding treatment is possible for tetragonal crystals [18]. For this system

$$a = \frac{\lambda}{2}\sqrt{\left/\left\{h^2+k^2+\frac{l^2}{(c/a)^2}\right\}\right.}\,\text{cosec }\theta \qquad (6.30)$$

For orthorhombic crystals two axial ratios are required and they can be treated similarly but with more difficulty—or otherwise by methods which are indicated below.

It has, however, been taken for granted that the indexing is already known. This may not always be the case and indeed the determination of the crystal system lattice type and unit cell may be required from a powder photograph.

6.4.3.4 Further General Interpretation: From this point of view the analysis of cubic crystals is simplest. Here the integer N increases with θ and from 5.2.6.2 and 5.2.6.5 we have:

(i) No cubic pattern can have N given by the forbidden numbers $4^m(8n+7)$, m and n are separate integers $(0, 1, 2, \ldots)$.
(ii) A primitive cubic structure will otherwise show most or all values of N except where the structure factor F is vanishingly small.
(iii) A body-centred cubic structure has $N=2, 4, 6, 8, \ldots$ (a consequence of the condition $h+k+l$ even).
(iv) A face-centred cubic structure has $N=3, 4, 8, 11, 16, 19$ ($4n$ or $8n+3$).
(v) A diamond cube has $N=3, 8, 11, 16, 19$ ($8n, 8n+3$).

The indexing, and in some cases also the lattice parameter determination itself, may be more readily achieved by the use of the reciprocal relationships such as equations (4.10)–(4.15) with the reciprocal form of the Bragg equation (5.7). Thus:

$$\frac{2}{\lambda}\sin\theta = \frac{1}{d}; \qquad \frac{4}{\lambda^2}\sin^2\theta = \frac{1}{d^2} = \frac{h^2}{a^2}+\frac{k^2}{b^2}+\frac{l^2}{c^2} \quad \text{orthorhombic}$$

or
$$\frac{4(h^2+hk+k^2)}{3a^2}+\frac{l^2}{c^2} \quad \text{hexagonal}$$

or
$$\frac{h^2+k^2}{a^2}+\frac{l^2}{c^2} \quad \text{tetragonal} \qquad (6.31)$$

Hence $\sin^2 \theta$, or the derived $1/d^2$, is treated as the square of a reciprocal vector and resolved into its components. The following relationships assist this process.

(i) Low angle lines again are likely to have low indices such as (100), (200), (110), etc.

(ii) Procedures of analysis can also use the fact that any $\sin^2 \theta$ or $1/d^2$ values which go in a sequence of multiples 1, 4, 9, ..., probably derive from one cell edge (two indices zero).

(iii) Subtracting each of the $\sin^2 \theta$ (or $1/d^2$) values from all the others may reveal quantities which occur frequently (within experimental limits). These are likely to be identical with, or simple multiples of, $1/a^2$ ($4/(3a^2)$ in the hexagonal case), $1/c^2$, etc. Numerical methods based on these principles have been proposed by Hesse [19] and Lipson [20] (and Ref. [1] of Chapter 5) whilst the general triclinic case was treated by Ito [21]. For further details the standard texts should be studied (Refs. [1] and [9] of Chapter 5). In hexagonal and tetragonal cases recognisable relationships of this kind may lead to estimates of a (lines with $l=0$) and of c (h and $k=0$), which can then be used to provide a value of c/a for direct use in (6.29).

Alternatively, for these systems in particular charts have also been constructed—those by C. W. Bunn [22] are perhaps the most widely used. (They can be obtained from The Institute of Physics.) These charts involve a principle derived directly from the basic equations and are simply illustrated here for the tetragonal case. From equation (4.13), the interplanar spacing can be expressed by

$$\frac{1}{d^2} = \frac{1}{a^2}\left\{h^2+k^2+\frac{l^2}{(c/a)^2}\right\}$$

Hence the ratio of two spacings d_1, d_2 corresponding to $(h_1k_1l_1)$ and $(h_2k_2l_2)$ gives an expression dependent on the indices and axial ratio but not the separate magnitudes of a and c. Therefore curves of

$$\log\left\{h^2+k^2+\frac{l^2}{(c/a)^2}\right\}$$

are drawn as functions of c/a. The distance between the two curves for $(h_1k_1l_1)$, $(h_2k_2l_2)$ is always $-2(\log d_1 - \log d_2)$ for a given axial ratio. Values of experimentally determined d's are plotted logarithmically on a strip of paper and moved over the chart until the points coincide with points on a sequence of curves at a fixed value of c/a. This indicates immediately the ratio and the indices. (As previously indicated, not all indices need be represented by actual lines on the film—a consequence of either systematic absences for which $F=0$ or accidental absences in which it is vanishingly small.) A number of other charts based on similar principles have been devised [23–25].

The information derived by the measurement of powder patterns and calculation of θ, d and/or lattice parameters, as outlined in these paragraphs, leads to results of varying degrees of accuracy according to factors which must be considered separately after the other powder methods have been described. The *achievement of the highest accuracy* and *the determination of accurate lattice parameters* by extrapolation methods is important in connection with many uses of the method (see section 6.4.7).

6.4.4 Flat Film Methods for Transmission, Side-Reflection and Back-Reflection

6.4.4.1   General: All these methods have already been indicated in Fig. 6.8. One may also refer to Fig. 6.3 for the Laüe method. If the single crystal specimen here is replaced by a thin layer of powder, a metal foil or a thin slide of any polycrystalline solid, then in Fig. 6.3(a) the diffracted ray CP would lie anywhere on a cone of semi-angle 2θ obtained by rotating CP about the axis XCO. For the back-reflection case a solid polycrystalline block can replace the single crystal specimen. The diffracted rays, now numerous, perhaps sufficiently so to give continuous rings, lie on a cone for which the semi-angle is $2(90 - \theta)$ as in Fig. 6.3(b).

6.4.4.2   Transmission case: The main requirement here is that the specimen should be thin enough so that the X-ray absorption is not too high. Otherwise the exposure time would be far too long. Alternatively a small specimen must be used. If the specimen film distance is r then the distance of a spot from the film centre, i.e. the ring radius is $r \tan 2\theta$ as before. If r is known the ring diameter D is used to give θ, e.g. $\tan 2\theta = D/2r$.

6.4.4.3   Side-reflection of glancing angle method: This technique is useful for solid specimens including metals which cannot reasonably or conveniently be made into rod or wire specimens for cylindrical cameras. Such a specimen is set in the path of the X-ray beam at an angle to it. Diffracted rays are cut off for values of 2θ below this angle, and so in order to get as much of the pattern in as possible the glancing angle should be small, but this tends to make the lines broader than is desirable. For many purposes an angle of 10–20° is, however, quite satisfactory. A curved film cassette can be used or otherwise a cylindrical camera can be adapted for this purpose. Alternatively a flat film holder is located in a suitable position (see Fig. 6.7).

The determination of θ in this arrangement may be assisted by the use of calibration if an adapted cylindrical camera is employed. If a separate film holder is used this can be provided with a scale graduated in angles, or a material of known spacings can be used for calibration separately or as an 'internal' standard (e.g. a standard substance coated onto the metal surface).

205

In some metallurgical applications the principal phase is a simple metallic constituent (e.g. body-centred or face-centred cube) and its lines can be used to determine the intervening lines for other phases such as intermetallic compounds, carbides, oxide layers, etc. The interplanar spacings should not be interpolated directly but angles θ can be, although less accurately with a flat film. It is noted that

$$\frac{1}{d} = \frac{2}{\lambda} \sin \theta = \frac{2}{\lambda} \left[\theta - \frac{\theta^3}{3!} + \frac{\theta^5}{5!} - \cdots \right]$$

with θ in radians and proportional to the linear distance along the curved film. Hence interpolation is easily achieved by plotting $1/d$ against distance along film. The separation between two points reasonably close together is:

$$\Delta\left(\frac{1}{d}\right) = \frac{1}{d_2} - \frac{1}{d_1} = \frac{2}{\lambda}(\sin \theta_2 - \sin \theta_1)$$

$$= \frac{2}{\lambda}[2 \sin \tfrac{1}{2}(\theta_2 - \theta_1) \cos \tfrac{1}{2}(\theta_1 + \theta_1)]$$

$$\Delta\left(\frac{1}{d}\right) = \frac{2 \, \Delta\theta}{\lambda}[\cos \theta_m] \quad \text{where} \quad \theta_m = \text{mean } \theta \qquad (6.32)$$

This expression justifies linear interpolation between such points up to differences in θ of several degrees, but is less justified towards higher absolute values of θ because of the increasing curvature of $\cos \theta$.

It is also evident that if a point for $\theta = 0$ is on the film and distances of lines from this point are given by S then $d \times S$ will vary very slowly with S and can be plotted graphically against S. Any value of $d \times S$ read off from the curve is then divided by S to give d.

6.4.4.4 Back-reflection case: Since the angle 2θ between incident and diffracted ray is now greater than 90°, the acute angle $2(90° - \theta)$ is taken. The *radius* of a back-reflection ring is then given by

$$r \tan 2(90° - \theta) = r \tan (180° - 2\theta)$$

(or $= -r \tan 2\theta$). In this case the thickness of the specimen is not critical. Also even if the radiation is filtered to remove the β component the $K\alpha$ radiation is now split into two components due to the pair of wavelengths α_1 and α_2. These are too close to be separated at low angles in normal cameras but the resolution increased with increasing θ—a point which receives further consideration in connection with the general accuracy of powder methods.

It will be apparent that the measurement of ring diameter on the film and a knowledge of the specimen to film distance r is sufficient to determine θ. r may be accurately measured by a mechanical device such as that suggested by Thomas [26]. Alternatively a powder substance of known standard spacing may be coated lightly onto the specimen. The two diffraction patterns then superimpose and the standard can be used to find a value of r appropriate not only to the experimental conditions, but also allowing for film shrinkage.

206

6.4.5 Focusing Methods: Including Monochromators

6.4.5.1 A disadvantage in the use of X-ray cameras with parallel collimated beams is that most of the X-radiation from the tube is not used. A relatively small fraction of the total emission emerges through one window and most of this is then cut off except for the tiny pencil that passes into the camera. Secondly, the intensity of a line or ring will naturally be limited by the amount of diffracting material or number of crystals and their size.

In focusing arrangements a larger fraction of X-ray intensity is first of all brought into use by having an X-ray tube with a line focus. The rays from this focus S (see Fig. 6.9(a)), over a considerable solid angle, fall on a curved crystal CC' which is arranged to reflect the rays again to a sharp point S'. The system employs the principle that if S, CC' and S' lie on an arc of a circle then angle CSC' = angle $CS'C$ so that the rays will be exactly focused at S' if this requirement is met. This condition is indicated simply in Fig. 6.9(a) and the curved crystal CC' constitutes a *monochromator* which in fact not only focuses the rays but does indeed only reflect and focus those of one wavelength. In practice a small range is focused, e.g. over a single characteristic line or including perhaps $K\alpha_1$ and α_2 but not β or much of the background on either side.

The monochromator uses Bragg reflection to focus the rays and so suitable strongly diffracting planes must be in the appropriate position. A ray path SCS' must involve the precise angle θ for incidence and diffraction. So likewise must a path $SC'S'$ or any in between. This result can be achieved firstly by cutting the crystal with the diffracting plane parallel to the surface and grinding it cylindrically to a radius R. The reflecting plane at C then makes an angle α at C with the tangent at C and thus the chord CC' subtends 2α at the centre of this (larger) circle. The crystal is now bent round a circle of (smaller) radius r so that it passes through SS'. The illustration here is for a symmetrical case. The reflecting plane PC still makes the angle α with the new tangent CT, and if PC is produced to P', $\angle P'CS' = \theta$, but if A is the mid-point θ is also the angle between the tangent at A and the chord $S'A$. Hence angle (in opposite segment) ACS' also $= \theta$ and so $P'C$ passes through A. Therefore angle $AS'C = \angle PCT = \alpha$ and angle $AOC = 2\alpha$ and CC' now subtends 4α at O. Hence the radius r must be half R for the same arc CC'.

The arrangement in Fig. 6.9(a) is due to Johansson [27]. An asymmetrical arrangement based on the same principles has been described by Guinier [28]. This places the focusing crystal nearer the tube focus S and so saves X-ray intensity (which decreases inversely as the square of the distance). An alternative system developed by de Wolff [29] bends a thin crystal round a logarithmic spiral for which all rays from S meet the curve at the same angle θ. Further details are given by G. W. Brindley [30]. It will be appreciated that the ground and curved monochromator is not adjustable for different radiations but the other type can be. It is also noted that parallel plate

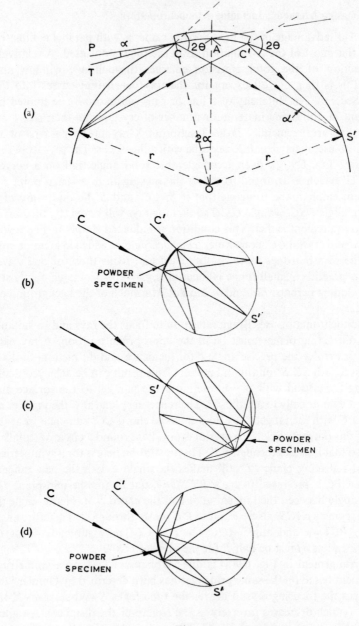

Fig. 6.9.

(a) focusing monochromator
(b) symmetrical transmission arrangement
(c) back-reflection arrangement
(d) asymmetrical transmission arrangement

monochromators have been devised for use with single crystals and other cameras with parallel beams.

6.4.5.2 The focusing cameras use the same principle of focusing on a circle but the circle has a smaller radius. The essential geometry is therefore illustrated in the remaining details of Fig. 6.9. It is noted that the powder specimen should ideally lie along the circumference of the circle, but, especially for the transmission case, a thin layer of powder can conveniently be spread over a flat film of thin transparent supporting material such as cellophane. Fig. 6.9(b) represents the transmission case—only one pair of diffraction lines is shown. In Fig. 6.9(c), the back-reflection case, the point S' lies on the small circle opposite the specimen. An asymmetric transmission arrangement is also shown. This virtually excludes all diffracted rays on one side of the main beam and allows a large fraction of the total circumference to take diffraction from $\theta = 0$ to (about) $45°$ (i.e. up to $2\theta \simeq 90°$ or slightly more). This gives very high resolution. An asymmetric arrangement using reflection rather than transmission has also been used.

A useful development of Fig. 6.9(c) is due to de Wolff [31]. Because of the length of the line focus S (Fig. 6.9(a)) the fullest advantage has been taken of this, by splitting the beam into four and having four powder specimens in a metal plate holder. The diffracted rays are separated in layers by partitions. The camera is thus like four separate focusing cameras placed one above the other. The advantages of this *quadruple focusing camera* are obvious—especially the economy in total exposure times for the same number of patterns and the high resolution.

In all these arrangements use is made of the fact that the angle between incident and diffracted rays is always 2θ, or $\pi - 2\theta$. In Fig. 6.9(b), for example, the incident ray is anywhere in the wedge between CS' and $C'S'$. This cuts through the (curved) specimen at a point on the circumference. The angle subtended at all such points by the chord $S'L$ is 2θ. Hence exact focusing occurs at L.

6.4.6 Powder Diffractometry

6.4.6.1 Microdensitometers: As a preliminary to the principles of diffractometry mention should be made of the use of a *microdensitometer* which is an instrument used in conjunction with a photographic record. A typical instrument uses a chart record to show the variation of blackening with position (or angle θ) along a film. Normally a narrow beam of light passes through the film and measures its intensity. Alternatively the intensity is automatically matched against the intensity transmitted through a standard photographic wedge—a standard strip exposed with progressively increasing blackening along its length. A chart record then shows the intensity along a film as a function of θ as the film is slowly traversed through the optical system. A marked change in transmitted light results when a line passes through the light beam and so the recorder chart plots out the line contour and so enables the measurement of its height, width or area under peak. This information can be useful for determining the intensities for calculations connected with the structure itself, with relative proportions of phases in a mixture, with strain or crystallite size measurement. These topics receive attention later. The experimental techniques of microdensitometry will not be considered in detail. It can be very useful but is limited by the available intensity range and non-linear response of photographic emulsions.

6.4.6.2 Diffractometers: There would be an obvious advantage especially for applications where the line contours and heights, etc., are needed, if the film could be eliminated and the diffracted intensity measured directly. This was the principle used in the earliest diffraction experiment using a goniometer fitted with an ionisation spectrometer. The photographic techniques developed more rapidly because of their relative simplicity and they still have many advantages. More recently the development of electronic radiation counting tubes and modern electronic recording systems have led to the establishment as an efficient laboratory instrument of the *counter diffractometer*. The geometry of such an instrument commonly follows one of the two arrangements in Fig. 6.10—the first being more general.

Fig. 6.10(a). This employs a flat powder specimen which receives a divergent beam of X-rays. The diffracted rays then converge and pass into the counter tube as indicated. The arrangement thus corresponds to the glancing angle or side-reflection system (6.4.4). The instrument mechanism is, however, arranged so that the specimen rotates at exactly half the speed of the goniometer arm which holds the counter and moves round an arc with centre at the middle of the specimen surface. It follows that when the specimen surface has rotated through an angle θ the counter has moved through 2θ. This means that the specimen is always symmetrically placed with respect to the incident and diffracted rays. It would only require to be slightly curved and then the X-ray focus, specimen surface and the slit in front of the counter tube would form a perfect focusing system. The use of a flat specimen makes no appreciable difference, except that the radius of the focusing circle gets smaller as θ increases (Bragg–Brentano focusing).

Fig. 6.10(b). This system has a curved specimen and employs exact focusing according to the principles of the powder cameras in Fig. 6.9 (sometimes called para-focusing or, like the camera systems, named after their originators, Seemann and Bohlin). The system clearly has some advantages but the counter tube is required to move round the focusing circle and this progressively increases its distance from the specimen. A more complicated mechanism is therefore required.

In both systems it is desirable to limit vertical divergence. The X-ray beam usually comes from a line focus which is normal to the plane of the diagram (Fig. 6.10) and the ray paths should all be parallel to this plane. This con-

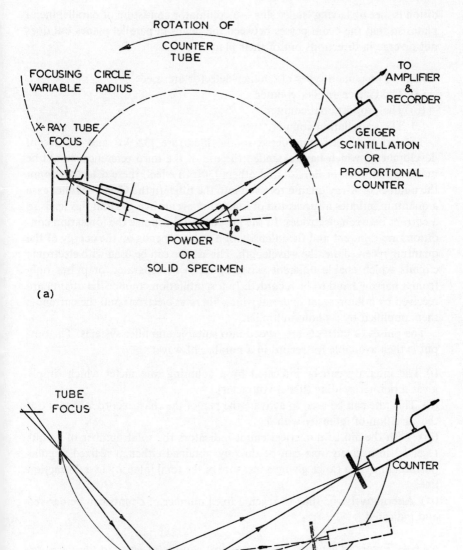

Fig. 6.10   Recording diffractometer arrangements.
(a) standard arrangement with powder specimen at centre of circle
(b) arrangement giving focusing (Seemann–Bohlin) (no monochromator)

211

dition is met by having 'soller slits'—a collimator consisting of parallel metal plates so that the beam passes between the plates in parallel planes but does not diverge in directions out of their plane.

6.4.6.3   Three main types of counter-detector are used:
(a) The Geiger–Müller counter.
(b) The proportional counter.
(c) The scintillation counter.

The principles are considered in the literature [32, 33] and for recent developments which have extended the use of the third reference should be made to the work of Parrish and others [34]. In effect, these detectors count the number of X-ray quanta received into the tube. In the Geiger–Müller case a quantum initiates a succession of ionisation events in a gas and this leads to a current between electrodes. In the proportional counter the ionisation conditions are different and the intensity of a pulse depends on the energy of the quantum received, i.e. the wavelength. This device can be used with electronic circuits which enable different wavelengths to be separated or pulses only from a narrow band to be recorded. In a scintillation counter the quanta are received by a fluorescent material which liberates electrons and the current is then amplified by a photomultiplier.

The pulses or currents are passed into suitable amplifier systems. The output is then available for record in a number of ways:

(i) The intensity can be indicated by a counting rate meter which simply gives a pointer reading (like a voltmeter).

(ii) This rate can be used to activate the pen of the chart recorder to plot out the variation of intensity with θ.

(iii) With the aid of numerical count indicators the total number of counts received in a given time can be directly obtained either at a fixed angular position or over a range giving a measure of the total intensity in a diffraction line.

(iv) Alternatively the time to reach a fixed number of counts can be derived and indicated.

6.4.6.4   Errors: There are several possible sources of error in the measurement of intensity by these instruments. These include:

1. Mains voltage fluctuations leading to fluctuations in incident X-ray intensity. This is usually met by having a stabilised generator.

2. Instability or drift of counters.

3. Drift or instability of electronic circuitry
(2 and 3 small in modern instruments).

4. Inconsistencies in specimen preparation. These may arise in solid specimens from different degrees of grinding or polishing or surface roughness. With powders, the relative particle sizes of different phases and the efficiency

of their mixing can make appreciable differences. Particular care is needed when the constituents have markedly different absorption coefficients.

5. Counting statistics. This source of error is very important because it represents an irreducible minimum which would only be approached under the best conditions. Thus with half-wave rectification and inadequate stabilisation the following relationships will not be very closely approached. With full wave rectification and stabilisation, the emission of quanta from the X-ray tube may be regarded as a truly random process. Thus, it is supposed the instrument is set at a fixed position and the number of counts N in a time t are recorded or measured. The experiment can be repeated a number of times giving different values of N or $n = N/t$ (counting rate). These values should vary according to a Gaussian distribution. The mean value of N is $\overline{N}$ and the standard deviation is $\sigma = \sqrt{N}$. Any single measurement deviates from $\overline{N}$ by $\Delta N = N - \overline{N}$ and the probability that the deviation will be this ΔN or less is denoted by p which can be deduced from the Gaussian formula [34, 35]. The so-called 'probable error' is the value of ΔN when $p = 50$ per cent and is given by, approximately

$$\Delta N_{50} = 0 \cdot 67\sigma = 0 \cdot 67\sqrt{N}$$

The corresponding *fractional* error in either counting rate or total numbers of count is:

$$\epsilon_{50} = \frac{\Delta n}{n} = \frac{\Delta N/t}{N/t} = \frac{\Delta N}{N} = \frac{0 \cdot 67\sigma}{N} = \frac{0 \cdot 67}{\sqrt{N}}$$

For example if it is desired that there should be a 50 per cent chance of an error of 1 per cent or less then $N = 4500$. Similarly 27 000 counts would make $\epsilon_{90} = 1$ per cent and 67 000 counts are required to make $\epsilon_{99} = 1$ per cent (virtual certainty!). The formulae required are thus:

$$\epsilon_{50} = \frac{0 \cdot 67}{\sqrt{N}}; \qquad \epsilon_{90} = \frac{1 \cdot 64}{\sqrt{N}}; \qquad \epsilon_{99} = \frac{2 \cdot 58}{\sqrt{N}} \qquad (6.33(a))$$

The total number of counts N obtained in this way includes 'background' radiation which can arise in a variety of ways, for example from any incoherently scattered radiation or fluorescent radiation from the specimen itself, from radiation scattered by parts of the instrument or coming from outside altogether. One could therefore count the background on either side of a diffraction line and interpolate if there was much difference. If N_B is the number of background counts this measurement alone will be subject to the same statistical relationships as the total. The standard deviation of the difference (or sum) of two separate measurements is given by $\sigma^2 = \sigma_1^2 + \sigma_2^2$. In this case

$$\sigma = \sqrt{[(\sqrt{N})^2 + (\sqrt{N_B})^2]} = \sqrt{(N + N_B)}$$

so that
$$\epsilon_{50} = \frac{0 \cdot 67\sqrt{(N + N_B)}}{N - N_B} \qquad (6.33(b))$$

(with corresponding expressions for ϵ_{90}, etc.). When a series of measurements are made across a line and a mean curve drawn through the points the errors on individual points tend to be 'smoothed out' and so the accuracy is higher than for isolated individual points.

6. Counting losses. When a detector receives a quantum, there is a finite time before it is ready to receive another. Hence the counting becomes less efficient at high rates, arising from high X-ray intensities. The error is largest with Geiger counters and in fact is negligible with both scintillation and proportional counters which have much smaller resolving times, e.g. $T = 0.25$ μs which can be compared with a typical 270 μs for the other counter. It is easy to estimate the relative effects on different relative counting rates, e.g. the true rate is related to the observed by the relation $n_t = n_0/(1 - \gamma n_0)$.

(*Note:* The general mathematical theory of diffractometry is covered in a book by Wilson [35].)

6.4.7 Intensity and Accuracy of Powder Diffraction Patterns

6.4.7.1 The techniques which have been described provide either a photographic or chart record of the diffraction pattern as a whole over the experimentally practicable range of θ. Secondly one may obtain a measure of the intensities of individual diffraction peaks and their shapes, areas or widths. Thirdly, the patterns can be used to derive interplanar spacings or lattice parameters to varying degrees of accuracy.

In order to develop the use of X-ray diffraction for some of its metallurgical uses it is therefore necessary to know the precise form of the intensity for a diffraction powder line (ring and in more advanced applications of a spot in a single crystal pattern). It is also necessary to appreciate the factors which determine the accuracy of spacings or lattice parameters.

Intensity of cylindrical powder patterns: In section 5.2.3 the various factors which affect the fundamental diffracted intensity were separately considered. In addition there were the 'geometrical' factors and the absorption factor. The full intensity is therefore finally evolved from the following terms:

1. A constant including the mass and charge of the electrons, the wavelength, etc.

2. A factor depending on the effect of speed of rotation on the *time* spent in the diffracting position. This is inferred from Fig. 6.2(c). (If desired this can be checked by a separate drawing with simple construction.) For the case under consideration ρ is in the region of 90° ($\pm 5/10°$). If the angular velocity round axis CV is ω the reciprocal lattice (on this scale) also rotates about OS at the same speed. When a reflection occurs OP, now close to OQ, cuts the sphere. P is moving with a linear velocity $OP \times \omega = 2r\omega \sin \theta$. Its component normal to the reciprocal sphere surface is $2r\omega \sin \theta \cos \theta = r\omega \sin 2\theta$. The velocity also alters with ρ but in this region it need not be considered—in any case intensities are usually measured on the equatorial line. The greater the

velocity the shorter the time in the reflection position. Hence the total intensity contains a time proportional to $1/(\sin 2\theta)$.

3. In Fig. 6.7(a), however, the diffracted intensity is spread all round a ring of radius $r \sin 2\theta$ which cuts the film in a constant length l. The intensity recorded thus varies as $l/(2\pi r \sin 2\theta)$ or $1/(2\pi r \sin 2\theta)$ per unit length.

4. Again in Fig. 6.7(a), assuming *random orientation* of the normals CN the probability of diffraction occurring must be proportional to:

$$\frac{\text{area of band of width } \Delta\theta}{\text{total area of sphere}} = \frac{2\pi r^2 \sin (90-\theta) \, \Delta\theta}{4\pi r^2} = \tfrac{1}{2} \, \Delta\theta \cos \theta$$

5. Although 2 alone was originally termed the 'Lorentz factor' it has become customary to combine it with 3 and 4 to give a factor

$$\propto \frac{\cos \theta}{\sin^2 2\theta} \propto \frac{1}{\sin^2 \theta \cos \theta}$$

6. Further combined with the polarisation factor $\tfrac{1}{2}(1+\cos^2 2\theta)$, this expression gives the Lorentz polarisation or angular factor for a cylindrical powder photograph, viz. $(1+\cos^2 2\theta)/(\sin^2 \theta \cos \theta)$. Tables of this function are available in standard references.

7. The number of normals CN under these conditions will also be proportional to the planar factor p. In this case p is the total multiplicity whereas in a rotation photograph this is not necessarily the case and the individual value must be deduced.

8. The above terms come into the expression for the *integrated intensity* of a diffraction line—equivalent to the total area under the peak. It also follows that this intensity will be proportional to the total volume irradiated or to the total number of unit cells $N = N_1, N_2, N_3$ as in section 5.2.7 where, however, it was shown that the *maximum* intensity was proportional to

$$N^2 = N_1^2 N_2^2 N_3^2$$

For practical applications it is usually sufficient to use the expression:

$$I \propto \frac{1+\cos^2 2\theta}{\sin^2 \theta \cos \theta} p|F|^2 AT \tag{6.34}$$

for a powder diffraction pattern using a cylindrical film and rotating specimen. (The last factor may also be omitted in some applications.) A, the absorption factor, is also dependent on θ (see below). T, the temperature factor, is a function of $(\sin \theta)/\lambda$ and so of θ (or d) for constant λ and, as noted in section 5.2.3, is given by $\exp\left[-(B \sin^2 \theta)/\lambda^2\right]$. Equation (6.34) will apply equally well to the diffractometer except that the absorption factor will be different. If it is derived to use the expression for a flat film the factor of $1/(\sin 2\theta)$ needs to be replaced by $1/(\tan 2\theta)$. In the single crystal rotation case the appropriate value of p is used and the angular factor modified for the layer lines and absence of

215

any randomness of the normals CN for individual planes (see Vol. II of *International Tables* (Ref. [9] of Chapter 4), Bunn [23], etc.).

6.4.7.2   The Absorption Factor: The absorption factor has been introduced in section 5.2.4 and specific cases of its calculation referred to. Full details are generally given elsewhere but it is useful to summarise here the cases of general interest in powder work and these are denoted in Fig. 6.11.

In these cases the various other factors represented in equations such as (6.34) are taken collectively, as equal to Q. Thus if I_0 is the intensity before diffraction and v the diffracting volume, then the integrated intensity after diffraction is QI_0v. The absorption factor A is then the factor $\exp(-\mu x)$ as in equation (5.26). It is necessary to find the total path for the different experimental arrangements and if this is not constant to take account of the fact. Thus:

(i) Fig. 6.11(a) represents the transmission case with normal incidence. By equation (5.26) the intensity along C_0C_x is reduced to $I = I_0 \exp(-\mu x)$ at a point C_x. A diffracting region of cross-sectional area (beam area) S and thickness δx therefore has volume $S\,\delta x$ and diffracted intensity $Q(I_0 \exp(-\mu x)S\,\delta x$. The remaining path from C_x to D_x is $(t-x)/\cos 2\theta = (t-x)\sec 2\theta$ and so the intensity is further reduced by a factor $\exp[-\mu(t-x)\sec 2\theta]$. The diffracted beam $C \to P$ will therefore contain all such contributions for paths along C_0C_t, t being the thickness. Hence the resultant intensity is given by

$$I = QSI_0 \int_0^t \exp[-\mu\{x+(t-x)\sec 2\theta\}]\,dx$$

$$= \frac{-QSI_0}{\mu(1-\sec 2\theta)}[\exp(-\mu t) - \exp(-\mu t \sec 2\theta)] \qquad (6.35)$$

Since the diffracting volume is St

$$A = \frac{1}{\mu t(\sec 2\theta - 1)}[\exp(-\mu t) - \exp(-\mu t \sec 2\theta)]$$

(ii) The general transmission case for a tilted specimen is given by A. Taylor (Ref. [2] of Chapter 4). One particular position which simplifies the calculation of A, and can be used with this intention, is shown in Fig. 6.11(b). The incident beam makes an angle of $90° - \theta$ with the surface so that all diffracting planes for this value of θ are parallel to C_0C_t. All paths then add up to a constant which is easily seen to be $t \sec \theta$, so that

$$A = \exp(-\mu t \sec \theta) \qquad (6.36a)$$

and $$I = QI_0vA = QI_0St \exp(-\mu t \sec \theta) \qquad (6.36b)$$

For these two and other *transmission* cases the intensity will increase with the thickness t due to the increasing volume of material irradiated but decrease

216

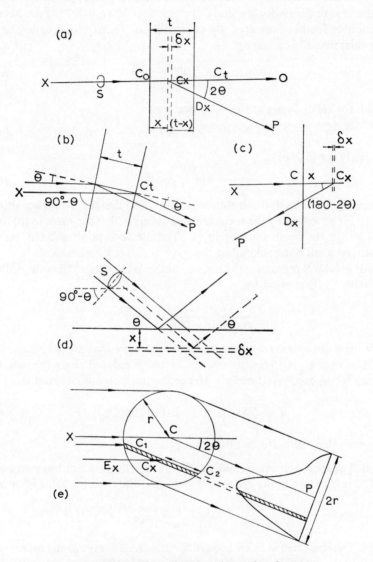

Fig. 6.11   Calculation of X-ray absorption factors.

(a) transmission
(b) symmetrical Laüe case
(c) back reflection
(d) symmetrical Bragg case
(e) cylindrical or powder specimen

because of the increasing absorption. It is therefore of value to be able to find the *optimum thickness*. This can be useful for giving an approximate estimate of the best thickness for a metal foil or powder layer as for the simple transmission or focusing cameras. By differentiating (6.35) and equating to zero, this thickness t^* is given by

$$\mu t^* = \frac{\ln (\sec 2\theta)}{\sec 2\theta - 1} \tag{6.37a}$$

which for small Bragg angles $\theta \to 0$ becomes

$$\mu t^* \to 1 \tag{6.37b}$$

Similarly for (6.36b)

$$\mu t^* = \cos \theta \tag{6.37c}$$

(which again $\to 1$ as $\theta \to 0$). It should be noted also that the X-rays are absorbed by the air as well as by the specimen and the air path is the same in (b) for any position in the beam but not in (a). The air absorption effect is generally relatively small but could affect the most accurate measurements.

(iii) Fig. 6.11(c) represents the simple back-reflection case. Here the diffracted ray path $C_x D_x$ is given by

$$\frac{x}{|\cos (180° - 2\theta)|} = x|\sec 2\theta|$$

(a positive distance is required). As for the transmission case the intensity after diffraction at C_x is $Q[I_0 \exp(-\mu x)]S \, \delta x$ and is reduced along the path $C_x D_x$ by the factor $\exp(-\mu x|\sec 2\theta|)$. Hence the integrated final result is

$$I = QSI_0 \int \exp [-\mu x(1 + |\sec 2\theta|)] \, dx \tag{6.38}$$

It is seen that there are two possibilities:

(a) The specimen can be sufficiently thin for the incident beam to emerge. The integration limits are then 0 and t and the effective value of A is

$$\frac{1 - \exp [-\mu t(1 + |\sec 2\theta|)]}{\mu t(1 + |\sec 2\theta|)}$$

(b) The specimen is thick enough for the second term in the numerator to approach zero. This is equivalent to integration from 0 to ∞, in which case t should be replaced by an arbitrary t' corresponding to the 'effective depth of penetration'. Thus the diffracting volume $= St'$ in the numerator, and so t' cancels out and is not required in the intensity calculation. Alternatively:

(iv) The symmetrical Bragg case in Fig. 6.11(d) is important for the usual diffractometer arrangement. Here the irradiated volume is always

218

$\{S\,\delta x/[\cos{(90°-\theta)}]\}=(S\cosec\,\theta\,\delta x)$ and the total path at depth x is $2x\cosec\,\theta$ so that

$$I = QI_0S\cosec\,\theta \int \exp{(-2\mu x\cosec\,\theta)}\,dx$$

As for the previous case, one may consider a specimen of thickness t, or sufficiently thick for the incident intensity to be reduced to negligible magnitude before emerging from the other side of the layer. Thus:

$$I = QI_0S\cosec\,\theta\left(\frac{1}{2\mu\cosec\,\theta}\right)[1-\exp{(-2\mu t\cosec\,\theta)}] \to \frac{QI_0S}{2\mu} \quad (6.39)$$

(v) The cylindrical rod or powder specimen is not so easy to treat but the essential features are illustrated in Fig. 6.11(e). It is assumed for simplicity that the X-ray beam is strictly parallel and non-divergent. The diffracted line then has a width $2r$ (r=radius of specimen). The intensity at a point along the film, over this line, will therefore depend on the part of the specimen which corresponds to it. Material from anywhere along the strip C_1C_2 can contribute but the paths followed by the X-rays are not simple. Thus a small element at C_x diffracts rays which enter the circular section at E_x. The total path is the sum of the two distances $(E_xC_x+C_xC_2)=(a+b)$, say. The equations have been integrated numerically [36, 37]. Taylor and Sinclair [37] have also provided a technique for evolving the line shape relationship indicated in (e).

It is worth noting that

(a) If the absorption is low the peak is near the true centre.

(b) If the absorption is high the peak will be near the outer edge towards higher θ. Sharp lines can be obtained by coating powders onto a fine lead glass fibre. The displacement is allowed for in the correction methods described below.

(c) The specimen should be uniform in shape and constitution.

(d) The total intensity from a powder line follows (based on Taylor's treatment, Ref. [2] of Chapter 4). Consider a small area of strip $d\sigma$ at C_x. Then the diffracted intensity provided by the material here is:

$$QI_0 \exp{[-\mu(a+b)]}h\,d\sigma$$

where h=height of specimen irradiated.

Integration along the strip would give the contribution of the whole strip, and then integration across the circle, that for the whole specimen. Therefore for the whole specimen

$$QI_0 \int \exp{[-\mu(a+b)h\,d\sigma]} = QI_0vA = QI_0\pi r^2hA$$

so

$$A = \frac{1}{\pi r^2 h} \int \exp{[-\mu(a+b)]}h\,d\sigma = \frac{1}{\pi} \int \exp{(-\mu rx)}\,ds \quad (6.40)$$

where $x=(a+b)/r$ which is dimensionless, and $ds=d\sigma/r^2$ also dimensionless.

Values of A as a function of θ and (μr) have been tabulated ([37]—and Ref. [2] of Chapter 4). The calculation or construction can be based on a circle of unit radius and then the only variable parameter is the effective absorption (μr) where $r =$ actual radius.

Powder specimens: The above considerations would apply to solid specimens in which all the volume is occupied by the diffracting crystals. In actual powder specimens allowance must be made for the fact that not all of the total volume is occupied by the crystalline material. If V and V' are the respective true volume of powder and the occupied volume ($V < V'$) and the whole compact has an apparent density ρ' then $\rho V = \rho' V' =$ weight. The linear absorption coefficient depends on this absorbing mass, so $\mu'/\rho' = \mu/\rho$. Therefore in the above formula μ' is substituted for μ, ρ' for ρ and V' for V. To find the actual density of powder specimen one can measure its dimensions carefully and then weigh it.

6.4.7.3 Accuracy of spacing and parameter measurements: Fig. 6.11(e) has already illustrated a point about absorption. The maximum intensity taken as the position of the line may not in fact be the true centre—the peak appears to be shifted. This and other factors give rise to an error in the measurement or determination of θ and so in the determination of d or lattice parameters.

From the Bragg law: $\lambda = 2d \sin \theta$, therefore

$$\Delta\lambda = 2 \, \Delta d \sin \theta + 2d \, \Delta\theta \cos \theta$$

Divide the left side by λ and the right by $2d \sin \theta$

$$\frac{\Delta\lambda}{\lambda} = \frac{\Delta d}{d} + \Delta\theta \cot \theta$$

or omitting $\Delta\lambda/\lambda$
$$\frac{\Delta d}{d} = -\Delta\theta \cot \theta \tag{6.41}$$

which gives relative errors in d in terms of actual errors in θ. It is a useful feature of this equation that as $\theta \to \frac{1}{2}\pi$, $\cot \theta \to 0$ so the same errors in θ give decreasingly small errors in d (or a, etc.) in this region. (The back-reflection region.)

With regard to the *wavelength* itself the literature contains spacings in both ångström units, Å, and kX. (It is possible to find older work stated to be in Å but actually in kX but this is unlikely to cause any confusion now.) The difference arose from the more accurate determination of basic standard wavelengths and constants (including Avogadro's number). It is advisable to check what units have been used in a publication and the authors should have stated their wavelengths.

To convert kX to Å multiply by 1·002 02. Thus the Å unit is a slightly smaller length than the kX.

For most practical purposes it is assumed that $\Delta\lambda$ is negligible, but care is needed in the most accurate work. For example a monochromator could affect the basic shape of the X-ray line actually used. The overlap of $K\alpha_1$ and $K\alpha_2$ can vary with the resolution or geometry. Equation (6.41) will thus be taken as sufficient. It remains to consider the possible sources of error in $\Delta\theta$.

(*Note*: In all accurate work care must be taken to keep the temperature of the specimen constant during the exposure due to the possible, slight but significant effect of thermal expansion.)

(a) Errors in line measurement or camera calibration—taking account of film shrinkage:

From equation (6.26) for the van Arkel camera

$$\Delta\theta = \left(\frac{\pi}{2} - \theta\right)\left(\frac{\Delta R}{R} - \frac{\Delta T}{T}\right) \tag{6.42}$$

and from equation (6.27) for the Bradley–Jay arrangement

$$\Delta\theta = \theta\left(\frac{\Delta\theta_k}{\theta_k} + \frac{\Delta S}{S} - \frac{\Delta S_k}{S_k}\right) \tag{6.43}$$

or if the radius is measured and $\theta = S/4R$

$$\Delta\theta = \theta\left(\frac{\Delta R}{R} - \frac{\Delta S}{S}\right) \tag{6.44}$$

In (6.42), $\Delta\theta \to 0$ as $\theta \to \frac{1}{2}\pi$ but in the other cases its small finite value is unimportant in comparison with the decrease $\to 0$ of $\cot\theta$ in equation (6.41). For the back-reflection case with flat film of ring diameter $D = 2r\tan(\pi - 2\theta)$, similar relationships can be found expressing $\Delta\theta$ in terms of errors in D or r or alternatively if r (specimen film distance) is found by calibration then the errors in D_s and θ_s' for the standard are included. (A useful exercise to determine these relationships.)

(b) Errors in eccentricity in cylindrical powder cameras. Reverting to Fig. 6.7(b) if the specimen is not at the centre of the camera C but displaced to a point C' the effect on S is to be found. Let $CC' = u$, and $\angle C'CO = \sigma$. Clearly the ray on one side is shifted laterally by the distance $u\sin(\sigma - 2\theta_1)$ and that on the other side by $u\sin(\sigma + 2\theta_1)$. These are the perpendicular distances from C' onto the undisplaced ray paths. The two arcs of the circle are closely similar to these distances and ΔS is their *difference*. That is,

$$\Delta S \simeq u\sin(\sigma + 2\theta_1) - u\sin(\sigma - 2\theta_1) = 2u\cos\sigma\sin 2\theta_1$$

The back-reflection formula is similar. Hence in equation (6.43), ΔS varies as $\sin 2\theta = 2\sin\theta\cos\theta$ and so equation (6.41) with this term inserted leads to: $\Delta d/d \propto \cos^2\theta$ for this error.

(c) Effect of absorption. From Fig. 6.11(e) it can be inferred that the relative shift of the point of maximum intensity decreases as $\theta \to \frac{1}{2}\pi$ and $2\theta \to \pi$. This is particularly easy to envisage for a specimen of high absorption when most of the diffraction is from the surface of the cylinder. Bradley and Jay [38] dealt with this case and others have given more general treatments. Again, the resultant angular displacement contains a $\sin 2\theta$ term and $\Delta d/d$ varies as $\cos^2 \theta$. In addition there is also a variation with $1/\theta$ or $1/\sin \theta$.

(d) Horizontal or general divergence of beam can be shown to vary in a similar way.

(e) Vertical divergence varies as $(\cot^2 \theta - 1)$ and so is a minimum at $45°$ and finite at $90°$. The use of fine pinhole collimators virtually eliminates this relatively small error. Apart from this last source of error all the errors diminish towards zero as $\theta \to \frac{1}{2}\pi$, i.e. $\cos^2 \theta \to 0$, $\cot \theta \to 0$. The uniform film shrinkage has been provided for by the general calibration procedure and the errors in the calibration or individual line measurements are then covered by equations (6.42)–(6.44). The errors in individual line measurements should be reduced by repetition of readings as when a film is being measured on a travelling microscope and scatter of points of the calculated lattice parameters over a range of θ values will diminish with increasing θ. One could thus calculate lattice parameters from equations such as (6.28) or (6.29) and plot the values of lattice parameter, e.g. a for a cubic crystal as a function θ or one of the following extrapolation functions, so that the correct value a_0 is obtained at $\theta = 90°$ or $\frac{1}{2}\pi$.

The functions are:

(a) $\cos^2 \theta$ proposed by Bradley and Jay [38]. This tends to give good extrapolations for $\theta > 50°$ where the curvature is negligible—the graph being virtually linear. An example is shown in Fig. 6.12(a).

(b) $\frac{1}{2}[(\cos^2 \theta)/(\sin \theta) + (\cos^2 \theta)/\theta]$ proposed by Nelson and Riley [39] and shown to give a straight line plot over nearly the whole range of θ—see Fig. 6.12(b). Taylor and Sinclair [37] also provided a study of this and similar relationships. The basis of it is that some of the error is represented as a function of the first term and some as a function of the second. The mean is closely correct for most practical cases. The main conditions are that absorption shall be the chief source of error and eccentricity vanishingly small (as it should be with accurately centred specimens) and the X-ray source should have a Gaussian intensity profile. This function is better than $\cos^2 \theta$ and since it can be used down to low angles is more successful for non-cubic substances for which the $(h00)$, $(0k0)$, $(00l)$ reflections can be used to determine a, b, c separately.

(c) For back-reflection cameras with only a few reflections (rings) available the most suitable extrapolation function is $(\frac{1}{2}\pi - \theta) \cot \theta$.

(d) A plot of spacings against $(\cos^2 \theta)/(\sin \theta)$ alone may often give a straight line from $\theta > 60°$. This is also of some interest for calculating and correcting interplanar spacings. For example, a cubic substance could be used as an internal standard to give $\Delta a/a = \Delta d/d$ as a function of $(\cos^2 \theta)/(\sin \theta)$ so that $(\Delta a/a) \sin \theta$ is plotted against $\cos^2 \theta$ to give a smooth curve extrapolating to a

222

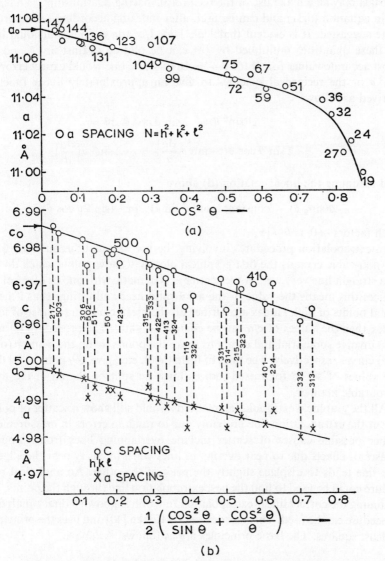

Fig. 6.12 Two extrapolation methods for powder photographs.

(a) extrapolation using $\cos^2 \theta$ (for a cubic phase M_6C)

(b) extrapolation using Nelson and Riley's function (for tetragonal SiO_2 at 208 °C (Cristobalite))

finite value at $\theta = 0$. On the other hand $\sin \theta \to 0$ and $\Delta a/a \to \infty$ as $\theta \to 0$ and so the actual error curve is hyperbolic in form. This result shows how spacing errors become increasingly large towards low angles.

(e) Procedures applicable to the calculation of lattice parameters of non-cubic

223

crystals may also make use of the reciprocal spacing relationships as referred to in equation (6.31) and can be used after indexing according to the procedure suggested. It is evident that $1/a^2$, $1/b^2$, $1/c^2$ for various individual lines or these quantities multiplied by $\frac{1}{4}\lambda^2$ can be plotted against $\sin^2 \theta$ and the most accurate values found for $\theta \to \frac{1}{2}\pi$. In this case one would expect errors in $\sin^2 \theta$ or the reciprocal spacings to give an approximately linear function derived as follows

$$\Delta(\sin^2 \theta) = 2 \sin \theta \cos \theta \, \Delta\theta =$$

$$2 \sin \theta \cos \theta \left\{ -\tan \theta \frac{\Delta a}{a} \right\} = -2 \sin^2 \theta \left(\frac{\Delta a}{a} \right)$$

and assuming the linear relation (d) above

$$\Delta(\sin_2 \theta) = \text{constant } (\cos^2 \theta \sin \theta) \quad \text{or} \quad (\sin 2\theta \cos \theta)$$

both factors $\to 0$ as $\theta \to \frac{1}{2}\pi$.

In extrapolation procedures involving the use of equations such as (6.29) for non-cubic crystals the first graphical plot may show points which do not fit a straight line very well. If the scatter is systematic with one or other of the indices this means the value of the axial ratio needs adjusting: thus if hexagonal points of high l index are further off the mean curve than those of low l index then the values of c/a can be changed slightly. The procedure (e) may also involve some trial and errors to make the points lie on the a and c (or a, b, c) curves respectively. The lines of form (h00), etc., give key points and then the values of $\sin^2 \theta$ for mixed indices must be split or shared in the most favourable manner.

All the graphical extrapolation methods would still show a scatter of points about the extrapolation line probably due to random errors in measurement. Other possible causes of scatter include overlapping lines from impurity phases or effects due to near overlap of lines when intensity near the edge of one line tends to displace slightly the peak of the other. An analytical procedure could be used to find the best extrapolation line through the points, but assuming the correct linear extrapolation function is known a direct analytical procedure can be used. This was devised by Cohen [40] and uses the principles of least squares. The basic principles are as follows. As above

$$\Delta(\sin^2 \theta) = -2 \sin^2 \theta \, (\Delta a/a)$$

and $\Delta a/a$ is then considered to be a linear function of one of the expressions indicated under (a), (b) and (d). These lead to

$$\Delta(\sin^2 \theta) = D \sin^2 2\theta \quad \text{or} \quad D \sin 2\theta \left(\frac{1}{\sin \theta} + \frac{1}{\theta} \right)$$

or $D \sin \theta \cos^2 \theta$ respectively where D is the slope summing up the systematic errors for the particular X-ray film. In general this function is $Df(\theta)$.

224

From relationships such as (6.31) one may express $\sin^2 \theta$ as a function ideally of the indices only and constants, e.g. $A = \lambda^2/4a^2$. Thus:

Cubic case: $A(h^2 + k^2 + l^2) - \sin^2 \theta = 0$

Hexagonal case: $A(h^2 + hk + k^2) + Cl^2 - \sin^2 \theta = 0$

Tetragonal case: $A(h^2 + k^2) + Cl^2 - \sin^2 \theta = 0$

To each such equation in practice the term $Df(\theta)$ must be added to make it approach a fit to the experimental results, but even so the random errors will be such that for each equation

$$A\alpha_i + C\gamma_i + Df(\theta_i) - \sin^2 \theta_i$$

is not exactly equal to zero, as it would be if the random errors were nil. Here $\alpha_i = $ the sum of indices squared or products $\gamma' = l^2$, etc., for a diffraction line at θ_i. According to the principle of *least squares* the best values of the constants A, C, D for a set of experimental points represented by θ_1, θ_2, θ_3, ..., θ_i, ... and the corresponding indices, are those for which the sum of all such differences is a minimum, i.e. the minimum value is required of

$$\sum (\Delta)^2 = \sum [A\alpha_i + C\gamma_i + Df(\theta_i) - \sin^2 \theta_i]^2$$

This minimum is found from

$$\mathrm{d} \sum (\Delta)^2 = \frac{\partial(\sum (\Delta)^2)}{\partial A} \, \mathrm{d}A + \frac{\partial(\sum (\Delta)^2)}{\partial C} \, \mathrm{d}C + \frac{\partial(\sum (\Delta)^2)}{\partial D} \, \mathrm{d}D = 0$$

and hence requires each partial differential to be simultaneously zero. For example

$$\frac{\partial(\sum (\Delta)^2)}{\partial A} = 2 \sum \alpha_i(A\alpha_i + C\gamma_i + Df(\theta_i) - \sin^2 \theta_i) = 0$$

or
$$A \sum \alpha_i^2 + C \sum \alpha_i\gamma_i + D \sum \alpha_i f(\theta_i) = \sum \alpha_i \sin^2 \theta_i \qquad (6.45)$$

with similar equations which are strictly analogous. In this case there are thus finally three simultaneous equations in A, C and D which can be solved in the usual manner. In the cubic case the equations reduce to two

$$\left. \begin{array}{l} A \sum \alpha_i^2 + D \sum \alpha_i f(\theta_i) = \sum \alpha_i \sin^2 \theta_i \\ A \sum \alpha_i f(\theta_i) + D \sum f(\theta_i)^2 = \sum f(\theta_i) \sin^2 \theta_i \end{array} \right\} \qquad (6.46)$$

The individual values of α_i ($= h^2 + k^2 + l^2$ or otherwise), $\sin^2 \theta_i$ and $f(\theta_i)$ are known for each value of θ_i and can easily be tabulated, and the products required in equations (6.45), (6.46) found and summed. (The process is much assisted by the use of a calculating machine or computer.)

6.5 ELECTRON DIFFRACTION PROCEDURES

6.5.1 General

6.5.1.1 The electron diffraction procedures which are now of most metallurgical interest are those which can be operated in an electron microscope, although separate electron cameras are used in some research laboratories. The essential principles are already outlined in Chapter 5 and experimental methods are essentially similar to those used in X-ray diffraction. (General Refs.: [4] of Chapter 5 and [41] of this chapter.)

In the case of transmission methods through thin foils there is an interconnection between the microscope images and the diffraction pattern. There are visual effects in the former which are explicable in terms of diffraction phenomena. The experimental conditions for and interpretation of these effects are deferred until in Vol. 2.

6.5.1.2 It is also necessary to consider a point arising from the fact that electrons have a much smaller penetrating power than X-rays. In section 5.1.2 it was seen that for diffraction to occur in a three-dimensional crystal, the three Laüe conditions must be satisfied simultaneously. If a foil, film or thin layer of crystals is placed in an electron beam, effectively, all the electrons will be absorbed and no diffracted beam emerge, unless the specimen is thin enough. If we take the three axes for convenience with c parallel to the incident beam and a and b perpendicular to it, then such a crystal or foil will have many unit cells in the a and b directions but few in the c direction. This means that the third factor in equation (5.38) will become broadened according to the principles of section 5.2.7.1. Therefore the first and second Laüe conditions are represented by sharp lines in Fig. 5.2(c) (q.v.). In the X-ray case diffraction would not normally occur because the sharp rings representing the third condition would not pass through the intersections of the first two sets of curves. In the electron case with a thin enough film or foil these circles would become broadened or smeared out into bands so they actually do pass through the points and so a diffraction pattern can occur for a very thin stationary specimen—a so-called cross grating pattern—as if it came from a two-dimensional crystal.

It is apparent that the higher the tube voltage, the shorter will be the electron wavelength (section 5.3.2) and the greater the thickness or depth of penetration or possible thickness for a foil. Secondly because of the short wavelength the angle of diffraction 2θ is very small (0–3 or 4° with 100 kV electrons). The planes giving rise to the diffraction are nearly parallel to the electron beam and perpendicular to the foil or crystal surface. It follows that if a thin film were to become progressively thicker (the circles in Fig. 5.2(c) getting correspondingly sharper) the pattern would tend to fade out, but would appear again in part if the specimen is tilted slightly one way or another. The appearance would then be of spots in curved bands where a part of one circle (still somewhat broadened) cuts a number of cross grating intersections on one side of the centre, thus forming the so-called Laüe zones which move in and out of view if the specimen is moved.

This point must also be considered in connection with another. Because of the short wavelength and the small values of θ, the reciprocal lattice points R

(on this scale) are much nearer to the diffraction spots P in Fig. 5.3(c). The effect of this on Fig. 5.2(c) is to make the curvature of the lines much less so that a pattern of this kind becomes more nearly rectangular and so resembles closely the plane in the reciprocal lattice itself. Alternatively one may say that the curvature of the Ewald sphere is so large ($\propto 1/\lambda$) and so it is nearly planar and can pass through several points in the reciprocal lattice plane. The curvature decreases as λ decreases with higher tube voltage. Hence the increasing thickness reduces the possibility that the three Laüe conditions will be simultaneously fulfilled due to the partial relaxation of the one normal to the plane, and so of the full three-dimensional pattern being obtained in a fixed position, but increasing the tube voltage makes it more easy for the reciprocal lattice sphere condition to be satisfied for more points (regarded as having a finite size in reciprocal space).

The first effect may be important for very thin films and layers. Many metal foils used in transmission work, however, have thicknesses from 100 to 1000 Å or more and the second effect is more important. Thus most metallurgical electron diffraction is genuinely three-dimensional, except where very thin platelets of precipitate are involved.

6.5.1.3   In Fig. 5.3(c), the effective camera length is L, then:

$$\left.\begin{array}{l} QP = L \tan 2\theta \simeq 2L\theta \\ QR = 2L \sin \theta \simeq 2L\theta \end{array}\right\} \quad \text{with small } \theta$$

and

If QP is taken as a ring radius $r = \frac{1}{2}D$ ($D =$ diameter),

$$r = \frac{D}{2} = L \tan 2\theta \simeq 2L\theta \simeq 2L \sin \theta = L\left(\frac{\lambda}{d}\right)$$

or
$$rd = \tfrac{1}{2}Dd = L\lambda \tag{6.47}$$

or
$$d^*(= 1/d) = r/L\lambda$$

Equation (6.47) is sufficiently correct for many practical purposes and $L\lambda$ is called the *camera constant*. Values of the interplanar spacings d are thus easily obtained from the diameters or radii, providing the constant is known. It can be determined by using a standard substance. Such calibration may reveal that $L\lambda$ varies slightly with distance chiefly due to the increasing error in the approximation used.

6.5.2   Single Crystal Patterns

6.5.2.1   Single crystal patterns are obtained when small isolated crystals are placed in the path of the electron beam or from thin films or metal foils. Metallurgical interest in these patterns has increased considerably with the use of the modern microscopes with selected area diffraction.

The patterns obtained will, in the first instance, appear like planes in the

227

reciprocal lattice for the reasons indicated above. If the specimen can be tilted various arrays will appear on the fluorescent screen and the operator will tend to photograph those that give the sharpest patterns. One such pattern from a known phase may give all the information needed. In the case of an unknown phase, however, photographs from more than one orientation can be used to assist in the identification. It is clear that a particularly useful application of such patterns will be to determine the crystallographic plane parallel to the surface of the foil. If two phases are present, this provides a means of determining the mutual orientation—usually described in terms of significant planes and directions which are parallel in the two phases. The following general treatment will therefore include an indication of how this may be done.

6.5.2.2   From section 6.5.1.2 it will be apparent that a single crystal pattern will resemble the network in Fig. 5.2(c) except that the hyperbolae will approximate more closely to straight lines. In that figure the pattern was assumed to be normal to the c axis in an orthogonal case. In the general case the beam direction normal to the plane is a zone $[uvw]$. Since the plane of the pattern is in effect a reciprocal lattice plane the points on it—at least those near the centre—will have values of hkl which satisfy the relation $hu + kv + lw = 0$ (equation (4.20)). The pattern is unlikely to be completely symmetrical about the centre and a number of principles may be used in the interpretation. This process is illustrated in a simple case in Fig. 6.13. The indexing, etc., will proceed on lines such as the following:

(a) It is useful to have a travelling microscope with two traverses at 90° or to use a standard enlarged print and measure with a ruler. Spots which go in pairs on either side of the centre spot can be used to locate the mid-point accurately. Distances of spots from the centre enable spacings d or their reciprocals d^* (directly proportional to the distances) to be found from equation (6.47).

(b) The process of indexing may be assisted by the recognition of prominent rows of spots through the centre such as that indicated by d_4^* in Fig. 6.13(a). These points represent successive orders or successive even orders from the same set of planes, e.g. in body-centred metals (such as iron) these spots are usually from {110} planes, a typical sequence being indexed $(1\bar{1}0)$, $(2\bar{2}0)$, $(3\bar{3}0)$, etc. The intensities are not necessarily equal on either side.

(c) It may assist to have a piece of tracing paper as an overlay if necessary large enough to enable a stereographic projection to be drawn. The pattern will then show the vectors denoted by $\mathbf{d}^*$ and the angles ϕ which are angles between the reciprocal vectors (or planes in real space) and can be represented as shown by points on the circle of projection. This is an illustration of the connections indicated in Table 4.2.

(d) The spacings and angles so found should fit those for the structure if known, and for cubic and other substances reference may be made to the

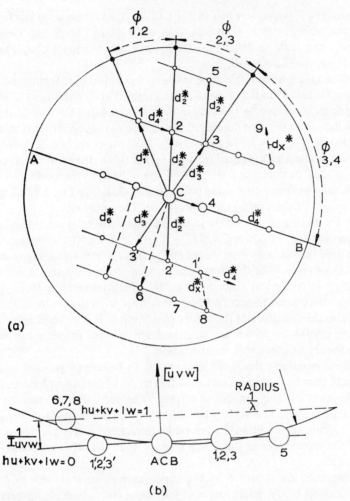

Fig. 6.13 Single crystal pattern from selected area diffraction.
(a) general
(b) showing occurrence of points from successive reciprocal layers and
tilt of principal zone axis

tables of angles. (In addition to the *International Tables* (Ref. [9] of Chapter 4), useful separate individual tables have also been published for cubic [42], hexagonal [43] and tetragonal [44, 45] structures.) The identification of unknown structures may require a search of standard interplanar spacings, but the matter is considerably easier if the lattice can be recognised from the angles and relative spacings.

(e) At an early stage spacings (particularly of known materials) can be allocated general indices $\{hkl\}$. The axis $[uvw]$, which may be normal to the plane of the diagram or slightly tilted as in Fig. 6.13(b), can be found by the use of equation (4.20) from any pair $(h_1k_1l_1)$, $(h_2k_2l_2)$, but these must be chosen from the available planes in their sets to give the correct angle $_1\phi_2$. Other planes $(h_3k_3l_3)$ can then be fitted into the system, and all must obey the relation (4.20), i.e. $hu+kv+lw=0$.

(f) When once a probable zone is found the indexing is further extended by adding vectors $\mathbf{d}^*$ in the plane of the diagram. This is equivalent to the adding of indices according to the system in section 4.4.2. Thus in Fig. 6.13 $\mathbf{d}_3^* + \mathbf{d}_2^* = \mathbf{d}_5^*$ where the spot numbered 5 is located as indicated (vector not shown), and $h_5=h_3+h_2$, $k_5=k_3+k_2$, $l_5=l_3+l_2$. The row of spots 1, 2, 3, is similar to the centre row and has equal separation. Indexing along this row is confirmed or extended by successive addition of the indices corresponding to $\mathbf{d}_4^*$.

(g) It may be found that some points, e.g. 6, 7, 8, 9 in Fig. 6.13(a), cannot be accounted for directly by equation (4.20). They may either

(i) Fit the same zone axis $[uvw]$ for non-zero layers, i.e. they obey the equation $hu+kv+lw=C=\pm1$, ±2, ±3, ±4, etc. (In some patterns from centred lattices only even values appear.) See Fig. 6.13(b). These are similar in some respects to the layer lines in X-ray rotation patterns and the separation in reciprocal space is given by equations (6.1) or (6.2), and (6.3). This effect tends to occur with spots further from the centre of the pattern, due to the curvature of the sphere of reflection. It is more likely to be found with crystals of higher lattice parameter. A scale drawing can be used if necessary to check this interpretation.

(ii) Alternatively the spots may appear to belong to another zone. These could then be slightly misoriented fragments of the same phase arising from sub-grain or polygonised structures. The second zone may be indexed independently. The angle ρ between two such zones $[u_1v_1w_1]$, $[u_2v_2w_2]$ may be calculated or obtained from tables (equation (4.21)). If this angle is only of the same order as the Bragg angle θ, then the other possibility is more likely.

Such cases are recorded by the appearance of vectors such as $\mathbf{d}_5^*$ which may connect to the origin and also join some pairs of points. Others like $\mathbf{d}_x^*$ appear only at a distance from the origin. There may thus be no point (h_x, k_y, l_z) in the original zone but it would fit the second zone $[u_2v_2w_2]$, or it joins points on two different layers of the reciprocal lattice represented by the single zone.

(h) The diagram in Fig. 6.13(b) gives a general indication of how the second layer of the zone may account for the spots nearer the edge. This is the view along BCA and also shows how the principal zone is likely to be slightly tilted and why patterns cannot be equally intense and symmetrical about the origin. The curvature and spot sizes are exaggerated here.

(i) If the stereographic projection is to be completed the projection circle is then a great circle corresponding to the zone axis $[uvw]$ or plane of same indices in a cubic case. Reference to a standard projection (such as Fig. 4.12) will enable the angles between the standard projection circle and the $[uvw]$

circle to be found, and by the relationships of Fig. 4.11 and a standard Wulff net the positions of other zones and the cell edges or faces can be located.

It follows from the above treatment that the direction [uvw] or the plane normal to it (same indices in cubic case) and a reciprocal direction (hkl) give the orientation of the pattern and so of the planes and/or directions on the crystal or foil from which it is derived. Patterns of two or more phases can be used to determine mutual orientations. Some care may be needed because a given zone axis [uvw] could sometimes be equally well replaced by another of the same set, e.g. [$\bar{u}vw$] makes an equal and opposite angle with [100] in the cubic system but the same angles with [010] and [001] so that the cell edges will appear in different but symmetrical positions in the two cases. It is usual to define these in terms of a plane in one phase parallel to a plane in the second and a direction in each plane.

6.5.3 Ring Patterns

6.5.3.1 Because of the small diffraction angles involved an electron diffraction pattern corresponding to the polycrystalline glancing angle or side reflection arrangement indicated in Fig. 6.7(b) will be obtained on a flat film perpendicular to the original beam direction (i.e. as in the X-ray transmission case). Such a pattern then shows almost complete semi-circular arcs. On the other hand a transmission pattern from a polycrystalline pattern will show completed rings. These two types of electron diffraction polycrystalline or powder patterns may be considered together.

Work of metallurgical interest involving the reflection type of pattern was done originally in a direct electron diffraction instrument—of a type which can be constructed or obtained commercially (but for which there is not a wide demand). The main point of interest is that the depth of penetration of the electrons is much less than that of X-rays, and since the Bragg angles are much smaller information is obtainable about very thin surface layers, the total path being a few hundred Ångströms (cf. with X-rays penetration is one hundredth of a millimetre or more). Thus to obtain the full pattern the angle at which the electrons enter the surface must be less than 2θ for the innermost spacing to be recorded. This angle is likely to be of the order of 1–3 degrees which means that a depth of penetration of ~ 500 Å represents a penetration normal to the surface of something between say 10 and 30 Å. The method is therefore valuable for the study of surface films formed by corrosion or oxidation when these are too thin for detection or adequate identification and study by X-rays. Surface finish, the effects of polishing and the existence of 'amorphous' layers, or alternatively surface roughness or the existence of phases which are left above the level of the matrix after chemical or other attack, can be studied.

The transmission type of ring pattern can be obtained from powders mounted suitably on a copper grid or as particles precipitated in a poly-

231

crystalline foil. In this case the foil or specimen must be thin enough to allow the electrons through. In the case of particles the path between may be of less dense material and diffraction then occurs in the glancing angle manner from their surfaces nearly parallel to the incident beam.

6.5.3.2 In both cases the patterns give interplanar spacings direct from equation (6.47). If it is reasonable to do so higher accuracy can be obtained by using the transmission relationship as for X-rays, i.e. $R = L \tan 2\theta$ and then the Bragg equation to find d corresponding to ring radius R.

In both cases the pattern may show concentrations of intensity at points round the ring indicating that the film, layer or powder contains a preferred orientation (general treatment in Chapter 7). If the number of crystallites is small the rings will in any case be spotty and in fact there can be any intermediate condition between the single crystal pattern and the fully continuous ring pattern. In some conditions this could provide an indication of crystallite size analogous to the X-ray methods (also to be described in Chapter 7).

References

1. BERNAL, J. D., *Proc. Roy. Soc. A*, 1926, **113**, 117.
2. DUNN, C. G. and MARTIN, W. W., *Trans. A.I.M.E.*, 1949, **185**, 417. (Also J. Leonhardt, *Z. Krist.*, 1924, **61**, 100.)
3. GRENINGER, A. B., *Z. Krist.*, 1935, **91**, 424.
4. GRENINGER, A. B., *Trans. A.I.M.E.*, 1935, **117**, 61, 75.
5. GRENINGER, A. B., ibid., 1936, **122**, 74.
6. DEBYE, P. and SCHERRER, P., *Phys. Z.*, 1916, **17**, 277.
7. HULL, A. W., *Phys. Rev.*, 1917, **10**, 661.
8. AZAROFF, L. V. and BURGER, M. J., *The Powder Method in X-Ray Crystallography*. McGraw-Hill, New York, etc., 1958.
9. TAYLOR, A. and KAGLE, B. J., *Crystallographic Data of Metal and Alloy Structures*. Dover Publications, New York, 1963.
10. VAN ARKEL, A. E., *Z. Krist.*, 1928, **67**, 235.
11. BRADLEY, A. J. and JAY, A. H., *Proc. Phys. Soc.*, (*London*), 1932, **44**, 563.
12. BRADLEY, A. J. and BRAGG, W. L., *J. Iron Steel Inst.*, 1940, **141**, 70.
13. STRAUMANIS, M. E. and IEVIŅŠ, A., *Z. Phys.*, 1936, **98**, 461.
14. WILSON, A. J. C., *Rev. Sci. Instrum.*, 1949, **20**, 831.
15. PARRISH, W., EKSTEIN, M. G. and IRWIN, B. W., *Data for X-Ray Analysis*, vol. I (Charts). Philips Tech. Library, New York and Eindhoven, 1953.
16. PARRISH, W. and IRWIN, B. W., *Data for X-Ray Analysis*, vol. II (Tables). Philips Tech. Library, New York and Eindhoven, 1953.
17. MASSALSKI, T. B. and KING, H. W., *J. Inst. Metals*, 1960–1, **89**, 169.
18. KING, H. W. and MASSALSKI, T. B., *J. Inst. Metals*, 1961–2, **90**, 486.
19. HESSE, R., *Acta Crystall.*, 1948, **1**, 200.
20. LIPSON, H., *Acta Crystall.*, 1949, **2**, 43.
21. ITO, T., *X-Ray Studies on Polymorphism*. Mamzen, Tokyo, 1950.
22. BUNN, C. W., *Chemical Crystallography*, 142–46, and Appendix 3 (see Ref. [11] of Chapter 4).
23. HULL, A. W. and DAVEY, W. P., *Phys. Rev.*, 1921, **17**, 47.
24. BJURSTRÖM, T., *Z. Phys.*, 1931, **69**, 346.
25. HARRINGTON, R. A., *Rev. Sci. Instrum.*, 1938, **9**, 429.

26. THOMAS, D. E., *J. Sci. Instrum.*, 1941, **18**, 135.
27. JOHANSSON, T., *Z. Phys.*, 1933, **82**, 507.
28. GUINIER, A., *C. R. Acad. Sci. (Paris)*, 1946, **223**, 31.
29. DE WOLFF, P. M., *Appl. Sci. Res.*, 1950, **31**, 119.
30. BRINDLEY, G. W., Monochromators and Focussing Cameras, Chapter 4 of *X-Ray Diffraction by Polycrystalline Materials* (see Ref. [9] of Chapter 5).
31. DE WOLFF, P. M., *Acta Crystall.*, 1948, **1**, 206.
32. KLUG, H. P. and ALEXANDER, L. E., *X-Ray Diffraction Procedures*. Wiley, New York, 1954.
33. ARNDT, U. W., Counter Diffractometers, Chapter 7 of *X-Ray Diffraction by Polycrystalline Materials* (see Ref. [9] of Chapter 5).
34. PARRISH, W. (ed.), *Advances in X-Ray Diffractometry and X-Ray Spectography*. Centrex Publishing Co., Eindhoven (especially paper by W. Parrish and T. R. Kohler, pp. 174–87), 1962.
35. WILSON, A. J. C., *Mathematical Theory X-Ray Powder Diffractometry*. Philips Tech. Library, Cleaver-Hulme Press, Ltd., London, 1963.
36. BRADLEY, A. J., *Proc. Phys. Soc. (London)*, 1935, **47**, 879.
37. TAYLOR, A. and SINCLAIR, H., *Proc. Phys. Soc. (London)*, 1945, **57**, 108 and 126.
38. BRADLEY, A. J. and JAY, A. H., *Proc. Phys. Soc. (London)*, 1932, **44**, 563.
39. NELSON, J. B. and RILEY, D. P., *Proc. Phys. Soc., (London)*, 1945, **57**, 16.
40. COHEN, M. U., *Rev. Sci. Instrum.* (a) 1935, **6**, 68; (b) 1936, **7**, 155.
41. MARSH, K. J. and QUARRELL, A. G., Chapter VI of *The Physical Examination of Metals*, ed. B. Chalmers and A. G. Quarrell. Arnold, London, 1960.
42. ANDREWS, K. W., DYSON, D. J. and KEOWN, S. R., *Interpretation of Electron Diffraction Patterns*. 2nd edition, Hilger, London, 1971.
43. PEAVLER and LENUSKY, *I.M.D. Special Report Series*, No. 8, A.I.M.E.
44. FROUNFELKER, R. E., SEITZ, M. A. and HIRTHE, W. M., *Nucl. Sci. Engin.*, 1962, **14**, 192.
45. FROUNFELKER, R. E. and HIRTHE, W. M., *Trans. A.I.M.E.*, 1962, **224**, 196.

Chapter 7

APPLICATIONS OF DIFFRACTION TECHNIQUES

7.1 SUPPLEMENTARY TECHNIQUES AND SPECIMEN PREPARATION

7.1.1 General

The practical methods which are of metallurgical use have been described in the previous chapter and are sufficient for most applications. Some applications will already have been apparent, but may require further description. Some of these and certain others will now be considered. It is also necessary to take account of the preparation of metallurgical samples in some detail for reasons which will appear.

7.1.2 Metal Crystals

Single metal crystals for X-ray study may be grown or developed by the methods of Chapter 3. Alternatively research into the mechanical or other properties of single crystals may require diffraction examination perhaps to establish or to confirm the orientation. It is also apparent that in the case of metals which have developed very large grains on recrystallisation—as for example in magnetic steel sheet (Fe–Si)—diffraction patterns can be obtained from individual grains which can be picked out from the etched surface. Such patterns could then be used for orientation determination.

Another procedure for single crystals which has been found useful in the investigation of complex phase diagrams employs electrolytic or chemical separation. The matrix is dissolved away and the particles of intermetallic phase isolated and removed for diffraction study. A separation method of this kind was used by Raynor and Wakeman [1, 2] in connection with aluminium alloys. The method has been used in connection with steels where however the isolated particles are not generally of sufficient size for any other than powder photographs. It may be that larger crystals obtained (as for the aluminium alloys) can still be adequately studied by powder methods, but alternatively it may be necessary also to obtain single crystal patterns. Some of the particles show definite crystallographic shapes and forms (e.g. needles) and this may assist their investigation and/or identification.

7.1.3 Powder Specimens

7.1.3.1 Preparation of metal filings: For the investigation of phase diagrams and for other metallurgical constitutional studies, powder photographs may be obtained from metal filings. In the preparation of filings note must be taken of the following:

(a) The file or cutting tool will probably be made of steel. Particles from it may contaminate the sample. Therefore magnetic separation may be required. This raises a difficulty if the specimen itself contains a ferromagnetic phase. Hard metal carbide tools provide an alternative or for the hardest materials diamond containing tools.

(b) The particles may easily oxidise in air. This requires a device for operating in an inert gas or liquid. The filing may for example be done with the specimen under liquid paraffin or in a closed vessel, through which an inert gas is passed. Apparatus for this purpose has been described by Hume-Rothery *et al.* [2] who also indicate apparatus for preparing filings in a complete vacuum.

(c) The operation of filing will cold work many metals and so the particles contain internal strains and in some cases a transformation may occur at the specimen surface (see section 7.1.4.1). These can be removed by heat treatment. A solid alloy containing two or more phases may have been heated to establish equilibrium at a specific temperature. The filings must therefore be heated at this same temperature since any other temperature might lead to a change in equilibrium. It is necessary to provide for the collection of the filings into a suitable glass, hard glass, or silica, capsule which can then be sealed off under vacuo so that it contains the particles in a form ready for annealing. Problems may arise here if the alloy contains a constituent which is highly volatile or which reacts with the glass.

(d) It may be that particles of a range of sizes are obtained and it is desirable to eliminate some of these—the coarsest fraction—by sieving. There is a possibility here that the finer particles are more representative of one phase than the other. This may arise if the metal has phases—such as a softer matrix and a harder secondary phase. The microstructure itself may assist at this point. The powder pattern would probably only be affected as regards the relative amounts of the phases but not their equilibrium composition or structure.

(e) Powders can generally be obtained from refractory materials, oxides, ceramics, etc., by direct crushing and grinding although this may require a percussion mortar for the hardest materials.

In any method of preparation there may be contamination in addition to the possibility of magnetic or other particles from the file or drill cutter or pestle and mortar. Particles of dust, small fibres from paper, soot and grease, etc., can be present. The use of suitable organic liquids for washing and cleaning is therefore desirable unless there is likely to be a reaction. It follows that the best methods of preparing filings, washing, drying and heat treating will depend on the type of alloys being studied and each case must be considered on its merits. In many cases the difficulties are not likely to be very serious and only seldom insuperable. The degree of care and precision can be less for a rapid survey of an alloy system—or part of a system—or the phases in an individual alloy. This could however be a dangerous attitude to adopt and it is always better to take rather too much care than too little.

7.1.3.2. Chemical and Electrolytic Isolation: The electrolytic isolation methods used in connection with alloy diagram studies (referred to above) have been used in the study of steels initially for the separation of non-metallic inclusions. This work owed much to the extensive investigations of Koch and co-workers and a survey of this work by Koch and Sundermann [3] should be consulted. Chemical methods—notably the solution of the metal in alcoholic bromine or iodine or sometimes gaseous chlorine—were also developed for the same purpose particularly for the study of oxide inclusions. The total or fractional determination of oxygen by vacuum fusion has tended to reduce the usage of the chemical solution methods for inclusion analysis. The chlorine method has been used to decompose and remove carbides or intermetallic phases from residues when they have been separated electrolytically or otherwise and when the main purpose of the separation is to obtain the oxide or silicate inclusions. Descriptions of these various procedures are to be found in the Iron and Steel Institute Special Reports on Steel Ingots [4].

In this work the possibility of using X-ray powder photographs of the separated particles was soon appreciated and thus the methods provide one means of establishing the identity and structural information about particles which represent far too small a fraction of the total to be studied by direct examination of the metal by diffraction methods although they may be seen under the optical or electron microscope.

Some of the same techniques could clearly be adopted or modified to provide particles of carbides or intermetallic phases from steels, nickel base alloys or other alloys. Here the particles are usually too small to be regarded as material for single crystal study. The methods have become increasingly important in this context in the development of high temperature alloys, and the study of low and high alloy steels. The electrolytic cells may be open or closed with neutral atmosphere (e.g. nitrogen or argon) and methods of separation, washing and handling depend on the specific type of alloy. The techniques described by Koch and Sundermann [3] are generally suitable. Other reviews including methods and applications are also available [5, 6]. In applying the methods to the isolation of phases for crystallographic study, care is required in the choice of electrolyte since some phases may be dissolved with the matrix, or the separation of phases may be incomplete and particles of matrix may be found in the residue. In general the more stable phases are

236

easy to separate without oxidation or decomposition. The alloys with the more stable corrosion resistant base will also be more difficult to dissolve. It has been shown that the use of potentiostats to control separation conditions can provide a means of closer control and sometimes a method of selective separation [7].

7.1.4 Solid Specimens

7.1.4.1 Cold Work during Preparation: The preparation of solid specimens for diffraction study may involve cutting, machining or grinding to obtain a flat surface or to remove oxide layers from the surface. This preparation is liable to cause distortion of grains, their fragmentation and perhaps the formation of a flowed layer. This cold worked condition of the surface would be indicated by the diffraction pattern but would be unrepresentative of the bulk of the material. Electropolishing or chemical etching may be necessary to remove the surface layer. On the other hand the constitution of a machined or cold-worked surface may be of interest as such and the condition of the surface—grain size and distortion—can therefore also be used as a measure of the amount of work.

It should be noted that for certain metals and alloys phase changes may be brought about in the surface layers due to preparation. Perhaps the commonest example is those stainless steels which are fully austenitic but undergo a martensitic transformation on cold working. This transformation can be brought about by machining, grinding or filing and so it may affect the preparation of lump specimens and of filings for powder analysis.

7.1.4.2 Enhancement of phases on surface: The process of etching for micro-examination depends on the different solution rates of the phases present. It may be useful to prolong the etching process in the solution or to etch electrolytically in a suitable solution—as might be used for complete electrolytic separation. In this way one phase may become considerably enriched at the surface so that a satisfactory side-reflection or back-reflection X-ray pattern can be obtained. This technique has been used for alloy and other steels when there is sufficient of the second phase, e.g. cementite in a medium/high carbon steel, sigma phase in a high alloy steel.

7.1.4.3 Oxide and other layers *in situ*: Extremely thin films may be found during the initial stages of high temperature oxidation, or on many ordinary metals at room temperature, and similar films can form by corrosion in solution or damp atmospheres. The thinnest films can be examined *in situ* in an electron diffraction camera. It may be necessary to take account of constitutional changes in such films whilst in the electron diffraction camera—either due to electron bombardment or to the vacuum conditions. Thicker films and layers may be examined by X-ray diffraction and this is an important application of the methods for solid specimens. In the case of iron and steel,

237

examination of the layers *in situ* may be followed by removal crushing to powder and supplementary powder photography. The identity and lattice parameters of the phases may be required. This procedure may not be necessary in the case of specimens scaled under laboratory conditions for short times. Ingots, billets, etc., heated in industrial furnaces may develop thicker layers and some of these eventually form double layers involving a sequence from the lowest oxide in contact with the metal to the highest at the surface. More complicated layers can be formed.

7.1.5 Thin Metal Strips, Foils, etc.

7.1.5.1 The transmission X-ray diffraction technique can be applied directly to any metal strip or foil that is already thin enough. In some cases further reduction by rolling (followed if required by annealing) can be employed to make the material thinner. Alternatively thinning by chemical or electrolytic methods can be adopted. In applications where these methods are unsuitable a thin piece of the metal can be cut off a solid lump and then reduced progressively by rubbing down on emery paper to a suitable thickness (perhaps 0·1–0·2 mm). A suitable simple holder is a steel block with a recess of the right depth into which the specimen with surface already smooth can be held by black vacuum wax or some other laboratory adhesive. Foils produced in this way are also used for micro-radiography (Chapter 8).

7.1.5.2 The use of foils for transmission diffraction is in effect an adjunct to the use for electron microscopy (Chapter 10 in Vol. 2). It is however useful to refer to it here since diffraction may still be the primary object. For this purpose the same initial procedures as mentioned above will enable foils to be produced down to say 25 μm (0·001 in). Further thinning by chemical or electrolytic means is still required. A number of simple devices have been constructed for this purpose. One such device places the specimen as anode between two cathodes. Electrolysis thus proceeds from both sides of the foil [8, 9]. Very thin foils may result and often these have holes and the operator has a choice of thickness in these regions and can choose regions which show diffraction or microscope images of interest.

7.1.5.3 Films or layers can be removed from the surface of a piece of metal. For example some oxide layers or corrosion films can be removed in this way and may be used for identification by electron or X-ray diffraction.

Similar in some respects are the 'extraction replica' techniques in which an etched specimen is coated with a polymer—or more satisfactorily with carbon (see Chapter 2 in Vol. 2). Further etching then leads to the film floating off, containing the particles which were in the surface. This method is used to obtain particles for electron diffraction or microscopy [8, 9]. It may sometimes be convenient to use the extraction replica produced in more bulk for X-ray diffraction which then complements the electron microscope/diffraction work [10].

Mention may also be made here of the direct surface replica techniques used in electron microscopy. The replica in this case provides only an image of the contours of the etched surface and so gives no diffraction pattern.

7.2 IDENTIFICATION OF PHASES

7.2.1 General

In Chapter 4 (section 4.3.2.2) some simple structures were described including some of metallurgical interest, and reference was made there and in section 6.4.1 to various tabulations and lists of structures which are available (Refs. [6], [7] of Chapter 4; [9] of Chapter 6). It is evident that although many compounds might have the same structure they are unlikely to give absolutely identical diffraction patterns—for example the lattice parameters may be different enough to enable discrimination. In general therefore diffraction patterns are a means of identifying phases and may be obtained primarily for this purpose.

Assuming the crystal system can be found or recognised from the pattern, the space lattice follows from the presence or absence of certain reflections. Other absences or intensity differences may narrow down the possibilities further and the lattice parameter(s) even more so. Patterns from simple structures like those from face-centred cubes, body-centred cubes, hexagonal structures with axial ratios in the region of $1 \cdot 5$–$1 \cdot 8$ (say) should be recognisable with a little practice (see also section 6.4.3.4). Alternatively the interplanar spacings alone can be used to provide an identification. The following further details are probably sufficient for practical purposes.

7.2.2 Use of Standard Patterns—Isomorphous Types, etc.

7.2.2.1 A practical procedure in a laboratory using diffraction techniques is to provide a collection of standard patterns taken under the conditions generally employed, i.e. the same camera(s) and radiations. A laboratory working on the identification of phases in complex alloys will for example build up a set of patterns of the relevant phases.

In some cases a structure may belong to the same type as a standard but not be otherwise identified with it. As examples one may refer again to the structures listed in section 4.3.2.2:

1. Diamond, silicon and germanium all have the same sequence of powder lines but different lattice parameters. As indicated previously these patterns have $N = h^2 + k^2 + l^2 = 8n$ or $8n + 3$.
2. Zinc blende has face-centred cubic lines with $N = 4n$ or $8n + 3$ but the $4n$ lines represented by odd n are weak.
3. Calcium fluoride again has face-centred cubic lines ($4n$, $8n + 3$ present) with some differences in relative intensities and a larger lattice parameter than the face-centred cubic metals or the two previous structures.

4. The sodium chloride structure likewise has an essentially face-centred cubic pattern but the relative intensities are different from those for the metals. There are many phases and compounds with this structure and lattice parameters can lie over a wide range. Care is needed in some work to discriminate. An example is the identification of the cubic carbides and nitrides formed in steels to which Ti, Nb or V have been added. Solid solutions or two (or more) phases of the same structure can be obtained in one specimen.

5. The body-centred cubic pattern only shows lines with N even. The pattern for the caesium chloride type structure derived from some body-centred cubic alloys which have undergone a disorder→order transformation on cooling, will however show weaker additional lines due to the diffraction for other values of N (always excluding 7, 15, 23, 28 etc., which cannot resolve into the sum of three squares). The reason for the extra reflections is given in section 5.2.6.2.

6, 7. Additional reflections appear for the same reason but in 6 the tetragonality is present and each cubic powder line is now represented by a double line.

7.2.2.2   The recognition of solid solution effects by slight differences in lattice parameters or interplanar spacings or by changes in the relative positions of lines where the axial ratio has changed (hexagonal, etc.), provides a valuable adjunct to simple identification and may in some cases be a necessary part of it (so in the cubic carbides and nitrides noted above). In this connection in particular the information collected by Pearson (Ref. [6] of Chapter 4) may be particularly useful. (It has other applications to be considered in section 7.3.)

7.2.2.3   In the case of electron diffraction a collection of ring patterns would serve as standards for polycrystalline work. The scale of the patterns is, however, small under the usual conditions and identification by direct comparison of standards without measurement is generally inadvisable. For spot patterns certain simple facts emerge. A square network of spots generally comes from a cubic phase, or tetragonal with c axis normal to the pattern. A hexagonal network of diffraction spots comes from planes parallel to the c axis in a hexagonal structure or parallel to $\langle 111 \rangle$ in a cubic structure. An example of the former is indeed shown in Fig. 4.8(b). Other cases can be calculated by reference again to equation (4.20). In this case if the pattern for a zone $[uvw]$ is required the equation may be used to find the smallest whole number indices that satisfy the equation. The angles between these directions in reciprocal space are then calculated and angles and vectors $\mathbf{d}^*$ are plotted as in Fig. 6.13. It is then easy to extend the pattern by vector or index addition, checking if desired by re-substitution in (4.20). It is a useful exercise to construct such networks for a few cubic zones, e.g. [110], [112], [321].

240

7.2.3 The Index of Powder Diffraction Data

7.2.3.1 Identification has been much assisted by the compilation of index cards, and an index book originally known as the *A.S.T.M. Index of Diffraction Data* and now as the *Powder Data File* [10]. This originated from earlier compilations of data for large numbers of compounds [11].

7.2.3.2 Identification would proceed by typical steps:

1. A powder pattern is obtained and the interplanar spacings and intensities tabulated in decreasing order of the former (increasing θ). The intensities are usually estimated for this purpose by eye.
2. The three strongest lines are selected and in addition the 'innermost' line, i.e. the one of smallest θ and largest d may also be noted.
3. The book accompanying the index is then consulted. The (apparently) strongest of those selected is taken first. The book lists spacings for groups of compounds between certain limits (e.g. a typical set has lines between 2·40 and 2·44 Å). Within this group the compounds are arranged in order of the second strongest line. Hence these lines are examined to see whether any of them coincide with either of the two other lines recognised as stronger than the others. The innermost line (which may or may not be one of the three) is also indicated.
4. If one of the compounds listed in the group appears to have spacings near to the observed, and is possible on other grounds, the card itself is taken out of the index. This card tabulates all the lines in order of interplanar spacing and indicates the indexing (*hkl*) when known, and the relative intensities. A full comparison may then be made. Two or more cards may require examination.
5. If the identification is not successful a fresh start may be made from one of the other strongest lines. If there is only one phase present this should lead to the same choice of compounds. Sometimes however the relative intensities are somewhat different if the radiation is not identical with that for which the standard pattern was obtained, or there may be two or more phases present.

7.2.3.3 It is worthwhile to point to some limitations of this method although for laboratories using X-ray diffraction as a tool for the identification of many different kinds of constituent in a variety of materials it can be of very great value. The index contains data for several thousands of compounds including metals, inorganic salts, organic salts and compounds, and many minerals. It is however still possible to meet compounds or phases which are not (yet) included. In metallurgical work particularly solid solutions may be present and so the interplanar spacings may be systematically shifted relative to those given in the index.

The point about the differences in relative intensity noted earlier is not often critical but it could affect the ease of identification especially when radiations of widely different wavelength have been used (e.g. Cu and Mo or even Cu and Cr). The data have usually been compiled from cylindrical powder photographs or diffractometer charts. The intensities are therefore affected by the angular factors (noted in section 6.4.7). If a focusing camera is used the relative intensities will be somewhat different because of the different geometry and unrotated specimen. For many purposes such differences are not critical in X-ray work. More care is needed in accepting X-ray standards for the identification of phases by their electron diffraction patterns since the differences in relative intensities may be much more marked due to a variety of causes including specimen texture.

The index cards also contain data about lattice parameters, refractive indices, density, etc., which may help in the identification or be useful otherwise.

7.3 LATTICE PARAMETER MEASUREMENTS

7.3.1 General

Methods for obtaining accurate lattice parameters and interplanar spacings and the various sources of error in these measurements have already been described in the previous chapter. The applications to be considered now are those which may be of value in the study of phase diagram relationships for pure or commercial alloys, or provide other information about the alloys. It is already apparent that such measurements will sometimes materially advance the precise identification of phases.

One application of such measurements particularly with solid specimens and using a back-reflection camera is for the determination of residual elastic stresses. These are the macro-stress systems which may be left in a piece of metal or component after casting, bending, welding, rolling, etc. This application along with other methods of stress measurement is considered in Volume 2. (There may be independent or superimposed micro stresses which are self-compensating over short distances and these are revealed by line broadening to be considered in section 7.4.)

7.3.2 Application to Phase Equilibria

7.3.2.1 It will be evident that if the phases in an alloy can be seen and recognised under a microscope, observations of this kind can be used to help to establish the phase relationships in say a binary or ternary alloy system. Two alloys close together in composition, one of which is single phase and the other two-phase together 'bracket' the phase boundary. Diffraction patterns could be, and are, used in the same way. The detection of small amounts of the second phase may however be difficult and the bracket may be in error unless the two-phase region is narrow so that a small composition difference leads to an appreciable change in the proportions of phases. A plot of proportions of two phases against composition in a binary system can sometimes be used to locate the boundary.

7.3.2.2 Another obvious way of locating the boundary is to obtain lattice parameter measurements of alloys in the single phase and two-phase region. Providing the atomic (or ionic) diameters of the components are not identical, which is unlikely, there will be a progressive change in lattice parameter in the one-phase region. In a binary two-phase region however the phases in equilibrium will have constant compositions corresponding to the ends of a tie line. Alloys in this region will thus show varying proportions of these two phases with constant lattice parameters. A minimum of two alloys will be used to confirm that the lattice parameter is constant. The resultant lattice spacing/composition curve therefore shows a sloping part and a horizontal part. The phase boundary is thus accurately located where the two intersect. In Fig. 7.1(a) a typical case is illustrated.

In a ternary or higher order system the compositions chosen for the study may happen to lie along a tie line when a constant lattice parameter will again be found in the two-phase region for these alloys. In general however this will not be the case and so it may be necessary to prepare a number of alloys of both one and two phase and to construct lattice parameter composition curves. The phase boundaries are located where these lines show inflections. An example is shown in Fig. 7.1(b) where the alloys are chosen on a line of constant ratio of two components (based on diagrams in Pearson's book, Ref. [9] of Chapter 4). In general a constant ratio of $B:C$ is taken and parameters plotted against $B+C$, or one may also take a line of constant A when compositions are plotted in terms of B or C ($B+C=100-A$). Alternatively compositions can be chosen at points over an area in the ternary system and used to construct lattice parameter contours. A notable example is the important phase diagram Fe–Cr–Ni of interest in connection with stainless steels and nickel-base alloys [13].

7.3.2.3 The same techniques can be used to define solubility limits in commercial alloys which have a complex base. For example several components may in fact be present although one or two predominate. The precipitation of phases, for example intermetallic phases or carbides, may be followed by constructing pseudo-binary sections with variation of one component. In such alloys it is also possible to find direct relationships between the lattice parameters of certain phases and the amounts or ratios of significant alloy elements in the alloy as a whole. For example in a steel containing molybdenum and vanadium the cubic carbide shows a (V, Mo)C variation of lattice parameter according to the ratio V(Mo + C) and reaches a solubility limit if this ratio is decreased to just below 0·6. Numerous examples of this kind are available in the work of Krainer [14].

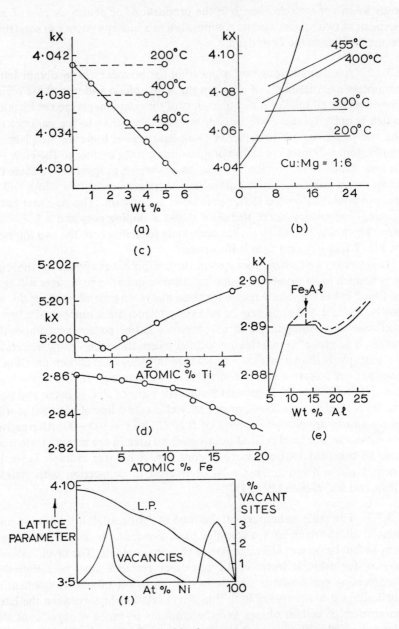

Fig. 7.1 Applications of lattice parameter measurements.

(a) binary system Al–Cu
(b) ternary system Al–Cu–Mg
(c) binary system MgTi (c axis)
(d) Fe–Si solutions
(e) Fe–Al alloys
(f) Au–Ni

(a) and (b) determination of phase boundaries

7.3.2.4   In work of this kind the precautions of ensuring that filings are properly representative of the compositions of lump alloys from which they are derived must be carefully observed (section 7.1). For some alloys it is obviously better or desirable to determine the lattice parameter by a back-reflection technique on the lump alloy. In constructing the lattice parameter composition curves in the one-phase region, high temperature equilibrium can be established first to ensure homogeneity and the alloys slowly cooled or quenched to room temperature. The binary alloys are then quenched to retain the equilibrium composition as in Fig. 7.1(a). The latter is desirable if the phase boundary moves in appreciably to lower compositions at low temperatures. In this case, however, it is necessary to examine the alloys carefully to see if they have undergone any transformation on quenching. Difficulties may arise if there is an increasing solubility followed by a decrease, if the inflection (due to a change in the second phase) is at a low enough temperature for the establishment of equilibrium to require long times. It is generally desirable to obtain some alloys in the one-phase region which have compositions near the phase boundary in order to avoid undue extrapolation.

These precautions and obvious difficulties do not detract unduly from the value of the method. It is perhaps useful to draw attention here to the value in some work of making lattice parameter/composition measurements at the equilibrium temperature itself (see section 7.7 on high temperature diffraction).

7.3.3   Other Applications to the Study of Alloys

7.3.3.1   It has been assumed in the previous paragraphs that the lattice parameter composition curves are straight lines or smooth curves, otherwise it would not be reasonable to use them for phase boundary determination. There are, however, a number of phenomena which give rise to changes in direction or irregularities and these may be used to help to interpret atomic or electronic factors which influence the formation of the alloys, and to detect such effects as ordering or vacancies. Only a brief account can be given and further reference may be made for example to Pearson (Chapter II in Ref. [9] of Chapter 4) and to a review by Massalski [15].

7.3.3.2   Relationship to atomic size factor: The slope of the lattice spacing curve is clearly an indication of the relative atomic or ionic sizes. Vegard's 'law' states that the mean interatomic distance in an alloy is a linear function of the distances in the crystals of the elements, e.g. in a binary system AB containing $x\%$ of B.

$$d_{\mathrm{m}} = d_A(1-x) + d_B x \tag{7.1}$$

One would *not* expect this equation always to be followed accurately for the following reasons:

i If there is a difference in crystal structure (co-ordination number) between

A and *B*. In this case there must be at least one two-phase region between the two primary solid solutions, for each of which a relationship like Fig. 7.1(a) would apply—lattice parameter decreasing or increasing with composition according to the relative size. Intermediate phases will show the same effects within a two-phase region at each end. In such cases, it is also unlikely that atomic diameters or volumes will lie on a smooth continuous curve joining the values for the pure components.

ii If there is a difference in valency there can be an effect depending on the square, viz. $(V_A - V_B)^2$ where the elements have the valencies V_A and V_B. This effect is additional to any direct size factor effect which might under ideal conditions lead to the Vegard relationship.

Another type of valency effect occurs in certain hexagonal solid solutions which depends on the axial ratio and the way it is affected by changes in electron concentration. The effect involves the Brillouin zone theory and is revealed by sharp changes in direction in the lattice spacing curve. Examples are magnesium base alloys with three- and four-valent solutes (e.g. Mg–Ti, see Fig. 7.1(c)). The system magnesium–cadmium shows an interesting effect which is also probably explained in relation to the zone theory. The *a* axis shows a marked deviation from Vegard's law, and the *c* shows a steep fall to a minimum near 50% Mg and then a less steep rise to 100% Mg. In this case there is no change in the electron/atom ratio since both elements have the same valency.

Into the same category one may place valency effects in alloys involving transition metals (of which iron and nickel are the most important)—either systems of transition metals with each other or with other elements. One example is the system Fe–Si illustrated in Fig. 7.1(d).

iii Not unrelated to relative valency is the possibility that a strong 'chemical attraction' between the elements may, without forming a compound, result in a shorter mean interatomic distance in the solid solution and a contraction relative to the Vegard relation. A mutual repulsion would give the opposite effect.

iv The existence of short-range order may be indicated by a significant but small contraction of the mean lattice parameter relative to the curve for a completely disordered solution. Long range order gives a more definite effect. A notable case is the system Fe–Al indicated in Fig. 7.1(e). In this simplified diagram the solid line represents the lattice parameters of alloys which are probably disordered up to the inflection but the horizontal solid line appears to correspond to alloys which are ordered in the region of Fe_3Al and beyond it where the curved region then corresponds to intermediate stages between Fe_3Al and FeAl. These are slowly cooled alloys. The broken lines for quenched alloys correspond to the larger lattice parameters of disordered alloys. For a fuller explanation see the original paper [16] (or Pearson, or books on alloy theory). This example serves to illustrate the possibilities. The additional diffraction lines are usually also present.

v A solid solution containing vacant lattice sites, e.g. vacancies which are quenched in or present for other reasons, would be expected to show a slightly lower lattice parameter than a solid solution of the same composition with all lattice sites occupied. Fig. 7.1(e) is largely diagrammatic but indicates a specific case—the system Au–Ni described by Ellwood [17]. The scale of vacancies is exaggerated and there are only small but definite deviations from linearity in the lattice parameter curve. This curve alone does not, however, provide all the evidence required and deviations in lattice parameter could clearly be due to other causes. Hence it is also necessary to determine the density. The unit cell and chemical composition can be used to find a theoretical density which is greater. The calculation is simple:

Let: d_0 = observed density

d_x = calculated (X-ray) density.

The percentage of vacant sites is $\left(\dfrac{d_x - d_0}{d_x}\right) \times 100$.

To calculate d_x the formula weight *per unit cell* must be multiplied by the atomic unit of mass ($1\cdot660\,35 \times 10^{-24}$ g) and divided by the volume, which in the cubic case is a^3 and in the tetragonal $a^2 c$, in the hexagonal $[\sqrt{(3)}/2](a^2 c)$, and in the orthorhombic *abc*.

This is an example of how interesting information may be obtained by the use of a simple additional physical measurement. It is, however, possible that some variation of density might arise from accidental flaws or cracks and care must be taken to check this point.

Further reference will be made to lattice parameters in the section on high temperature X-ray cameras.

7.4 GRAIN OR CRYSTAL SIZE

7.4.1 Basis of Methods

The methods to be considered here depend on the size or number of diffraction spots and usually concern crystal sizes greater than 10^{-4} cm. They are distinct from the methods which depend on line broadening to be considered in the next section. They arise from two simple facts. If there are not enough grains or particles to give smooth diffraction lines or rings individual spots will appear. Secondly the size or shape of a spot will depend on the size or shape of the grain from which it derives and of which it is in fact an image. Methods of determining grain size by spot counting are therefore analogous to the microscopical (or macroscopical) methods such as point counting which are referred to in Chapter 1 of Volume 2.

7.4.2 The 'Pinhole' Transmission Method

The first method arises from the simple use of a beam of unfiltered radiation containing the characteristic and white wavelengths. A collimator with a straight cylindrical hole is used, the diameter determining the area of specimen irradiated. The arrangement is usually like the Laüe transmission method and can be employed with a thin metal film or thin slice of refractory or rock, or with a flat powder compact. If desired a series of standard collimators can be kept. It can also be used with side-reflection or back-reflection arrangements.

The sequence of patterns for particles of different sizes arises in the following manner. If a single crystal is in the path of the beam a Laüe pattern is obtained. With a coarse-grained specimen there may be two or three, or a small number of irradiated grains giving their superimposed Laüe patterns. More crystals give a greater confusion of Laüe spots which gradually vanish into a continuous background. At the same time, however, more and more grains will be in a position to diffract the characteristic wavelengths and so spotty diffraction rings emerge corresponding to the ordinary transmission powder rings. If the grain size is even smaller these rings may be continuous and the background loses all traces of spottiness.

Applications of this method have been described in the literature [18, 19, 20].

7.4.3 Measurement of Spot Size

The shape and size of a diffraction spot depends on the shape of the grain, its orientation relative to the beam, the beam divergence and the angle at which the pencil of rays defining the image intersects the plane of the film. A linear relationship exists at least between 10 and 100 μm. A notable example is the work of Clark and Zimmer [19] illustrated in Fig. 7.2 which was established for brass and then shown to apply to steel, silicon carbide, silica and other materials.

The linear relationship between the average grain diameter, independently determined by microscopical methods, and the image length is clearly established.

The method will be most easily applied to powders with undistorted grains, sufficient in number to give a fair statistical sample, but not enough to give many overlapping images.

7.4.4 Spot Counting Methods

In the region where there are reasonably defined Debye–Scherrer rings, spot counting may be used as an index of crystal size. Such was the procedure developed by Stephen and Barnes [21] for back-reflection and for transmission by Schdanow [22].

Let a = diameter of irradiated area (may be determined by a direct exposure)
R = specimen film distance.

d=thickness of transmission specimen *or* depth of penetration in back-reflection.

then $\qquad\qquad V =$ irradiated volume $= \frac{1}{4}\pi a^2 d$

Also let N=number of spots in a ring and v=average volume per grain.

As in the determination of the complete intensity formula for powder patterns (section 6.4.7 and Fig. 6.7) it is necessary to consider the probability

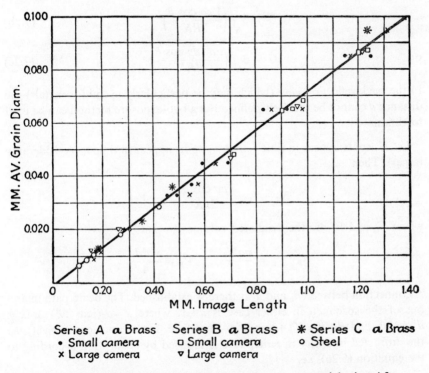

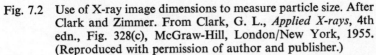

Series A a Brass   Series B a Brass   ✳ Series C a Brass
  • Small camera   □ Small camera   ○ Steel
  × Large camera   ▽ Large camera

Fig. 7.2   Use of X-ray image dimensions to measure particle size. After Clark and Zimmer. From Clark, G. L., *Applied X-rays*, 4th edn., Fig. 328(c), McGraw-Hill, London/New York, 1955. (Reproduced with permission of author and publisher.)

that a single grain will be so oriented as to contribute to the diffraction ring. The spherical belt is assumed to have a width approximately equal to a (diameter of collimator)* and so an area $2\pi Ra \cos \theta$ which on division by the area of the sphere $(4\pi R^2)$ and multiplying by the factor p for the planar multiplicity gives:

$$\frac{1}{2}\frac{pa}{R} \cos \theta$$

* In this case $a = r\varDelta\theta$ in Fig. 6.7.

Now v is the average volume per grain so that $V/v=$number of grains irradiated and so the number N of spots must be this number of grains times the probability, i.e.

$$N = \tfrac{1}{2}p\,\frac{V}{v}\,\frac{a}{R}\cos\theta \qquad (7.2a)$$

If the grains are approximately cubic of mean edge G, $G=3\sqrt{v}$, i.e.

$$G = 3\sqrt{\left(\frac{Vpa\cos\theta}{2RN}\right)} \qquad (7.2b)$$

or $$G = 3\sqrt{\left(\frac{\pi a^3\,dp\cos\theta}{8RN}\right)} \qquad (7.2c)$$

In the back-reflection case (which may be particularly useful for metals) the distance d cannot be precisely defined but a two-exposure method can be used to eliminate it:

Let N_1 spots appear in time t_1 and N_2 spots appear in time t_2 (considerably larger). Then

$$N_2-N_1 = \frac{pa\cos\theta}{2vR}\,[V_2-V_1]$$

where V_2 and V_1 are the irradiated volumes so that

$$V_2 = \frac{\pi a^2 d_2}{4}, \qquad V_1 = \frac{\pi a^2 d_1}{4}$$

A connection between d_2 and d_1 is thereby established. The beam path in and out of the specimen in either case is $d+x$, where $x=d/(|\cos 2\theta|)$ or $x=d|\sec 2\theta|$ and $d+x=d(1+|\sec 2\theta|)$. Therefore, for an incident intensity I_0, the diffracted intensity in either case is reduced by absorption according to the equation (5.26), i.e.

$$-\log_e\frac{I}{I_0} = \mu(d+x) = \mu d(1+|\sec 2\theta|)$$

Therefore $$V_2-V_1 = \frac{\pi a^2}{4}\,(d_2-d_1) = \frac{\pi a^2}{4\mu(1+|\sec 2\theta|)}\log_e\frac{I_1}{I_2}$$

The effective depth of penetration in each case obviously corresponds to grains from which the reflections are only just visible, whilst I_1 and I_2 correspond to the intensities of radiation leaving the specimen to produce these diffraction spots. The same blackening requires $I_1\times t_1=I_2\times t_2$, and so one may put

$$r = \frac{I_1}{I_2} = \frac{t_2}{t_1}$$

The final form of equation (7.2) is then

$$G = 3\Big/ \sqrt{\left\{\frac{(\pi a^3 p \cos \theta)\log_e r}{8NR\mu(1+|\sec 2\theta|)}\right\}} \qquad (7.2d)$$

Precautions in the use of the method are:

1. In the back-reflections case care must be taken when the α_1 and α_2 rings are separate to count all the spots in the α_1 ring only and avoid including α_2 spots, or to count all spots and divide by two. Spots may however be displaced somewhat from the mean ring position and ambiguity can arise.
2. Twinned crystals will contribute spots as if they were extra crystals so the mean grain size will appear smaller.
3. The effective range of grain size for which the method is most effective—grains not too coarse at one limit or too numerous to give separate spots—is rather limited with typical practical arrangements. The range is between 2 and 8×10^{-2} mm. The micro-beam technique mentioned below extends this range to lower limits.
4. The assumption that the width of the spherical belt, in the derivation of the formula, is equal to the collimator diameter is not completely justified. The spread of diffracting plane normals over an effective angular range Δ (or $\Delta\theta$) is more appropriate. A method for eliminating this not easily determined quantity has been suggested [23]. Two exposures are taken with the specimen oscillated through two different angles ψ_2 and ψ_1. Here $\frac{1}{2}a\cos\theta$ is replaced by $(\psi/\pi + \frac{1}{2}\Delta)\cos\theta$ so that

$$N_2 - N_1 \propto \frac{1}{\pi}(\psi_2 - \psi_1)$$

and the other factors are the same.
5. For coarse grained specimens there is a tendency for the mean grain size to be affected by the surface grains which are not truly representative of the whole material. In the transmission case two surfaces can be involved. A method of allowing for this has been described [23, 24].
6. One method of eliminating possible sources of error is to have a standard curve for a given experimental arrangement—plotting number of spots against grain size—the latter having been determined separately by another method, in practice by the optical microscope. Stephen and Barnes [21] show a typical case.

7.4.5 Micro-Beam Techniques

With ordinary cameras and experimental arrangements, e.g. with beam diameters between $\frac{1}{2}$ and 1 mm, continuous diffraction rings are formed with mean crystal size in the region of $2-8 \times 10^{-2}$ mm. The spot counting techniques are not able to indicate crystal sizes in the region $10^{-2}-10^{-3}$ mm (one micron). The micro-beam techniques developed by Hirsch and Kellar

[25, 26] have, however, enabled this range to be covered. With the smallest collimators available grain sizes as low as 0·6 μm have been indicated (12 μm collimator). Crystal sizes of 1 μm have been determined for a back-reflection aluminium ring with collimator diameters in the region of 16 μm. The collimators are made from lead glass capillaries and obviously accurate alignment is needed. The intensity is raised by using a micro-focus X-ray tube which gives a fine concentrated 'point source'.

7.5 LINE BROADENING: CRYSTALLITE SIZE AND LATTICE STRAIN

7.5.1 Introduction

A powder diffraction line has a finite width due to two main types of cause:

(a) The instrumental or experimental factors.
These include finite size of X-ray focus, specimen size and beam diameter, finite slit width in a diffractometer and grain size of film. Whatever collimation is used there will be angular divergence, although in a focusing arrangement this is taken advantage of to give a theoretically 'sharp' line and can eliminate the specimen size effect. For normal work the radiation used normally contains $K\alpha_1$ and $K\alpha_2$ components both of which have a natural finite spread and overlap at low Bragg angles. A monochromator does not usually separate these two lines (and if it did would leave an unsymmetrical line).
(b) Intrinsic or crystallographic factors.
These are factors which are determined by the state of the crystals giving the diffraction. The chief causes of broadening to be considered are:

  (i) Small crystallite size, and related
  (ii) Stacking faults or formation of Guinier–Preston zones, etc., and
  (iii) Lattice strains.

The first set of factors (a) can be found by experiment and are obviously usually kept within satisfactory limits in the typical camera and diffractometers. They are otherwise difficult to calculate. It is however useful to take note of the following property. The notional finite spread of a wavelength is an irreducible factor. Let this be given by $\Delta\lambda$. This could be, for example, the width of the (natural) peak at half the maximum height (see below). From the Bragg equation, as in the calculation of errors for lattice spacing measurement but considering here the relation between $\Delta\lambda$ and $\Delta\theta$, one has the result

$$\Delta\lambda = 2d\,\Delta\theta\,\cos\theta$$

or
$$\Delta\lambda = \left(\frac{\Delta\lambda}{\lambda}\right)\tan\theta \qquad (7.3)$$

252

It follows that at low values of θ the spread from this cause will be narrow so that not only will the spread be small but the $\alpha_1\alpha_2$ components will be much closer together and the total width will be less than any spread due to the instrumental causes. With increasing θ however it is clear from equation (7.3) that the $\alpha_1\alpha_2$ separation will increase and the spectral width of each line (or $\Delta\lambda/\lambda$) will give rise to an increased angular broadening. On the other hand, the broadening due to some of the factors associated with the experimental geometry decreases. Therefore towards higher values of θ the minimum line broadening is due to the natural spectral width of the X-ray line. Any effects due to small crystallite size or lattice strains are additional to this width.

The following discussion relates chiefly to X-ray powder lines, but the effects due to stacking faults and zones are also relevant to electron diffraction techniques.

7.5.2 Line Broadening due to Crystallite Size

7.5.2.1 Since there is an increasing line broadening with smaller crystallite size below 10^{-4} cm, the technique of measuring this broadening to determine the size is complementary to the techniques for larger grain sizes outlined in section 7.4.

It has been shown in section 5.2 that the peak intensity and shape of a diffraction spot (or line) contains a factor representing the number of unit cells in the diffracting crystal. This factor is represented by equation (5.38), one term of which is given by

$$I \propto \frac{\sin^2 \pi N_1(h+\Delta h)}{\sin^2 \pi(h+\Delta h)} = \frac{\sin^2 \pi N_1 \Delta h}{\sin^2 \pi \Delta h} \tag{7.4}$$

(the maximum value as $\Delta h \to 0$ is N_1^2).

To find the effect on a particular diffraction from a plane of spacing d one may first assume that the axes have been transformed so that, in (7.4), h becomes h' in the direction normal to the diffracting planes (and so $k' = l' = 0$). It is then true that the crystallite size ϵ in this direction is given by $\epsilon = Nd$ where N is the number of 'unit cells' of edge d in this direction ($N = N_1$ in (7.4)). It was seen in section 5.2 that the maximum value of (7.4) would be N^2 and that the intensity on either side fell off to zero as Δh increased from zero to points given by $\Delta h = \pm(1/N)$ (width $2/N$). This distance is not a useful measure of width because it is between two points of 'zero' intensity which are not sharply defined. The practical width may be defined in either of two ways the first of which to be considered is the *half-peak width* denoted by β in Fig. 7.3(a), i.e. the width at half the maximum intensity (N^2). This arises when:

$$\frac{1}{N^2} \frac{\sin^2 \pi N \Delta h'}{\sin^2 \pi \Delta h'} = \frac{1}{2}$$

providing only the genuine crystallite size broadening is represented—this is separated from the actual broadening by methods described below. Since $\pi\,\Delta h'$ is very small $\sin \pi\,\Delta h' \simeq \pi\,\Delta h'$ and so

$$\frac{\sin \pi N\,\Delta h'}{\pi N\,\Delta h'} = \frac{1}{\sqrt{2}}$$

This is of the form

$$\frac{\sin x}{x} = \frac{1}{\sqrt{2}}$$

The solution can be found graphically from tables (Ref. [9], Chapter 4). It is

$$x = \pi N\,\Delta h' = 1\cdot 39 \text{ radians} \quad \text{or} \quad \Delta h' = 0\cdot 44\frac{1}{N}$$

which is somewhat less than half the total width given by $\rightarrow(1/N)$. The corresponding value of $\Delta\theta$ in the diffraction pattern corresponding to $\Delta h'$ can be

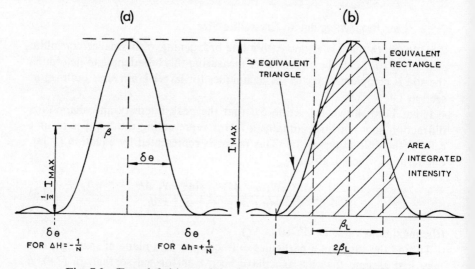

Fig. 7.3   Two definitions of line broadening due to crystallite size
(a) half-peak breadth
(b) Laüe integral breadth

deduced simply as follows. It is only necessary to consider that the cell to which h' refers is rectangular. The reciprocal form of the Bragg equation (5.9) and equation (4.12) then give

$$\frac{4}{\lambda^2}\sin^2\theta = \frac{1}{d_n{}^2} = \frac{h^2}{a^2}+\frac{k^2}{b^2}+\frac{l^2}{c^2} = \frac{(h')^2}{d^2}$$

or

$$\frac{2\sin\theta}{\lambda} = \frac{h'}{d}$$

and generally

$$\frac{2\,\delta\theta\cos\theta}{\lambda} = \frac{2\delta h'}{d}$$

254

Here d_n is the reflection of the nth order and n is a common factor in h, k, l. Therefore $n = h'$ and as already indicated $k' = l' = 0$. It follows that

$$\frac{2\,\Delta\theta}{\lambda} \cos\,\theta = \frac{2\,\Delta h'}{d}$$

for the specific widths $\Delta h'$ and $\Delta\theta$. Therefore

$$\beta = 2\,\Delta\theta = \frac{2\lambda}{\cos\,\theta} \left(\frac{\Delta h'}{d}\right) = \frac{2\lambda}{\cos\,\theta} \left(\frac{1{\cdot}39}{\pi Nd}\right)$$

which since $\epsilon = Nd$ gives the result

$$\beta = \frac{0{\cdot}885\lambda}{\epsilon \cos\,\theta} \tag{7.5}$$

or generally

$$\beta = \frac{k\lambda}{\epsilon \cos\,\theta} \tag{7.6}$$

where k is the 'Scherrer constant'. It does in fact differ somewhat according to the way the crystallite size and broadening are defined. Thus for cube-shaped crystals Scherrer [27] obtained

$$k = \frac{2\sqrt{(\log_e 2)}}{\pi} = 0{\cdot}94$$

and Bragg [28] derived a value of $0{\cdot}89$.

The alternative definition of β is indicated by β_L in Fig. 7.3(b) and is the Laüe [29] *integral breadth*. This is defined by the total intensity or area under the peak divided by the height, i.e. $\int [(I\,dx)/(I_{max})]$ or its equivalent in radians. This can be obtained by an integration procedure but a simple approximation gives the same result. The line defined by (7.4) has a width in terms of $\Delta h'$ given by $\Delta h'$ from $-(1/N)$ to $+(1/N)$. The height is proportional to N^2 and so if the total area can approximate to an isosceles triangle the area is $\frac{1}{2}(2/N)N^2$ and the integral breadth $1/N$ extending over the $\Delta h'$ range is

$$\pm\frac{1}{2N} = \pm 0{\cdot}5\,\frac{1}{N}$$

which is somewhat larger than the factor for the half-peak breadth (constant $0{\cdot}44$).

Therefore the angular spread β_L (radians) on the pattern is given by:

$$\beta_L = \frac{\lambda}{\epsilon \cos\,\theta} \tag{7.7}$$

and total width $= 2\beta_L$.

Hence in this case the constant k is unity. In view of the possible differences in the definition of ϵ it is usual to take equation (7.7) as general and ϵ to represent a *mean apparent crystal size*. Thus, if t is the cube root of the mean grain volume one may set $\epsilon = t/k$ in the direction normal to the planes giving

255

the diffraction line at θ, where k is the appropriate constant for the conditions. Further details of the methods for deriving the essential relationships are given in the literature noted above [27, 28, 29] and other papers [30, 31]. For a full discussion, and more advanced treatments see also Chapter 14 of Taylor's book (Ref. [2], Chapter 4).

7.5.2.2 For the *practical determination* of β (or β_L) it is necessary to be able to separate this component of breadth from the others which contribute to the total breadth and which have been outlined in section 7.5.1. The separation is more difficult for the smallest crystal sizes since the breadth is then wide enough (a few degrees) for the angular factors to vary appreciably across it. In general the separation can be regarded as depending on the total observed broadening B from which it is required to separate the instrumental broadening b.

The relation
$$\beta = B - b \qquad (7.8a)$$
was proposed by Scherrer but is not good enough.

$$\beta^2 = B^2 - b^2 \qquad (7.8b)$$

has also been used. A combination of these two expressions (actually the geometric mean) is given by

$$\beta = [(B-b)\sqrt{(B^2-b^2)}]^{1/2} \qquad (7.8c)$$

This expression would apply to a single line at low angles (α_1, α_2 almost coincident) or to separated $\alpha_1 \alpha_2$ lines towards high angles, but would not apply where there was any appreciable broadening so that normally separated lines overlap. Other overlapping lines should obviously be avoided.

An experimental method for determining b was devised by Jones [30] and involves the use of an internal standard of large crystallite size so that its line breadths are essentially due to the instrumental factors only. The values of b vary with θ but can be interpolated for the lines from the material under investigation. For the purpose of determining the line shapes a micro-densitometer or diffractometer would be used. (Overlapping lines must again be avoided.) The separation of B from b according to the above formulae can then be made.

Without considering the full details here, it is possible to give some appreciation of the reasons for the choice of equation (7.8c) and to indicate more clearly the significance of B and b. The diffraction broadening alone would be of the form shown in Fig. 7.3. The unbroadened standard lines would have a not dissimilar (but not identical) shape due to the instrumental conditions. Let the β broadening correspond to a function $f(y)$ and the b broadening to $g(x)$ where x or y are lengths along the film or record. The broadening due to b is then assumed to operate on each element of $f(y)$, spreading it out into a shape of the same form as for the standard lines alone. By finding the

256

resultant area and height this gives the broadened line whose width is B. The expressions obtained are more easily expressed as β/B or b/B. Suitable expressions for $f(y)$ are derived from equation (7.4) or similar expressions, whilst the $g(x)$ function could, for example, be Gaussian. Alternatively, both could be of this form without appreciable error. In the latter case

$$\left(\frac{\beta}{B}\right)^2 + \left(\frac{b}{B}\right)^2 = 1$$

which is thus an alternative form of (7.8b), and is the equation of a circle. Other forms lead to the line $\beta/B + b/B = 1$ which is a straight line as (7.8a). Experimental results are found to follow closely the curve corresponding to the geometrical mean of the two values of β in these equations and this is given by (7.8c) which in terms of the ratios is

$$\left(\frac{\beta}{B}\right)^2 = \left(1 - \frac{b}{B}\right)\left\{1 - \left(\frac{b}{B}\right)^2\right\}^{1/2}$$

so that the curve can easily be interpolated between the other two. These curves are shown in Fig. 7.4.

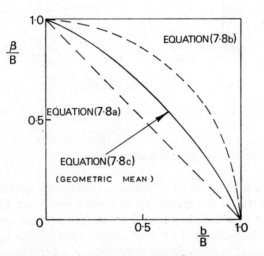

Fig. 7.4   Curve(s) for determining β from b and B.

In order to obtain accurate results when $\alpha_1 \alpha_2$ doublet overlap is likely to affect the measurement of B (or b) it is necessary to correct for this separation. By using equation (7.3), or otherwise, for any radiation of wavelengths $\lambda\alpha_1$, $\lambda\alpha_2$ one can find $\Delta\theta$ corresponding to $\lambda\alpha_2 - \lambda\alpha_1$, and express this in the same units as b, β, or B by the quantity d (*not* interplanar spacing here).

The overlapping lines have a resultant observed breadth B_0 which derives from the spread of intensities resulting from the addition of the superimposed α_1 and α_2 contributions. The ratio

$$\frac{B}{B_0} = \frac{\text{true breadth}}{\text{observed breadth}}$$

is plotted as a function of

$$\frac{d}{B_0} = \frac{\alpha_1\,\alpha_2 \text{ separation}}{\text{observed breadth}}$$

as in Fig. 7.5 which has a typical form. The curve enables B to be found from B_0 and d and applies equally well to the standard or unbroadened lines represented by b and b_0.

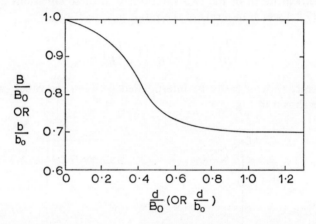

Fig. 7.5   Correction for $\alpha_1\alpha_2$ overlap (particle size determination).

The information in Figs. 7.4 and 7.5 is thus sufficient to enable the experimental determination of crystallite size.

7.5.2.3   Because of the effect of the $\alpha_1\,\alpha_2$ overlap the question arises as to whether the extent of the separation of the two peaks can be used directly, since the point at which they appear to be just resolved corresponds to a definite size. Alternatively the overall shape of the double line as indicated by a diffractometer or microdensitometer chart is a direct indication—at least over part of the particle size range for which the broadening occurs. These shapes can be calculated for a given wavelength and value of θ and estimates of intermediate cases made by interpolation.

7.5.2.4   Line broadening due to stacking faults or 'mistakes' as in hexagonal cobalt or certain face-centred cubic metals and alloys can be interpreted in similar terms but may require a detailed consideration of the structure factor to determine which lines are broadened. The size of domains in ordered

258

structures require similar treatment. For a detailed account reference should be made to Wilson [32].

Stacking faults are a cause of streak formation in electron diffraction patterns from metals such as iron manganese alloys or stainless steels and the interpretation is again essentially similar. The streaks are in a direction normal to the planes containing the faults and obey the relation $\Delta h \simeq 1/N$.

7.5.3   Line Broadening due to Lattice Strains

In connection with the determination of accuracy of measurement of interplanar spacings and lattice parameters it was seen that a small relative change in spacing is related to $\Delta\theta$ by the equation obtained from the Bragg law:

$$\frac{\Delta d}{d} = -\Delta\theta \cot\theta \quad \text{(equation (6.41))}$$

If a metal crystal is strained elastically one would expect the lattice parameter to change and so the diffraction line to be displaced according to the relation. In a cold worked metal—whether in the form of filings or a solid lump—however the grains or sub-grains are not equally strained. Indeed there must be a system of mutually compensating or balancing microstresses. In the lump specimen there may also be long range residual stresses which shift the diffraction line as a whole according to equation (6.41), but the microstresses will lead to a range of lattice parameters or interplanar spacings on either side of the mean. A line which has a very small breadth in the completely unstrained condition becomes broadened. Hence, for any point on the broadened line, there is an elastic strain given by

$$e = \frac{\Delta d}{d} = -\Delta\theta \cot\theta$$

according to (6.41), where d is the mean spacing. There is then a 'mean strain' $\bar{e}$ represented by some measure of Δd or the broadening itself. Let this mean strain broadening be β (radians) so that the mean strain itself depends on $\pm\frac{1}{2}\beta$. For a camera radius R the distance along the film is given by $x = 2R\theta$ and $dx = 2R\,d\theta = R\beta$, so

$$\Delta\theta = \frac{\beta}{2} \quad \text{and} \quad \bar{e} = \tfrac{1}{4}\beta \cot\theta \tag{7.9a}$$

The actual stress corresponding to this strain is given by

$$\bar{\sigma} = E\bar{e} \tag{7.9b}$$

where E is Young's modulus. This is not however the mean Young's modulus for the polycrystalline material (effectively isotropic) but varies with crystallographic direction.

The value of θ (or d) corresponds to specific planar indices hkl. Then for this set of planes in a cubic crystal

$$\frac{1}{E_{hkl}} = s_{11} - 2[(s_{11} - s_{12}) - \tfrac{1}{2}s_{44}]\frac{h^2k^2 + k^2l^2 + l^2h^2}{h^2 + k^2 + l^2} \tag{7.9c}$$

In this equation the quantities s_{ij} are fundamental coefficients of elasticity (or 'compliances'—in effect reciprocal moduli). The relationship will be referred to again in connection with stress measurement which is considered in Chapter 5 of Vol. 2.

Strain broadening was studied by Stokes and Megaw [33] and several others (e.g. Ref. [34]. See also A. Taylor, Ref. [2] of Chapter 4.

7.5.4 Combined Effect of Small Crystallite Size and Strain Broadening

The question naturally arises as to whether in a specific case the broadening is due to either of these causes or to both causes acting at once. It has for example been found that, for cold worked metals the line broadening, due to lattice strains, has largely accounted for the whole and the particle size contribution has been relatively unimportant. It is not necessary to discuss here the results of investigation of this subject which are important to any attempt to interpret the structure of cold worked metals. It is however necessary to appreciate the essential principles by which one could discriminate between the two causes. Thus for particle size from (7.6) with $k \simeq 1$,

$$\beta = \frac{\lambda}{\epsilon \cos \theta}$$

whilst for strain incorporating (7.9) and the directionally dependent E_{hkl},

$$\beta = \frac{4\bar{\sigma}}{E_{hkl}} \tan \theta$$

These expressions imply that plots of β against $\sec \theta$ and $\tan \theta$ should give a direct indication of the cause of the broadening. It is obviously safe to assume in the first instance *both* causes can be present if the two effects are additive:

$$\beta = \frac{\lambda}{\epsilon \cos \theta} + \frac{4\bar{\sigma} \tan \theta}{E_{hkl}}$$

or on rearranging $\qquad \dfrac{\beta \cos \theta}{\lambda} = \dfrac{1}{\epsilon} + \left(\dfrac{4\bar{\sigma}}{\lambda E_{hkl}}\right) \sin \theta \tag{7.10}$

Hence if one ignores any variation of E_{hkl} a plot of $(\beta \cos \theta)/\lambda$ against $\sin \theta$ will give a straight line. The intercept at $\theta = 0$ then gives $1/\epsilon$ and the slope a quantity $(4\bar{\sigma})/(\lambda E_{hkl})$ from which $\bar{\sigma}$ can be calculated. If points are included for which the values of E_{hkl} are sufficiently different this will show up with more scatter and it might be desirable to incorporate equation (7.9c)—in effect plotting $(\beta \cos \theta)/\lambda$ against $\sin \theta(\sum h^2k^2/\sum h^2)$ for each line. The assumption of simple additivity is only an approximation. A more satisfactory approach is the method developed by Averbach and Warren [35] and using Fourier analysis. In effect, the shape of the line is carefully measured and the mathematical procedure then separates or distinguishes between the two causes of broadening.

7.6 DETERMINATION OF ORIENTATION AND STUDY OF TEXTURE

7.6.1 Orientation of Single Crystals or Grains in an Aggregate

7.6.1.1 X-ray Methods: It may be required to determine the orientation of a single crystal or of a crystal in an aggregate where the grains are coarse enough to irradiate separately. Grains can then be chosen and separate diffraction patterns obtained from each. Such, for example, could be a grain in a piece of silicon transformer steel sheet, or large columnar grains in an ingot.

The X-ray method which can be most conveniently used for this purpose is the back-reflection Laüe technique using the Greninger chart. For thin sheets of metal, foils or small crystals the transmission method can equally well be used. These methods have been described in section 6.3, and the necessary information for interpreting photographs is given there and in general references. It was noted that it may be useful to establish a collection of patterns for various orientations. Also, angles determined from a chart or by direct measurements on a film followed by calculation can be inserted in a stereographic projection—as indicated in Fig. 6.6(c). From such a projection of the actual crystal giving the pattern and by reference to a standard projection, the location of crystal axes or other significant directions or planes can be indicated. Some of these may correspond with the direction of the X-ray beam or the plane or directions normal to it. When applied to large grains in an aggregate this constitutes one method of determining orientation texture, for which other methods are considered in section 7.6.2.

There are obvious applications of X-rays to the determination of the mutual orientation of phases such as a precipitate and matrix. The problem might be difficult to solve because of the presence of two diffraction patterns superimposed. It could be assisted by use of a fine collimator, or even a microbeam. It would probably require more than two determinations in different sections—more useful therefore for coarse grained metals.

7.6.1.2 Electron Diffraction Methods: The information given in section 6.5.2.2 should be sufficient to determine the orientation of a crystal in say a metal foil. This may be part of what is virtually a much larger single crystal or the foil could have been reduced down from a piece of sheet or strip or metal aggregate in some form. It may be of interest to know the orientation of the crystal for this reason, or, if two phases are present, to study the mutual orientation relationship. In general the determination of the orientation is likely to be incidental to the use of the pattern even if it is not primarily intended for this purpose (as when identification is the primary aim).

(*Note*: Methods for determining orientation by optical examination of etch pits are indicated in Chapter 1, Vol. 2, and the use of the torque magnetometer in Chapter 3, Vol. 2.)

7.6.2 Orientation Textures (X-ray Methods)

7.6.2.1 If a completely randomly oriented polycrystalline aggregate is subjected to a process of continuous plastic deformation (below the recrystallisation temperature) as in typical metallurgical processes such as wire drawing or rolling, the individual grains are not only distorted and broken down into smaller sub grains, they also tend to turn progressively into certain specific orientations. These are determined by the crystallography of the deformation processes in the grains—the available slip planes and directions and the angles they make with the principal directions of stress. The further the deformation process goes the more pronounced will this orientation texture become. The following facts are of interest here:

(a) The orientation is seldom complete or perfect and so the patterns are intermediate between those for single crystals and random powder patterns.
(b) Cast metals usually have some orientation and in fact any cast or wrought metal is unlikely to have completely randomly oriented grains.
(c) When the material is annealed, and recrystallisation occurs, the orientation texture may remain unaltered or a new texture may develop.
(d) Hot deformation will also produce textures which may be connected with these annealing textures.
(e) Different textures may appear in the same material due to different modes of deformation (e.g. rolling, drawing, pressing, etc.).
(f) Two (or more) well-defined orientation textures may appear together even after a simple deformation process.

The effect of texture on properties can be very important as for example in causing anisotropy of yield behaviour leading to 'ear' formation in deep drawing or pressing of sheet metals. The effect on magnetic properties is important in transformer sheet steels.

This very brief summary indicates the need for adequate techniques for determining the orientations. These are essentially the experimental methods described in Chapter 6 modified where necessary for the particular requirements. It is also useful to have a diagrammatic method of representation and for this purpose the pole figure is commonly used. This is an application of the stereographic projection. It can best be understood by reference first of all to a fibre orientation, the determination of which is now described.

7.6.2.2 *Fibre diagrams.* Textures found in wires or rods which are axially symmetrical would be expected to have a significant crystallographic direction parallel to the wire axis. The texture is thus similar to that for a fibre—including organic or synthetic fibres.

If an ordinary cylindrical powder camera with a strip film is used to obtain a diffractional pattern of such a wire the presence of a preferred orientation is indicated by the fact that the relative intensities of the lines do not match those of a polycrystalline randomly oriented specimen (e.g. of filings of the same metal)—the theoretically calculated intensities can be used as a reference. The strip film does however give limited information because the intensity may in fact be concentrated in directions on the diffraction cone which do not cut the film at all (line or ring apparently absent). For this reason it is an advantage to have either a larger flat film or to set the wire up like a rotating crystal and use the camera arrangement with a deep cylindrical film. Thus the sharper the orientation the more nearly will the pattern approach that obtained by rotating a single crystal of the same material. The principles of interpretation are similar and are based on Fig. 6.2 but the calculation can be considerably simplified according to Fig. 7.6(a).

In Fig. 7.6(a) it is shown how the scatter of the orientation about an ideal position extends the diffraction spot about an arc of the Debye–Scherrer ring. Alternatively the progressive development of an orientation results in intensity which would otherwise be evenly spread round the (2θ) diffraction cone being concentrated in this arc. One also notes that as the wire rotates diffraction occurs from the same plane in four positions. These correspond to two positions of the angle δ and 'opposite sides' of the plane, viz. $(hk\bar{l})$ and $(\bar{h}\bar{k}\bar{l})$. These facts are indicated for the flat film geometry in Fig. 7.6(a).

From the value of θ which is known *a priori* (known material) or easily ascertained from the ring diameter, and from the value of δ, the angle ρ can be found by using equation (6.10), viz.

$$\cos \rho = \cos \delta \cos \theta$$

This angle is sufficient to determine the angle between a given plane (hkl) and the wire axis. For small values of θ, $\rho \simeq \delta$.

If a cylindrical film is used, it is easiest to read ρ directly from one of the charts referred to. Alternatively if only a (ζ, ξ) chart is available $\tan \rho = \xi/\zeta$ (equation (6.18)) whilst these co-ordinates can be calculated from the (x, y) co-ordinates of a point on the cylindrical film (equations (6.6) and (6.8)).

In some experimental work it may be convenient or desirable to have the wire or fibre axis tilted at an angle to the X-ray beam. This is known as the method of 'diotropic planes' and is due to Polanyi. The spot pattern is not

263

symmetrical and two values of δ are obtained, viz. δ and δ' as in Fig. 7.6(b). In this case equation (6.20) is replaced by

$$\left.\begin{array}{l} \cos \rho = \cos \delta \sin \beta \cos \theta + \cos \beta \sin \theta \\ \cos \rho = \cos \delta' \sin (180-\beta) \cos \theta + \cos (180-\beta) \sin \theta \end{array}\right\} \quad (7.11)$$

and/or

(These equations can be derived by similar methods to the derivation of (6.20) or by direct application of formulae from spherical trigonometry.)

The knowledge gained from a single photograph gives ρ according to (6.20) or the angle between one set of planes (hkl) and the wire axis. Clearly this information alone is not sufficient. One could locate the pole of the plane in

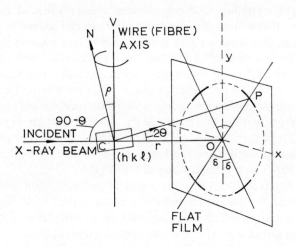

(a)

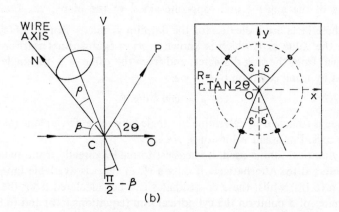

(b)

Fig. 7.6 Determination of preferred orientation in wire.
(a) flat film geometry: fibre axis normal to X-ray beam
(b) wire (fibre) axis tilted at angle β with X-ray beam

a standard stereographic projection as in Fig. 4.12 for the cubic case. The axial direction is then on a (small) circle ρ degrees from this pole. Consequently a second value of ρ is required for some other plane. This will already be available on the same film. Thus in a body-centred cubic metal the two innermost rings have the indices (110) and (200), and in the face-centred cubic case (111) and (200). In either case these represent two corners of the elementary spherical triangle in Fig. 4.12. (The entire projection can be developed solely from the triangle—corners (100) (110) and (111)—see section 4.5.3.3.)

In the two cubic cases referred to the solution may be easy because the fibre axis often happens to coincide with one of the directions normal to the planes stated (in b.c.c. $\langle 110 \rangle$ and in f.c.c. $\langle 111 \rangle$ or $\langle 100 \rangle$). Otherwise, given two values of ρ, ρ_1 and ρ_2, which the wire axis makes with the poles of two planes $(h_1k_1l_1)$, $(h_2k_2l_2)$, we require to find the directions corresponding to the intersections of two small circles of radius ρ_1, ρ_2 drawn round the respective poles of the two planes. This involves a principle not referred to previously— the small circle formed on the surface of the sphere by the intersection with the cone generated by lines making angle ρ with the pole of a plane becomes a circle in the plane of projection. Its centre does not coincide exactly with the pole of hkl, but can be found directly from the Wulff net (see section 4.5.3.3) by placing a diameter of the net through the pole of (hkl). A distance ρ is then marked off on the stereographic scale in opposite directions along this diameter. The two points are opposite ends of the true small circle diameter required.

If the intersections correspond to two different permutations of one set of zonal indices $\langle uvw \rangle$ the problem is solved. Otherwise it may be necessary to obtain a third measurement of ρ. Again, for either of the two cubic cases with the first two reflections stated the cubic symmetry makes two values of ρ sufficient since they give complementary answers. Reference to Fig. 4.12 confirms this, e.g. 122 and 212 are both equidistant from 111 and likewise from 001; 112, 121 are equidistant from 111 and from 011. These angles can be checked on a Wulff net or fitted into equation (4.21).

For all cubic cases tables of angles based on (4.21) can be applied since (see section 4.4.3.3) the directions always coincide with the normals to planes of the same indices. Otherwise the equation for angles between planes and zones for the appropriate crystal system must be used. Tables of angles or calculation can be used to confirm or calculate the exact values of ρ for possible indices $[uvw]$ of the wire axis.

It has been pointed out that if the method of equation (7.11) is used and photographs are taken with two or more different values of ρ one may happen to get a diffraction spot in the vertical plane. For this case $\beta = 90 - \theta$ and so the particular diffracting plane is normal to the wire axis, and the zone axis is found directly. Diffraction intensity may be found at these positions in any case from textures which are not strongly developed—hence note must be taken of the scatter about the ideal position.

7.6.2.3 *Pole Figures*. Pole figures are of most value for the representation of textures in sheet or strip material. They are not strictly necessary for the representation of wire or fibre textures since only one piece of information sums it up, viz. the indices of the direction (or directions) which align themselves parallel to the wire axis. Sometimes, however, a sheet metal deformed in certain ways will have a texture that is axially symmetrical and resembles a wire texture. This does for example occur in certain parts of brass cups stamped out of sheet. The pole figure concept is most easily introduced by reference to such cases first.

In the derivation of the stereographic projection given previously (sections 4.5.3.2/3) it was seen that the normal to a plane on a zone axis making an angle ρ with a vertical reference axis (OZ) (see Fig. 4.10(b)) could be projected as a point on a reference circle. Hence, if we consider all such directions which make a constant angle ρ with OZ, these would project on to a circle—giving a polar diagram as in Fig. 7.7(a). (The circle has radius $r \tan \frac{1}{2}\rho$ as explained in section 4.5.3.2.) Considering the plane normals in Fig. 7.6, for a given orientation, the location of this circle of constant ρ (replacing ρ_m in Fig. 4.10(b)) with OZ parallel to CV in Fig. 7.6, could thus be used to indicate the orientation. Alternatively one can turn OZ into the plane of projection so that the curve of constant ρ becomes an arc like the curves designated by constant ϕ in Fig. 4.11. Otherwise OX in the plane of projection is the wire axis and these curves are now regarded as constant ρ curves for the angle between plane normal and OX (rather than zone axis and OZ). This gives an alternative axial pole figure in Fig. 7.7(b). Two additional points now arise:

1. The orientation may not be ideal and so, corresponding to the spread of intensity in Fig. 7.6(a), the ideal pole figures are spread into bands. It is common practice to divide these into two, three or four regions shaded according to the intensity of reflection and so the density of poles CN and their spread about a mean value of ρ.

2. The angle ρ only applies to one particular plane. Hence the pole figures are simplest for planes of lowest indices where there are only a few possible values of the angle ρ. This is particularly so with the simple cubic cases referred to. The reader may examine Fig. 4.12 and confirm that if the wire axis is along the direction [100] (X direction), the pole figures for the first few reflections would have the following characteristics:

(a) (100) (absent in f.c.c. or b.c.c.), (200), etc., ρ can only be ideally 0 or 90° and so the ideal is an equatorial circle in Fig. 7.7(a), with a point in the centre; or, a horizontal diameter in Fig. 7.7(b) with a point at each end of the vertical diameter.

(b) (110) (present in b.c.c., not in f.c.c.), (220), etc. Equatorial intensity arises from (011) and (0$\bar{1}$1). Otherwise $\rho = 45°$ (in Fig. 7.7(a) or (b)) for $\pm$($\bar{1}$10) or (110)

266

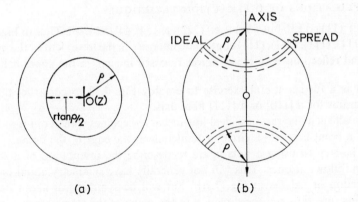

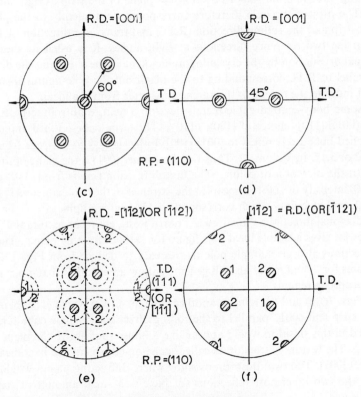

Fig. 7.7 Introduction to pole figures.

(a) axis at centre   (b) axis in plane of pole figure

(a) and (b) texture of a wire or fibre or a sheet metal with symmetry.
About one direction only

(c) pole figure for {110} planes   (d) {001} planes ((200) diffraction)

(c) and (d) a 'cube-on-edge' (110) [001] texture (b.c.c.)

(e) {111} planes (possible spread indicated)   (f) {001} planes ((200)
    diffraction)

(e) and (f) (110) [112] 'twin orientation' (f.c.c.)

(c) (111) (f.c.c. but not b.c.c.) (222). $\rho = 54° 44'$ (54′ 74°). Thus in Fig. 7.7(b) (111), (1$\bar{1}$1), (11$\bar{1}$), (1$\bar{1}\bar{1}$) contribute intensity in the lower half of the diagram and reflections with these indices reversed in sign, in the upper half.

It is a further useful exercise to employ Fig. 4.12 to construct the ideal diagrams for a [110] or a [111] fibre axis.

In all such diagrams an ideal line or curve becomes spread out into a band and a point becomes a circle or semicircle at the edge of the figure.

The step to using a pole figure to describe the orientation of a sheet or strip follows at once. This will not generally have symmetry about only one direction or axis as in Fig. 7.7(b) but will be symmetrical about both the rolling direction and the direction in the plane of the strip at right angles to it. The projection circle therefore corresponds conveniently to the plane of the strip and the rolling direction (R.D.) and transverse direction (T.D.) are then the two reference directions at right angles. It is usual to specify the actual orientation by the crystallographic plane in the sheet and the direction parallel to R.D. corresponding to the ideal, but the pole figures themselves will indicate the extent of the spread or scatter from the ideal.

Some body-centred cubic metals develop a texture on rolling represented by (001), [110] (choice of {100}, ⟨110⟩). The term 'cube-on-face' might have applied but strictly refers to (001) [100]. Either is easy to envisage from Figs. 4.12 or 3.17, in Volume 2. Thus for the former (001) is the plane of the strip. [110] the direction of rolling. The first diffraction ring is from {110} planes. (110) intensity therefore appears at the extremes of the two diameters R.D. and T.D. and at points which correspond to 45° (stereographic) from the centre in diagonal directions ('north east', 'north west', etc.) The texture (001) [110] likewise gives a simple ideal pole figure for {110} diffracting planes. The (200) reflections also give simple pole figures easy to deduce from Fig. 4.12. Pole figures for actual materials will generally show areas rather than points in the appropriate regions.

Figs. 7.7(c) and (d) refer to another cube texture in which {110} plane is in the strip and ⟨001⟩ parallel to the rolling direction. A single cube (unit cell) tilted in this position would appear like a house roof above the plane of the strip. The texture is sometimes called 'cube-on-edge' and can be indicated by (110) [001]. The two illustrations show where diffracting planes would occur for the two innermost reflections (of b.c.c.)—a small amount of scatter is indicated.

In the above examples both R.D. and T.D. contain crystallographic mirror planes (m) of symmetry normal to the surface of the strip and so the ideal pole figures have symmetry about these two directions. As stated in terms of planes and direction some preferred orientations do not appear to have this symmetry. The pole figure does nevertheless show it. An example is a rolling texture found in several f.c.c. metals denoted (specifically) by (110) [1$\bar{1}$2]. In this case a plane ($\bar{1}$11) is contained in R.D. normal to (110). It is not a mirror

268

plane in the (untwinned) f.c.c. lattice. The pole figure thus exhibits mirror symmetry about this plane because the strip would otherwise have developed an unsymmetrical plastic flow pattern which it could not do and remain straight. One way of regarding this symmetry is to say that from any given starting point (random orientation or previous texture) any part of the material has an equal chance of approaching one of two stable conditions. In either case let (110) be the plane of projection. If [$1\bar{1}2$] is the rolling direction, [001] (cube edge) lies 35·26° on one side, whereas if [$\bar{1}12$] is the rolling direction it is on the other side. This situation should again be confirmed by reference to Fig. 4.12. These two positions are exactly mirror images of each other through the ($\bar{1}11$) (or ($1\bar{1}1$)) plane. In fact the twin plane in a f.c.c. structure is of the {111} type. This is therefore called a twin orientation, although it is not necessarily assumed that exactly half the material is related to the other half by twinning as such. The same symmetry could arise with other directions along R.D. which do not place a twin plane normal to T.D. The pole figure should still show a double texture of this kind, and such pole figures are thus harder to interpret. Figs. 7.7(e) and (f) illustrate the above case by showing the positions of the poles (with a small spread) for (111) and (200) (i.e. second order from {100}) diffraction. The two sets of twin-related spots are indicated by 1 and 2 respectively. If the spread is large some of these spots may merge into each other and the pole figure then contains extensive shaded areas which have the symmetry about the two axes R.D. and T.D.—'except in special circumstances' (see section 7.6.2.5).

The problem of interpretation may be complicated by the rather large spread about the ideal position which is usually found under all but extreme conditions (illustrated by dotted lines in (e)). Again some pole figures suggest the presence of two or even more basic orientations. In the case of Figs. 7.7(e) and (f), for example, a typical pole figure for rolled copper showed regions of pole density which also contained the ideal (twin) orientation denoted by (112) [$11\bar{1}$] (see Richards [36]).

7.6.2.4 *Determination of Pole Figures and Orientation in Sheet Metal.* In the case of fine grained materials it is usual to take Debye–Scherrer photographs or diffractometer records with the sheet at varying angles to the incident X-ray beam. Coarse grained materials are probably best investigated by using the Laüe techniques to determine the orientations of several individual grains in order to obtain a statistical estimate of the texture—especially necessary if more than one orientation is present. For specimens of intermediate size oscillation in the plane of the specimen is sometimes used. The correction between the diffraction patterns and the pole figure is now considered. It is customary to represent intensity of diffraction equivalent to pole density by two, three or even four grades of shading. In the case of a diffractometer investigation contours of equal counting frequency can be plotted.

Providing the sheet is already thin enough or can be chemically thinned the transmission method can be used, and pole figures determined by this method can be constructed as follows. Firstly a photograph is taken with the incident beam normal to the sheet. The result is a diffraction ring which is symmetrical. One can infer this from Fig. 7.6 (or Fig. 6.2) in which the axis CV is replaced by a plane (equivalent to rotating CV about CO). As before the diffracted rays always make an angle 2θ and the normals and $90° - \theta$ with the incident beam direction. Consequently all these poles on the stereographic projection are $90° - \theta$ from the centre or θ from the circumference as in Fig. 7.8(a). This circle may thus be drawn in on the pole figure and the regions of high and low intensity distinguished by suitable marking. The positions round the circle naturally correspond to the angle δ which locates the plane containing incident and diffracted ray as well as the normal CN.

Further photographs may now be taken by turning the sheets through successive values of the angle α indicated in Fig. 7.8(b). The beam geometry remains the same but the cone of normals at $90° - \theta$ to XC is no longer symmetrically placed with regard to the strip surface. The situation may be explained by reference to a hemisphere placed over the circle of projection. In Fig. 7.8(a) the cone cuts this sphere in the (small) circle of radius equivalent to $90° - \theta$. On rotating through an angle α about the vertical direction which is conveniently taken as R.D. the circle moves round over the surface of the hemisphere and so across the plane of projection. If α is less than θ (α_1 in the figure) it is still entirely within the projection circle. If α is greater than θ (α_2 in the figure) the cone now intersects the sphere behind the strip as well as in front. These intersections constitute the two parts of a circle on the surface of the sphere which project stereographically to the curves shown (arc of circles). Diffracted intensity which appeared in the position shown on the diffraction ring and transferred firstly to the $90° - \theta$ circle, would thus come from poles along this circle. Since rotation is about R.D. the angle ρ is constant for angles between R.D. and any given plane normal. All that is necessary is to mark off the corresponding length along the arc by moving along lines of latitude through the angle α. A series of photographs for different values of α thus give arcs which represent the spread of the normals giving appreciable intensity. The areas enveloping these are then filled in to complete the pole figure. Thus the diffracted intensity for any value of α falls initially onto the $(90 - \theta)$ circle. The intensity along this circle immediately derives from the range of values of δ from which (see Fig. 7.8(b)) the range of values of ρ can be indicated and located on the circumference of a Wulff net, and hence the appropriate constant ρ curves to be followed in order to transfer the intensity through the angle α. Use may also be made of equation (6.20), with θ = constant and so also cos δ/cos ρ=constant.

Charts giving the positions of the arcs for different values of α were devised by Wever [37] and several have been published since. These are available in the literature (standard refs.). The charts show that there are regions near the 'north' and 'south' poles which are not covered. These can be investigated by one or two rotations with T.D. as the axis instead of R.D. The charts thus provide arcs of constant α and latitude lines which are in fact directly indicated for constant δ rather than ρ. These are sometimes denoted by $\phi = 90° - \delta$. When $\delta = 0$ or $\phi = 90$, $\cos \rho = \cos \theta$ ($\rho = \theta$) and so this defines the blank area. Such charts refer to a fixed value of θ, i.e. to diffraction from a particular plane of a given material.

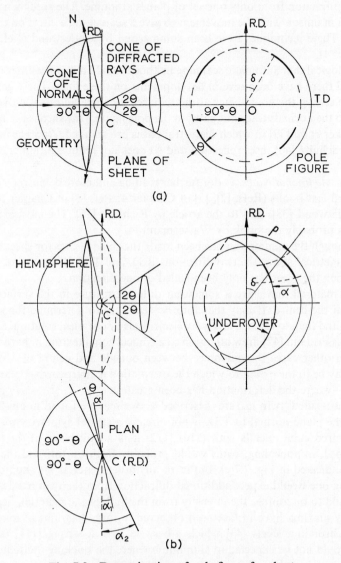

(a)

(b)

Fig. 7.8 Determination of pole figure for sheet.
  (a) X-ray beam perpendicular to sheet
  (b) X-ray beam at angle to sheet

271

In this method for determining pole figures the absorption is equal for all points round the ring only when $\alpha = 0$. Hence, in materials with strong absorption it will be necessary to make a correction. Photographic methods do not however, lead to very accurate estimates of intensity and pole figures are qualitative or semi-quantitative. Other variants using films, include procedures where the film is moved past a slot in a metal screen which is set to allow diffraction from only one set of planes at a time. The specimen is then rotated in unison with the movement to give a scan along a circle on the pole figure. These techniques have been summarised by Underwood (Ref. [2] of Chapter 4).

It is logical that automatic scanning methods should rely on counter recording and there have been several developments on these lines. These generally involve setting the goniometer at the required 2θ angle and then moving the sheet so the normals trace out paths in the pole figure. One such is the method of Decker *et al.* [39] in which diametral paths are scanned. Another method, due to Schultz [40], uses reflection and so does not require a thin specimen.

7.6.2.5 *Additional Notes.* 1. For further reading and more detail refer to the standard text books (Refs. [2], [3] of Chapter 4; Ref. [3] of Chapter 5, etc.), to Underwood [38], and to the article by Richards [36]. The most extensive work is probably the book by Wassermann [41].

2. Although the assumption has been made that pole figures for sheet should be symmetrical about the two directions R.D. and T.D. this may not be the case when the sheet has only been rolled in one direction.

3. The material may have a somewhat different texture in the surface than it has in the centre. It may therefore be necessary to determine the surface texture and then to remove layers by electrolysis or chemical solution. Schmid and Wassermann [42] drew attention to a conical zonal texture in hard drawn wire. In other cases the difference between outside and centre of a wire or strip may be in the extent to which the orientation is sharpened—there is less scatter—where the deformation has been greatest.

4. The so-called 'twin texture' described above may be found in any metals where the plane normal to T.D. is not already a crystal symmetry plane. In face-centred cubic metals with (110) [1$\bar{1}$2] it is a twin plane. If the texture is retained on annealing, twins would probably form on this and the other planes indicated in Fig. 7.7(e). In this or any other material with appreciable twinning one would expect additional diffraction from the twin bands which might add to or confuse the intensity from the bulk of the material. Separate intensity maxima have in fact been observed in fibre diagrams of austenitic nickel–chromium steels [43] which otherwise have a strong [111] texture. They would not be expected in aluminium where the stacking fault energy is high. In metals such as silver, copper and gold, however, there is a considerable admixture of a cube-edge [100] texture and this spreads out the minor components of intensity arising from any twin bands in crystals with either alignment.

5. Important technological applications have been indicated above in section 7.6.2.1. In addition orientation textures are found in electro-deposits and other metal coatings. The techniques in one form or another may be used to determine the orientation of oxide or other layers.

7.7 DIFFRACTION AT HIGH OR LOW TEMPERATURES

7.7.1 An Extension of the Metallurgical Usefulness of Diffraction

Diffraction methods to high temperatures have many metallurgical applications and have much enlarged the useful scope of the techniques especially in fundamental and background studies. Applications to industrial research problems are less common, but a valuable contribution will sometimes be made to the solution of a technical problem—providing the applicability of the diffraction technique is properly understood. As will be seen high temperature involves a number of problems which vary in their importance according to the material being studied and variables such as temperature and pressure. This may account for the comparative rarity of high temperature techniques outside basic research laboratories. In the following section a general account is given of the essential principles of high temperature cameras. It is not intended to describe any in detail. Some metallurgical applications are then indicated. Low temperature techniques are then briefly considered but have few metallurgical applications at present.

Apart from the summaries in some of the standard X-ray texts reference should be made to papers and articles by Goldschmidt (Refs. [44], [45] and [46] (bibliography)) for the high temperature cameras. For low temperature diffraction see Steward [47], and Post's bibliography [48].

7.7.2 High-Temperature Cameras

7.7.2.1 Elementary Principles: The earliest high-temperature cameras were natural developments from the cylindrical powder cameras, particularly the types described in section 6.4.3. The tendency has therefore been to retain the rod, wire or powder specimen in the centre of the cylinder. This is heated by
  (i) small furnaces
  (ii) induction
  (iii) internal resistance (applicable to wire or conducting specimen).
It is necessary to make provision for a number of circumstances:

(a) The specimen must be kept hot but the film cassette, any vacuum seals, bearings, etc., must be cold. In cameras where the film is placed round cylindrical flanges these now take the form of hollow annuli through which cooling water circulates. There may, in addition be inner wa￼ ￼r cooled rings,

273

which surround the furnaces and protect the bearings and thermocouple connection (see (d) below).

(b) There must be a gap wide enough to let the diffracted rays out (and of course a collimator system for the incident X-rays). The gap would mean an open space from the inner furnace to the film and the outer atmosphere. A thin strip of some low absorbing but vacuum- and gas-tight material is often used to ensure that the furnace chamber is completely enclosed. It must have a sufficiently low absorption for the X-rays so that it does not excessively reduce the intensity of the diffracted beams. Aluminium foil is commonly used.

(c) The specimen itself may tend to oxidise in air, and for this reason or otherwise needs to be protected from change during the exposure. Sometimes a specimen is sealed off in a glass or silica capillary in which case the container must be sufficiently transparent to X-rays. If it is exposed to the furnace conditions directly either a vacuum or inert gas atmosphere is required.

(d) It is necessary to be able to control and measure the temperature. A thermocouple is generally employed. It may be advantageous to have separate couples for measurement and control, in which case one would expect the leads to be taken out at opposite ends of the axis of the camera so that any heat conducted away is removed symmetrically.

Variants of the above arrangements have been devised for special applications. Particular interest attaches to the use of a heating furnace system attached to a diffractometer for following metallurgical or other transformations. Examples are those of Van Valkenberg and McMurdie [49], and Heal *et al.* [50] (see also bibliography of Ref. [46]). The latter was applied to the isothermal transformations in steels—both need a focusing system.

7.7.2.2 Some details and limitations: It happens that some of the requirements of a high temperature camera are difficult to fulfil and may be incompatible with other requirements. Some conditions may be suitable for one application but not for another. The various requirements have been considered by Goldschmidt (Refs. [8], [9] of Chapter 4). The following notes summarise the main points:

1. Temperature. (a) This should be as high as possible or needed, but many problems increase in series with increasing temperature.

(b) It has been found that there may be differences in temperature between the specimen and a controlling or measuring thermocouple. The only certainty is to have the measuring couple in direct contact with the specimen.

(c) It is necessary to examine for temperature gradients and to examine whether the distribution is uniform.

(d) The thermocouple must be stable.

(e) The furnace or thermocouple must not evaporate, change or react during the exposure. Possible evaporation of the metal winding, or the couple or the specimen and therefore of interaction or change must be known or taken into account.

2. The specimen. (a) The last point is part of the general requirement that the specimen should remain stable with respect to its immediate environment. For example the need to prevent oxidation is clear and requires vacuum or inert gas (unless the oxidation is being studied).

(b) The specimen might react with a capillary tube if it is contained in one (e.g. Mg and its alloys can react with silica).

(c) The specimen could have a sufficiently high vapour pressure to evaporate (for example the vapour pressure of iron becomes appreciable above 1200 °C).

(d) If an oxide it may dissociate at higher temperatures (e.g. sub-oxides of silicon and aluminium can form).

(e) Arrangements for rotation or oscillation must usually be provided.

3. Atmosphere. (a) Apart from the question of oxidation the gas must not react in any other way with the specimen.

(b) An inert gas must have a low X-ray absorption (low atomic number).

(c) This favours a vacuum but the vacua obtained with ordinary rotary and diffusion pumps still contain oxygen at sufficient vapour pressure to oxidise (slowly) some metals and alloys. An inert gas flow (under the right conditions) would considerably reduce the oxygen partial pressure.

(d) Powder specimens sealed in capillaries have their own restricted space atmosphere (see 2(b) above and section 7.1 on specimen preparation).

4. X-ray optics and camera geometry. These conditions must be similar to those for room temperature cameras.

(*Note*: The existence of a continuous air-, gas-filled or 'vacuum' space between the hot furnaces and specimen and the cold water-cooled parts of the camera means that any evaporated metal or compounds from specimen, windings or couples will condense again on to the cold parts and so there is no possibility of setting up equilibrium even though any changes may often take place slowly enough for practical stability to be assumed. In some cases, e.g. the evaporation of a pure metal, or an oxidation study with air flow the effect will not invalidate the experimental results.)

These observations indicate the measure of care and discrimination required for the successful use of high temperature diffraction. The value of the techniques is then in no way diminished if they are subject to these considerations.

7.7.2.3 Some Applications: 1. Thermal expansion is an obviously useful application. The lattice parameter is measured at suitable temperature intervals. This provides a very accurate means of determination. There may be slight differences from the expansion determined by macroscopic means and these could depend on the presence of grain boundaries and voids, in the

275

macro specimen. The lattice parameter is the mean for the grains themselves. Examples in the literature include the expansion of iron, copper, aluminium and many other metals. Graphite, and several oxides and nitrides have also been investigated.

2. Phase changes in a pure metal or alloy can also be studied. The simplest method is to bracket the transformation temperature by taking photographs (or diffractometer records) on either side. Exact lattice parameter measurements in both phases can be used to determine the volume change involved. The progress of transformation can be followed if it occurs over a range of temperature. Typical applications include iron and some of its alloys.

3. An extension of 1 which may also involve 2 is the study of multiphase alloys. Other methods would give the bulk expansion only. The X-ray method —providing the spacings or parameters can be measured for each phase separately—gives the individual phase expansions. The results for these phases may show similar or markedly different thermal expansions. In the latter case thermal stresses are more likely to arise (see Chapter 5, Volume 2). The presence of a phase change with a marked expansion or contraction will provide an additional source of stress. Again iron and steel provide typical examples. Investigations of high speed steel [51] and of heat resistant steels [45] have been reported.

4. Under suitable conditions it is possible to follow the oxidation of a metal. The qualitative determination of the constitution of an oxide film or layer is useful in itself. The increasing thickness of the scale and so the relative decrease in intensity of the diffraction pattern from the metal is indicative of the oxidation rate and in favourable cases use can be made of the relative intensities of lines from each phase. If one component is oxidised more rapidly than another the lattice parameter of the metal may change progressively due to preferential removal of this component from solid solution (e.g. Ni–Cr heat resistant alloys).

5. Phase diagram determination is materially advanced if phases which are stable at high temperatures can be identified especially if they decompose on cooling. Equally important is the determination of the ranges of composition and temperature over which ordered phases exist. The presence of ordering may also be indicated by lattice parameter irregularities or apparent anomalies in physical properties, such as specific heat (section 2.2.4) and electrical resistance (Chapter 3, Vol. 2). The appearance of the superlattice lines on the diffraction patterns (see section 5.2.6) gives a positive indication that ordering is present, although the lines may be very faint if the atoms have closely similar scattering factors for the X-ray wavelength(s) used (atoms close together in atomic number). An extension of the use of lattice parameters to determine phase boundaries follows in two ways. The room temperature methods (see Fig. 7.1) can be repeated at higher temperatures. Alternatively the phase change method 2 above can be applied. The phase regions can be more accurately fixed by the change in direction of the lattice spacing

temperature curve as it crosses the phase boundary. The necessary care and limitations of the X-ray methods applied in this way are considered by Hume-Rothery *et al.*, pp. 214–21 [2].

7.7.3 Low Temperature Methods

7.7.3.1 General Principles: The principles are in some respects similar to those for high temperature diffraction. Chemical reactivity and oxidation are unlikely to present any problems, but condensation may occur. The following conditions should be met.

1. The diffraction record should be unimpaired.
2. The camera should be easy to use.
3. The temperatures required must be provided for and controlled accurately.
4. The specimen should not be damaged by the coolant.
5. The instrument should not be extravagant in the use of cooling media such as liquid gases (He, H, N, O or air).

A freezing mixture can be used to produce a standard fixed temperature. Such mixtures are listed in standard references (e.g. Ref. [52]). Alternatively a refrigerating system can be incorporated. Liquefied gases provide for the lowest temperatures. It is necessary to see that any air in the camera is dry in order to avoid icing from water vapour. Diffraction geometry then follows one of the usual patterns and both photographic and counter arrangements have been used. A cryostat for diffractometry from 4 to 300 K has been described by Mauer and Bolz [53]. Heating and cooling attachments described by Kellett and Steward [54], and originally used in studies of graphite, are applicable to temperatures from -196 to 2500 °C. (Several cameras and diffractometer arrangements are listed in the bibliography [48].)

7.7.4 Applications

Applications of low-temperature diffraction devices have been generally confined to structural investigations. A few of these are of metallurgical interest (see Barrett [55]). Industrial research applications are not known, but it is possible to envisage circumstances in which the techniques might be useful. For example in connection with materials intended for use in refrigeration plants, the thermal expansion coefficients and any structural changes possibly of the martensitic type may require investigation.

References

1. RAYNOR, G. V. and WAKEMAN, D. W., *Proc. Roy. Soc. A*, 1947, **190**, 82.
2. HUME-ROTHERY, W., CHRISTIAN, J. W. and PEARSON, W. B., *Metallurgical Equilibrium Diagrams*, Institute of Physics, London, 1952.
3. KOCH, W. and SUNDERMANN, H., *J. Iron Steel Inst.*, 1958, **190**, 373.

4. Various Authors: Iron and Steel Institute. *Reports on the Heterogeneity of Steel Ingots.* Nos. 16, 1937, and 25, 1939.

5. GURRY, R. W., CHRISTAKOS, J. and STRICKER, C. D., *Trans. A.S.M.*, 1957, **50**, 105.

6. ANDREWS, K. W. and HUGHES, H., *Iron and Steel*, 1958 (Feb.), **31** (2), 43.

7. GENSCH, C. I., *Arch. Eisenhüttenwesen*, 1960, **31**, 97.

8. KAY, D. (ed.), *Techniques for Electron Microscopy* (especially Chapter 5 by D. E. Bradley, pp. 82–137). Blackwell, Oxford, 1961.

9. HAINE, M. E. and COSSLETT, V. E., *The Electron Microscope*. Spon, London, 1961.

10. *Powder Diffraction File.* Amer. Soc. for Testing Materials, Philadelphia, 1965.

11. HANAWALT, J. D. and RINN, H. W., *Ind. Eng. Chem. and Ed.*, 1936, **8**, 244 and with L. K. Frevel, ibid., 1938, **10**, 457. (See also L. K. Frevel, ibid., 1944, **16**, 209.)

12. FREVEL, L. K., ibid., 1942, **13**, 109; 1953, **25**, 1697, and with H. W. Rinn and H. C. Anderson, ibid., 1946, **18**, 83.

13. REES, W. P., BURNS, B. D. and COOK, A. J., *J. Iron Steel Inst.*, 1949, **162**, 325; A. J. Cook and B. R. Brown, *J. Iron Steel Inst.*, 1952, **171**, 345.

14. KRAINER, H., *Arch. Eisenhüttenwesen*, 1950, **21**, 33; 1950, **21**, 39; 1950, **21**, 119.

15. MASSALSKI, T. B., *Metal. Rev.*, 1958, **3**, 45.

16. BRADLEY, A. J. and JAY, A. H., *Proc. Roy. Soc. A*, 1932, **136**, 210.

17. ELLWOOD, E. C., *J. Inst. Metals*, 1951, **80**, 217, 605 and with K. Q. Bagley, ibid., 1951, **80**, 617.

18. ROOKSBY, H. P., *J. Roy. Soc. Arts*, 1942, **90**, 673.

19. CLARK, G. L., *Applied X-rays*, 4th edition (Fig. 328(c)). McGraw-Hill, London, New York, 1955.

20. CHESTERS, J. H., *Steelplant Refractories* 2nd Edition (pp. 91, 92). The United Steel Cos. Ltd., Sheffield, 1957.

21. STEPHEN, R. A. and BARNES, R. T., *J. Inst. Metals*, 1937, **60**, 285.

22. SCHDANOW, H. S., *Z. Krist.*, 1935, **90**, 82.

23. ANDREWS, K. W. and JOHNSON, W., *Br. J. Appl. Phys.*, 1959, **10**, 321.

24. HIRSCH, P. B., *Br. J. Appl. Phys.*, 1954, **5**, 257.

25. HIRSCH, P. B., *Micro-Beam Techniques*, Chapter 11, pp. 278–97 of Ref. [9] of Chapter 5.

26. HIRSCH, P. B. and KELLAR, J. N., *Acta Crystall.*, 1952, **5**, 162; P. B. Hirsch, ibid., 1952, **5**, 168, 178.

27. SCHERRER, P., *Nachr. Ges. Wiss. Göttingen*, 1918, **2**, 98.

28. BRAGG, W. L., *The Crystalline State*, vol. 1, Bell, London, 1933.

29. VON LAÜE, M., *Z. Krist.*, 1926, **64**, 115.

30. JONES, F. W., *Proc. Roy. Soc. A*, 1938, **116**, 16.

31. WILSON, A. J. C., ibid., 1943, **181**, 360.

32. WILSON, A. J. C., *X-Ray Optics* 2nd Edition. Methuen, London, 1962.

33. MEGAW, H. D. and STOKES, A. R., *J. Inst. Metal*, 1945, **71**, 6.

34. STOKES, A. R. and WILSON, A. J. C., *Proc. Phys. Soc. (London)*, 1944, **56**, 174.

35. AVERBACH, B. L. and WARREN, B. E., *J. Appl. Phys.*, 1949, **20**, 1066. (Also chapters in *Modern Research Techniques in Physical Metallurgy*, A.S.M., 1952 and *Progr. Metal Phys.*, 1959, **8**, 147.)

36. RICHARDS, T. LL., *Progr. Metal Phys.*, 1949, **1**, 281. (Preferred Orientation in Non-Ferrous Metals.)

37. WEVER, F., *Trans. A.I.M.E.*, 1931, **93**, 51.

38. UNDERWOOD, F. A., *Textures in Sheet Metals*. Macdonald, London, 1961.

39. DECKER, B. F., ASP, E. T. and HARKER, D., *J. Appl. Phys.*, 1948, **19**, 388.

40. SCHULTZ, L. G., *J. Appl. Phys.*, 1949, **20**, 1030.

41. WASSERMANN, G. and GREWEN, J., *Texturen Metallischer Werkstaffe*, 2nd edition. Springer, Berlin/Göttingen/Heidelberg, 1962.

42. SCHMID, E. and WASSERMANN, G., *Z. Phys.*, 1927, **42**, 779; *Naturwiss.*, 1929, **17**, 321.

43. ANDREWS, K. W., *J. Steel Inst.*, 1956, **184**, 274.

44. GOLDSCHMIDT, H. J., *High-Temperature Methods*, Chapter 9, p. 242 of Ref. [9] of Chapter 5.

45. GOLDSCHMIDT, H. J., *Advances in X-Ray Analysis*, ed., W. M. Mueller, **5**, 191. Plenum Press, New York, 1961.

46. GOLDSCHMIDT, H. J., *High-Temperature X-Ray Diffraction Techniques*. Bibliography I. International Union of Crystallography, 1964.

47. STEWARD, E. G., *Low-Temperature Methods*, Chapter 10, p. 265 of Ref. [9] of Chapter 5.
48. POST, B., *Low-Temperature X-Ray Diffraction*. Bibliography 2. International Union of Crystallography, 1964.
49. VAN VALKENBERG, A. and MCMURDIE, H. F., *J. Res. Nat. Bur. Stds*, 1947, **38**, 415.
50. (a) HEAL, H. T. and SAVAGE, J., *Nature Lond*, 1949, **165**, 105; (b) HEAL, H. T. and MYKURA, H., *Metal Treatment*, 1950, **17**, 129; (c) HEAL, H. T., GILLAM, E. and COLE, D. G., *J. Sci. Instrum.*, 1952, **29**, 380.
51. GOLDSCHMIDT, H. J., *Iron Steel Inst.*, 1957, **186**, 68, 79.
52. *Handbook of Chemistry and Physics*. 53rd Edition, The Chemical Rubber Co., Cleveland, Ohio, 1972.
53. MAUER, F. A. and BOLZ, L. H., *J. Res. Nat. Bur. Stds*, 1961, **65**, 225.
54. KELLETT, E. A. and STEWARD, E. G., *J. Sci. Instrum.*, 1962, **39**, 306.
55. BARRETT, C. S., *Advances in X-Ray Analysis*, ed., W. M. Mueller, **5**, 33. Plenum Press, New York, 1961.

Chapter 8

RADIATION TECHNIQUES (Other Than Diffraction Techniques Already Covered)

8.1 SUMMARY OF THESE TECHNIQUES

The techniques to be described in this chapter fall into three groups.

Firstly: X-ray spectrography strictly comes within the category of a physical method of analysis, but there are also quantitative, semi-quantitative and qualitative applications which can be of direct interest for physical-metal-lurgical research and investigations. The electron probe micro-analyser is then a development which provides a valuable technique complementary to the microscopes (electron and optical) and to diffraction.

Secondly: There are X-ray techniques which have been described as X-ray 'microscopy' and microradiography, which give a visual image, with contrast, as in an optical image, but arising from other physical causes.

Thirdly: There are some applications of radio-isotopes. One of these—autoradiography—again gives a visual image of a metal (surface, grain structure, etc.). Some would be described as tracer applications. Other applications pass from laboratory to plant and provide useful information in other aspects of metallurgy.

8.2 X-RAY SPECTOGRAPHY

8.2.1 Basic Principles

8.2.1.1 The Nature of X-rays: Reference should be made again to section 5.3.1 where an outline has been given of the nature of X-rays in connection with their applications to diffraction. Fig. 8.1(a) shows the relationship of the K, L and M shells to the atomic nucleus and to the *characteristic* wavelengths indicated by these letters. This is *schematic* and should not be taken as any more than a reminder of the electronic shell structure. The energy-level type of diagram in Fig. 8.1(b) represents the relationship between the energy levels of the electrons and the X-ray energies. It shows that if an electron is removed by one of the available means (see below) it will receive an amount of energy $E = h\nu$ (where ν can be the frequency of a quantum of incident radiation). If this is sufficient to take it completely out of the atom then E is greater than the ionisation energy for say the K level (as indicated)

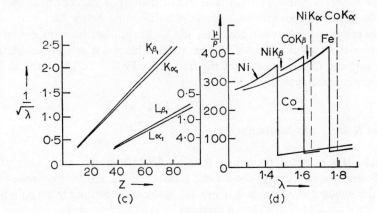

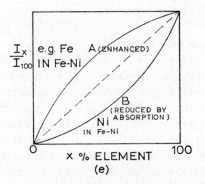

Fig. 8.1 Essential background to X-ray spectography.

(a) schematic relationship between energy levels and spectra
(b) energy level diagram (simplified)
(c) relationship between wavelength and atomic number
(d) essential absorption relationships
(e) inter-element effects

and the electron leaves the atom with a kinetic energy E_{kin}, equal to the difference. The vacancy created in the K shell is then filled by an electron which 'falls into' this shell from the L or M level. These transitions give the $K\alpha$, $K\beta$ lines (α_3 and β_2 are very weak).

As in optical spectrography there is some 'splitting' of energy levels or lines, but the number of lines is generally much smaller. The use of the characteristic X-ray lines to recognise the atoms themselves and to make a quantitative analysis of the relative amounts of different atomic species therefore follows and has this particular advantage—the simplicity of the spectra emitted. The absorption of the wavelengths in air or by thin films of light metals, cellophane, etc., increases rapidly as the wavelength increases, i.e. as the atomic number decreases. This makes practical spectrography more diffi-cult for elements of lower atomic number than say 22 (Ti) and exceptionally so for elements below 11 (Na). The use of helium or vacuum path spectro-graphs improves the detection especially in the range below 22.

The relationship between characteristic X-rays and atomic numbers known as Moseley's Law is indicated by Fig. 8.1(c). This is a nearly linear relation-ship between the square-root of the frequency, i.e. $\nu^{1/2}$ or $1/\sqrt{\lambda}$ and Z. The relation is represented by

$$\lambda^{-1/2} \propto \nu^{1/2} = K(Z-\sigma) \tag{8.1}$$

where K and σ are constants.

8.2.1.2 It was noted previously (section 5.3.1) that the mass absorption coefficient μ/ρ varied with the atomic number of the absorbing element (Z) and the incident wavelength λ (from any element) according to an equation,

$$\frac{\mu}{\rho} = \text{const. } Z^4\lambda^3 \tag{8.2}$$

In this case a single relationship only exists for a given atom (Z constant) between two absorption edges. If $\log(\mu/\rho)$ for this element is plotted against $\log\lambda$ a straight line results. At an absorption edge there is a step and a fresh straight line section follows (with a new constant). The absorption edges occur for succeeding elements in the periodic table in a regular sequence. The occurrence of the characteristic wavelengths of other elements in the vicinity of the absorption edge is such that for many typical elements the $K\alpha$ is critically absorbed by the element with atomic number two less, e.g. the line for Fe with equation (8.2) has a step at about 1·75 Å and a new branch starts. The Ni$K\alpha$ is at approximately 1·65 Å on the high side. This situation and the corresponding position for neighbouring elements is indicated in Fig. 8.1(d) (linear scale). The diagram indicates the absorption curves for three metals, iron, cobalt and nickel, and shows the positions of the $K\alpha$ and $K\beta$ lines for both Co and Ni. From this diagram it is seen that:

(i)   Co and Fe strongly absorb the Ni$K\beta$ line.

(ii) Fe strongly absorbs the Co$K\beta$ line but (as might be expected) Co does not absorb its own $K\beta$ (or $K\alpha$) strongly.

(iii) As already seen (Fig. 5.6(a)) Ni absorbs Cu$K\beta$ ($\lambda = 1\cdot39$) but not Cu$K\alpha$ ($\lambda = 1\cdot54$).

Similarly Ni$K\alpha$ is heavily absorbed by Fe but not by Co and so on.

If radiation is strongly absorbed this may be taken to mean that the atoms of the absorbing element have received energy as in Fig. 8.1(b). It will therefore emit its own radiation. It follows that this radiation will add to the radiation already excited from this element. Hence if the intensity from some element is decreased by absorption there is a corresponding enhancement of the intensity of others. In general there will tend to be departures from a purely linear relationship between intensity and the amount of an element present if the components of the alloy, compound, or mixture under analysis, have appreciably different absorption coefficients. In the neighbourhood of absorption edges, however, there is a marked departure from linearity. This is illustrated by Fig. 8.1(e) which can also be related to Fig. 8.1(d). In a Ni–Fe alloy, Fe absorbs both Ni$K\alpha$ and $K\beta$ and so will be enhanced—more than if it only absorbed one of these wavelengths. A fraction of this absorbed radiation is then converted to Fe$K\alpha$ radiation. There will be a different relative effect for Fe–Co where only the Co$K\beta$ is strongly absorbed.

Effects of this kind in binary alloys were reported by Koh and Caugherty [1] and other aspects are considered in, for example, a book by Liebhafsky et al. [2]. The physics of X-ray spectra is extensively covered in the classic work of Compton and Allison (Ref. [10] of Chapter 5). The first textbook on analysis by this means was by von Hevesy [3]. A more recent monograph should be consulted for a modern viewpoint [4]. (The inter-element effects are considered further in section 8.2.3.)

8.2.2 Techniques

8.2.2.1 Alternatives: The techniques fall into two categories in two different ways—in regard to (a) method of excitation and (b) method of detecting and recording spectra.

Thus the choices are:

(a) The excitation can be by electron beam as described for the production of X-rays for diffraction. This method was used in some of the early spectrography but has the disadvantage that successive specimens have to be introduced into the high vacuum to be brought into position under the electron gun. It has, however, recovered in importance in the development of the *electron probe micro-analyser* (section 8.3). Or:

The excitation can be produced by X-rays of sufficient energy, i.e. exceeding the amount of energy required to eject an electron from the corresponding level as in Fig. 8.1(b). This is the process of fluorescence, and the *X-ray fluorescence spectrograph* is based on this principle. The enhancement effect is really fluorescence produced within a sample.

(b) Photography can be used to record the spectra as for diffraction. Or:

The counter techniques can be used. These provide for speed and accuracy and it is the development and improvement of the counting tubes which have greatly extended the use of X-ray spectrographs in the past ten to twenty years.

8.2.2.2 Photographic Methods: Two simple arrangements are shown in Figs. 8.2(a), (b). These are focusing cameras. The primary X-ray source is an X-ray tube—generally with a tungsten or gold target. These arrangements can be used to demonstrate the principles and give satisfactory spectra for reasonably simple alloys.

The primary X-rays impinge on the specimen which then emits the characteristic X-rays (and some white radiation) corresponding to the atoms present in it. These different wavelengths are either reflected by or transmitted through the analysing crystal. This crystal is cut in either case so that a prominent crystal plane diffracts the spectral wavelengths according to the Bragg Law. It is, however, necessary to oscillate the crystal through a large enough angle so that the diffracting plane is brought successively into the right position for the relation $n\lambda = d \sin \theta$ to be satisfied for each value of λ in turn with d constant. (In diffraction λ is generally constant and d different.) The films from such cameras will then show a series of lines corresponding to the various wavelengths. The separation by the analysing crystal is indicated in Fig. 8.2(c).

8.2.2.3 Counter Spectrographs: It would be expected that the counter spectrograph would be somewhat similar to the diffractometer—the diffracting crystal would be at the centre of the instrument as in Fig. 6.10(a) or on the circumference with a focusing arrangement as in Fig. 6.10(b). This is the case, except that the crystal or powder specimen is now replaced by the analysing crystal as in Figs. 8.2(d) and (e). The specimen undergoing analysis is then placed so that the fluorescent X-rays which are emitted in all directions from the specimen surface are restricted to either a parallel beam as in (d) or a focused beam as in (e). The parallel Soller slits in (d) prevent sideways scatter and divergence.

Fig. 8.2 Arrangements for X-ray fluorescence spectrography.

(a), (b), (c) are photographic methods
(a) reflection by crystal
(b) transmission through crystal
(c) separation of wavelengths for (a)
(d) and (e) are counter methods
(d) general arrangement
(e) focusing arrangement

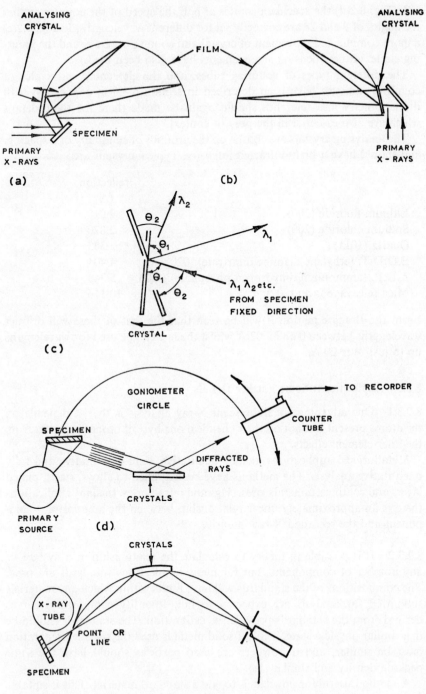

In Fig. 8.2(d) the specimen rotates at half the speed of the counter so that the angles of θ and 2θ are correctly set for different wavelengths. In Fig. 8.2(e) a more complicated movement of crystal and counter occurs round the focusing circle. (Other focusing arrangements have also been used.)

The different types of counting tubes, and the elementary principles of counting statistics have been described in section 6.4.6 in connection with diffractometry and reference should again be made there for these details which are also essential in the present context.

The analysing crystals are chosen on the grounds of suitability of d spacing, and should have a high diffracted intensity. Typical crystals are:

	reflection (Å)
Lithium fluoride (200)	2·01
Sodium chloride (200)	2.820
Quartz (10$\bar{1}$1)	3·343
E.D.D.T. (ethylene diamine ditartrate) (020)	4·404
A.D.P. (ammonium dihydrogen phosphate) (202)	5·348
Mica (cleavage face) (002)	10·116

From the Bragg equation it will be seen that the first of these will diffract wavelengths between 0 and 4·02 Å whilst the last can be used for wavelengths up to just over 20 Å.

8.2.3 Applications: Some Further Details

8.2.3.1   One advantage of fluorescent X-ray analysis is that it depends on the atoms present but not on their chemical or physical properties apart from the inter-element effects.

Metallurgical applications include qualitative, semi-quantitative and fully quantitative analysis. The methods have been applied to alloys, steels, plated layers and coatings, minerals, ores, slags and refractories. In many applications there is an approximately linear relationship between the amount of a component and the recorded X-ray intensity.

8.2.3.2   It is possible in theory to calculate the relationship in a system of any number of components, but for many applications standards are used. The compositions of the standards will have limits within which any materials submitted for analysis are expected to fall. Intensity–composition curves derived from the standards provide the calibration. The standards should be in a similar physical state. Thus, if solid metal is used the surface preparation must be similar, and if powders are used particles should have the same packing density and similar size.

A method useful for powders is to add a standard material. This is suitable for analysis of elements which form a small proportion—10 per cent or less—

286

of the total. The standard element can be chosen to be next to the element to be determined so that it has nearly the same absorption and enhancement properties. The amount of the element undergoing analysis is then determined by the relative intensities of corresponding lines for this element and the standard. Naturally this method can only be used when the samples are in powder or solution form and can be thoroughly mixed with any added component.

Another variant consists of adding a known amount of the element to be determined so that the intensity from this element is increased by a known relative amount. Thus if the percentage of an element is X and the known amount S is added

$$\frac{I_x}{I_{x+s}} = \frac{X}{X+S} \tag{8.3}$$

Solution of solid specimens can be a useful step in the analysis of some complex alloys and other materials. The standards can then be built up from accurately prepared solutions containing standard amounts of the separate components. This method can be useful when only a small amount of a material is available, but could not be used for elements of low atomic number where a vacuum path is desirable. It can be applied to alloy phases separated from the matrix by electrolysis.

8.2.3.3 Calculation Methods. A brief account of methods for calculating fluorescent intensities will now be given. This is based largely on the work of Birks and others [4, 5, 6] but with some simplification of notation. Consider the paths of rays in the case illustrated in Figs. 8.2(d) or (e) and let ϕ_1 be the angle of incidence of the primary beam and ϕ_2 the angle at which the fluorescent beam emerges. The absorption effects are calculated by methods similar to those used for absorption calculations in connection with diffraction (section 6.4.7.2). The primary X-rays at a depth x will thus have an intensity given by

$$I = I_0 \exp\left\{-\left(\frac{\mu}{\rho}\right)\rho x \operatorname{cosec} \phi_1\right\} = I_0 \exp\left(-\mu x \operatorname{cosec} \phi_1\right) \tag{8.4}$$

in which I_0 is the incident intensity and $\mu/\rho=$ mean *mass* absorption coefficient for the matrix for the wavelength(s) in the primary beam.

(*Note*: In some work (e.g. Ref. [4]) μ is taken to be the mass coefficient, but it is simpler to take it as the linear coefficient since μ then applies in particular conditions and μ/ρ is a fundamental property of the element.)

The incident intensity at the depth x will be further absorbed on passing through a further depth dx. A fraction of this absorbed radiation will be converted to the fluorescent radiation of one of the elements—i, say—present in the sample. The intensity of this fluorescence radiation will depend on an excitation constant $Q_i=$ ratio of excited intensity to primary intensity for the

287

pure element—dilution in a compound, mixture or alloy, must therefore be taken account of. If the density of this element in the sample (=weight of element i in unit volume) is ρ_i the intensity of fluorescent radiation is thus

$$dI_i = Q_i\rho_i I \, dx$$

This intensity on reaching the surface at an angle ϕ_2 will undergo absorption according to the expression

$$dI_i \exp\left(-\mu'_i x \operatorname{cosec} \phi_2\right) \tag{8.5}$$

where μ'_i is the linear absorption coefficient for the fluorescent wavelength from element i. The final intensity is thus given by integration combining (8.4) and (8.5)

$$\int \exp\left(-\mu'_i x \operatorname{cosec} \phi_2\right) dI_i$$

$$= Q_i\rho_i I_0 \int \exp\left(-\mu x \operatorname{cosec} \phi_1\right) \exp\left(-\mu'_i x \operatorname{cosec} \phi_2\right) dx$$

The limits of integration are from 0 to a finite depth x or, if the specimen is thick enough, to ∞. Also since for a given element the analysis employs ϕ_1 and ϕ_2 as constants one may write

$$\left(\mu \operatorname{cosec} \phi_1 + \mu'_i \operatorname{cosec} \phi_2\right) = \mu''_i$$

and the equation for intensity becomes

$$I_i = \frac{Q_i\rho_i I_0}{\mu''_i}\left\{1 - \exp\left(-\mu''_i x\right)\right\} \tag{8.6}$$

Hence

$$I_i = \frac{Q_i\rho_i I_0}{\mu''_i} \quad \text{for} \quad x \to \infty \tag{8.7}$$

This expression thus allows for the excitation of the wavelength of i by the primary beam only, but the enhancement effect may also increase the intensity I_i, and decrease the intensity of other wavelengths from other elements, or vice versa. The simplest way of dealing with this is to let μ''_i include both absorptions and enhancements (a kind of 'negative absorption'). This is a concept that is supported sufficiently closely by experiment [5], so that one can use an equation like (5.28) to find μ''_i, i.e.

$$\left(\frac{\mu''}{\rho}\right)_i = \frac{w_1}{100}\left(\frac{\mu''}{\rho}\right)_{i1} + \frac{w_2}{100}\left(\frac{\mu''}{\rho}\right)_{i2} + \cdots + \frac{w_j}{100}\left(\frac{\mu''}{\rho}\right)_{ij} \tag{8.8}$$

where $w_1, w_2, \ldots, w_j$ are the weight percentages, and $(\mu''/\rho)_{ij}$ are the composite absorption coefficients. The relation between the weight percentages and the densities ρ_i is that

$$\frac{w_i}{100} = \text{weight fraction} = \frac{\rho_i}{\rho} = \text{density fraction}$$

Equation (8.8) can thus be re-written:

$$\mu''_i = \left(\frac{\mu''}{\rho}\right)_i \rho = \left(\frac{\mu''}{\rho}\right)_{i1}\rho_1 + \left(\frac{\mu''}{\rho}\right)_{i2}\rho_2 + \cdots + \left(\frac{\mu''}{\rho}\right)_{ii}\rho_i + \cdots + \left(\frac{\mu''}{\rho}\right)_{ij}\rho_j$$

or
$$\mu''_i = a_{i1}\rho_1 + a_{i2}\rho_2 + \cdots + a_{ii}\rho_i + \cdots + a_{ij}\rho_j \tag{8.9}$$

where a_{ij} is the mass absorption/enhancement coefficient for the element j as it affects the intensity from the element i, including its effect on the absorption of the primary X-ray beam. μ''_i is still a linear coefficient (as it should be). Equation (8.7) can be written:

$$\mu''_i = \frac{Q_i I_0 \rho_i}{I_i}$$

and so from (8.9):

$$a_{i1}\rho_1 + a_{i2}\rho_2 + \cdots + a_{ii}\rho_i + \cdots + a_{ij}\rho_j = \frac{Q_i I_0 \rho_i}{I_i} \tag{8.10}$$

This is a useful result since it shows that the reciprocal of the intensity is a linear function of the composition. On multiplying through by $100/\rho$ the weight percentages appear again instead of the ρ's.

For the pure element i

$$a_{ii}\rho_i = \frac{Q_i I_0}{I_{pi}}\rho_i$$

where the suffix pi refers to the pure element i by the definition

$$I_{pi} = Q_i I_0$$

and so
$$a_{ii} = 1$$

On substituting the weight percentages (or weight fractions if desired) and combining the term on the right-hand side with the a_{ii} term in (8.10), one obtains

$$a_{i1}w_1 + a_{i2}w_2 + \cdots + \left(1 - \frac{I_{pi}}{I_i}\right)w_i + \cdots + a_{ij}w_j = 0 \tag{8.11}$$

$$\left(\text{since} \qquad a_{ii}w_i - \frac{Q_i w_i I_0}{I_i} = \left(1 - \frac{I_{pi}}{I_i}\right)w_i\right).$$

Equation (8.9) is for the intensity of one element i. Corresponding simultaneous equations exist for all the other elements. Clearly also

$$w_1 + w_2 + w_3 + \cdots + w_i + \cdots + w_j = 100$$

(or unity if weight fractions are used).

The simultaneous equations as (8.11) cannot be solved for an actual analysis unless the coefficients are all known. Furthermore it is not strictly

289

correct to regard them as constants since each a_{ij} depends on the other elements present. The equations can, however, be applied to systems of a few components and a method of determining the coefficients is to use the pure elements (i–j) and a binary alloy to determine the coefficient a_{ij}. This can be either near the equiatomic composition or with the elements in approximately the same proportions as they are expected to occur. Coefficients will be required for every element pair present and so for example if ternary alloys are being analysed, three pure components and three binary alloy standards are required. For n components the numbers are n phase samples and $^nC_2$ binary systems. The coefficients so determined can then be used with equations (8.11) to analyse unknown samples. This is regarded as a first approximation. A better approximation follows if ternary standards are used.

Sherman [6] has suggested another way of dealing with the equations. In a fluorescent spectrograph the intensity is often measured as the number of counts in a fixed time. Instead one may measure the time to reach the same number of counts from each element. (If necessary, elements which are present in only small amounts can be counted for shorter times and then the times scaled up but this would reduce the statistical accuracy.) If N is the fixed number of counts

$$\frac{N}{t_i} = I_i, \quad \frac{N}{t_{pi}} = I_{pi}$$

for each element. Thence

$$\left(1 - \frac{I_{pi}}{I_i}\right) = \left(1 - \frac{t_i}{t_{pi}}\right) = \frac{1}{t_{pi}}(t_{pi} - t_i)$$

On multiplying through equations (8.11) by t_{pi} (constant) one obtains a series of constants ($t_{pi}a_{ij}$) which can be replaced by b_{ij}. The equations then become:

$$\left.\begin{aligned}
(b_{11} - t_1)w_1 + b_{12}w_2 + b_{13}w_3 + \cdots &= 0 \\
b_{21}w_1 + (b_{22} - t_2)w_2 + b_{23}w_3 + \cdots &= 0
\end{aligned}\right\} \tag{8.12}$$

$$w_1 + w_2 + w_3 + \cdots = 100 \quad (\%)$$

(Sherman used as instead of bs but it is necessary to show that these are different from the coefficients in (8.11).) Equations (8.12) are now completely linear and the times inserted give a set of simultaneous equations. The possibility of determining the coefficients by a least squares statistical method also arises. Enough standard samples would be required, and the formulae then used only for materials within the standard ranges. A development of this kind has been reported for analysis of stainless steels [7] and a treatment for steel furnace slags also provided [8] which considers the limitations of the method.

290

8.2.3.4 Another result is derived from equation (8.6) when x is small and equal to t (thickness).

$$I_i = \frac{Q_i \rho_i I_0}{\mu''_i} \{1 - (1 - \mu''_i t)\}$$

$$= I_{pi} \rho_i t \tag{8.13}$$

This equation shows that the intensity under these conditions is directly proportional to the thickness t and amount of element and is not then affected by the other elements present—a useful result for the analysis of thin films including stripped oxide films, plated layers removed from substrate, etc.

A further conclusion (from calculation) affects the use of solutions—acid solutions of metals or alloys and fused salt solutions of slags, minerals, or oxides; e.g. borate glass discs can be prepared in this way. These solutions have an advantage over powders in that they distribute the atoms much more evenly through the actual sample under analysis and avoid effects due to grain size, packing and surface irregularity. In both cases the actual sample under analysis is diluted considerably by the solvent, i.e. water containing acid radicals or other constituents, or borate glass. Two (limiting) possibilities arise although intermediate situations are possible:

(i) There is interaction between atoms in the solvent and the sample under analysis. This must be taken care of as far as necessary, e.g. by using the various equations (8.6)–(8.11).

(ii) If there are no inter-element effects between solvent and solute atoms and the solution may be regarded as sufficiently dilute then examination of (8.8) shows

(a) that μ''/ρ will be composed of the terms for the solvent for which μ'' is low but w_i relatively large. Since the same concentration of solvent should always be present and its effects on the variable amounts of atoms in the solute constant and negligible, these terms are effectively constant.

(b) The remaining terms may involve interactions between solute atoms from the sample but the concentrations (ρ_i or w_i) are small so that the effect of variations in these amounts on the magnitude of μ'' is relatively small. Therefore:

(c) The variation of μ''_i with solute composition should be negligible and (8.7) becomes a linear equation between I_i and the density of atoms i on their weight percentage,

$$I_i = \left(\frac{Q_i I_0}{\mu''_i}\right) \rho_i = \text{(const.)} \, w_i \tag{8.14}$$

If μ''_i does not vary for w_i between 0 and 100 per cent in the sample (but not the solution) this is an ideal relationship.

8.3 ELECTRON PROBE MICRO-ANALYSIS

8.3.1 General

It has been indicated that the instruments considered in this section are in fact X-ray spectrographs. The excitation is, however, provided by an electron beam and not by primary X-rays and so this is a more direct means of excitation. The beam is focused so that it impinges on a small area of the specimen and so the analysis recorded is the composition at this 'point'. By moving the beam or specimen a point by point analysis is possible.

Such an instrument is probably most accurately described as an electron probe micro-analyser but terms such as electron probe or micro-probe analyser, and others are used.

It will be appreciated that many of the spectrographic principles and the method of recording the spectral intensities are the same as described in the first part of this chapter for ordinary X-ray spectrography (section 8.2.2). The primary beam absorption effects are now replaced by electron absorption effects and other problems arise (section 8.3.3).

The reason for considering such instruments here is that they have provided a valuable addition to the techniques of physical metallurgy in recent times. Many of the instruments are elaborate and expensive but they are found with increasing frequency in metallurgical research laboratories and educational establishments. The brief description below is intended to indicate the principles and some metallurgical uses. (General Refs.: [9], [10] and some chapters of [11]–[14].)

8.3.2 The Instruments

8.3.2.1 Instruments for micro-probe analysis were made with electron beam and lens systems similar to those for electron-microscopes. Advances in the latter field therefore helped the evolution of the probe. The first patent was in 1947 [15], and other early developments were due to Castaing and Guinier [16], and a Russian, Borovskii [17]. The first instrument to become commercially available was the 'Cameca' based on Castaing's design. Later instruments have embodied various principles which extend the versatility and range. These will not be described in detail but it is useful to have an appreciation of the essential features.

8.3.2.2 The instrument should contain:

1. A source of electrons—usually a hot filament.
2. A high voltage system to accelerate the electrons from the source—the cathode—to the anode (1 and 2 constitute the 'electron gun').

3. An electron optical lens system to focus the electrons at a point on the specimen.

4. An optical microscope system to enable visual observations and positioning of the sample.

5. One or more X-ray analysers or spectrographic attachments.

6. In addition, provision may also be made for the lateral movement of the electron beam so that the specimen can be scanned and a large number of points, i.e. an area, surveyed continuously.

For details of the electron optics reference should be made to the literature but an outline of some of the systems which have been used is shown in Fig. 8.3.

Fig. 8.3(a) shows the essentials of an electron gun. There is a source of electrons in the hot filament, and a cathode shield which, by means of a suitable biased voltage through variable resistance concentrates the electrons into a nearly parallel beam. The electrons are then accelerated by a high voltage between the cathode and anode and achieve a high velocity which carries them through a hole in the anode, and an aperture into the lens system.

Fig. 8.3(b). In briefest detail this represents the outline of Castaing's design.

Fig. 8.3(c). This is an arrangement in which the electron gun is vertically below the lens system and specimen. A horizontal spectrometer system can then be used.

Fig. 8.3(d). Here the X-rays are taken off from the specimen back through the objective lens. This makes a high 'take off angle' possible (see below).

Fig. 8.3(e). If supplementary coils are provided the fields in these coils can be varied synchronously so that the electron beam is bent to and fro in the manner indicated and so scan the specimen.

8.3.2.3 The optical microscope systems are not shown but these usually employ lenses and mirrors or prisms to provide the light paths and focusing and are built in so that the precise location of the electron beam can be pre-selected to give information required about particular features or constituents. In the scanning instruments the choice may be made of an area over which a survey is made.

8.3.2.4 The X-ray spectrometers may take advantage of the geometrical principles already considered in connection with fluorescence analysis. In particular the focusing system enables advantage to be taken of the radiation over a wider solid angle and so more of the X-ray intensity reaches the counting system. Some instruments are now provided with two or three spectrometers with different crystals to provide resolution for different wavelength ranges. Alternatively a proportional counter can be used without a diffracting crystal since the actual quantum energy is measured and not the number of

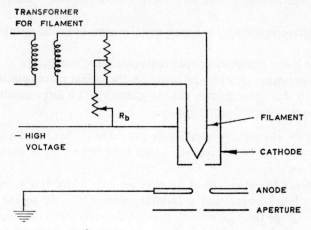

(a)

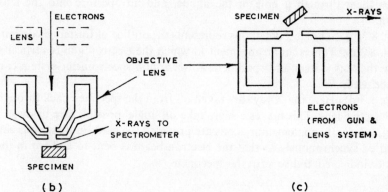

(b)

(c)

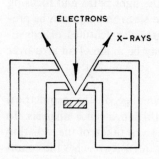

(d)

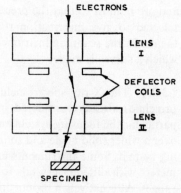

(e)

294

quanta (per unit time). Problems arise in connection with the resolution of pulses from adjacent atoms in the periodic table. The electronic circuits convert the X-ray intensity or quantum energy which can be recorded by counting tubes or a chart recorder. Alternatively scanning instruments have a cathode-ray oscilloscope on which images of the specimen can be recorded.

In the scanning circuits provision is made to synchronise the X-ray intensity record on the oscilloscope screen with the scanning path over the specimen. The image is retained (like a television image) and shows lighter areas in proportion to the intensity of the X-rays emitted. The instrument can record this intensity for one wavelength at a time, i.e. for one element. Hence a light patch signifies a concentration of the element in that area. Another element might be found to be depleted here and concentrated elsewhere. In addition the electrons are not all absorbed but some are scattered back from the specimen. This scattering increases with the atomic number of the element. Hence a detector which records these electrons is a useful adjunct. The modern instrument can therefore record:

1. An image of an area of the specimen for each element in turn (above a minimum atomic number) showing regions enriched or depleted in the particular element.
2. An image of the area showing the regions of higher and lower atomic number—providing these are not entirely uniformly mixed.

In addition, by confining the scanning to any chosen straight line over this area:
3. A chart record of the intensity variation for each element along this line.
   By confining the beam to a single point:
4. An accurate analysis of the relative proportions of atoms at this point. The metallurgical applications referred to below involve one or more of these facilities.

8.3.3 X-ray Intensities

8.3.3.1 The factors relating X-ray intensity and composition are (cf. L. S. Birks [9]):

1. Factors which are related to the physical processes involved:
   (a) Electrons which are back-scattered from the surface of the specimen are clearly not available for X-ray excitation within it. It would be expected that in those regions where more are scattered, the overall excited intensity

Fig. 8.3   Some details of electron micro-probes (based on diagrams due to L. S. Birks).

(a) electron gun
(b), (c), (d) some arrangements of specimen and objective lens and (e) with scanning
(For description see p. 293.)

would be less because less electrons have entered the specimen. In practice it appears that this effect is small and it is probably not important.

(b) Electrons are excited and ejected from the inner shells of atoms when electrons from the incident beam are absorbed. This absorption of electrons is proportional to the density and atomic number divided by atomic weight, i.e. $\rho Z/A$ and with Z/A between 0·4 and 0·5 the variation on this account is small but may be allowed for when desirable.

(c) Absorption of X-rays by matrix.

(d) Enhancement of X-rays by secondary excitation.

((c) and (d) as for fluorescence in X-ray spectrographs).

2. Factors depending on the geometry.

(a) Angle between incident electron beam and surface.

Penetration is greatest for normal incidence.

(b) X-ray take-off angle—usually denoted by ψ and equivalent to ϕ_2 in equations (8.5), etc.

8.3.3.2   The calculation of intensity based on these factors is not as straightforward as for fluorescence. The relation between the X-rays excited at a depth x from the surface, cannot be calculated accurately, but suitable curves have been obtained by experiment. The excitation at any point depends (as one would expect) on the density as well as the distance, and so the X-ray intensity on $\phi(\rho x)$, a function of the penetration ρx (units of mg cm^{-2}). According to the incident electron energy as determined by the accelerating voltage, these curves rise at first and then fall off with penetration. They vary only slightly with atomic number (Z/A being only slightly variable) and so a single standard curve is sufficient—see Fig. 8.4(a). The empirical curve thus gives intensity of excited X-rays as a function of depth. The intensity can then be calculated by dividing into a series of layers of thickness $\Delta(\rho x)$ to a depth below which the intensity is negligible. The X-rays which emerge from the surface at the angle ψ are reduced by the amount,

$$\exp\left\{-\left(\frac{\mu}{\rho}\right)' \rho x \operatorname{cosec} \psi\right\}$$

as in equation (8.5). $(\mu/\rho)'$ is then the mass absorption coefficient for the whole specimen as before. The intensity from an element is thus proportional to its thickness, to $\phi(\rho x)$, to the weight fraction or percentage and to the absorption factor. If all such elements are added up then:

$$I_i = w_i \sum \phi(\rho x) \, \Delta(\rho x) \exp\left[-\left(\frac{\mu}{\rho}\right)' (\rho x) \operatorname{cosec} \psi\right] \qquad (8.15)$$

(Note: $(\mu/\rho)'$ is used here, as in the treatment of fluorescence, for the mass absorption coefficient. μ' alone is used by Birks and others but the use of μ' for linear coefficient is preferred for consistency.)

If the electrons are not incident at 90°, but at some other angle ψ_0 say,

$\phi(\rho x)$ and so the right-hand side of (8.15) must be divided by sine ϕ_0 (or multiplied by cosec ψ_0). Equation (8.15) corresponds to the integration for (8.6), but a summation is required because the equation for $\phi(\rho x)$ is not known.

Since for a given incident electron energy the function $\phi(\rho x)$ is similar for all elements, equation (8.15) shows that I/w then depends solely on $(\mu/\rho)'$ cosec ψ. Hence one could plot I/w as a function of this absorption factor to

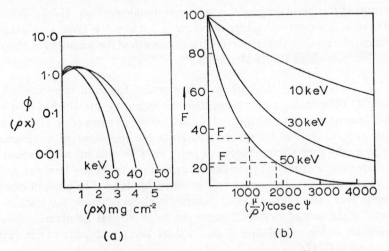

Fig. 8.4   Functions for intensity determination.
(a) electron excitation function $\phi(\rho x)$ (voltage of electron gun in keV)
(b) function for calculating intensities

give a series of curves for constant electron energy. A standard series of curves for general use has been obtained by plotting

$$F = \frac{I}{w \sum \phi(\rho x)\, \Delta(\rho x)}$$

which is equal to the right-hand side of (8.15) excluding $w\,(=w_i)$ and divided by the simple $\phi\Delta$ sum. Alternatively

$$F = \frac{\sum \phi(\rho x)\, \Delta(\rho x) \exp\left[-(\mu/\rho)'(\rho x)\cosec \psi\right]}{\sum \phi(\rho x)\, \Delta(\rho x)} \qquad (8.16)$$

which can be multiplied by 100 to give a percentage scale—immaterial because ratios of Fs are involved. Such curves are shown (approximately) in Fig. 8.4(b). Their use follows. The values of $(\mu/\rho)'$ cosec ψ can be calculated for two different compositions under the same conditions of excitation, e.g. 50 keV in Fig. 8.4(b). The two values of F are F_1, F_2. From the previous equations

$$\frac{F_1}{F_2} = \frac{I_1}{w_1}\Big/\frac{I_2}{w_2} \quad \text{or} \quad \frac{F_1 w_1}{I_1} = \frac{F_2 w_2}{I_2}$$

297

For 100 per cent of an element i one may find $(\mu/\rho)'$ for this pure element and from it

$$F = (F_i)_p$$

Thus

$$\frac{I_i}{(I_i)_p} = \frac{F_i w_i}{(F_i)_p 100} \tag{8.17}$$

where w_i is the percentage of i in a sample under analysis. If fractions are preferred 100 is omitted. Where the error in reading F is large a calibration curve may be needed. Otherwise (8.17) gives a method for quantitative analysis using the pure element as a standard.

8.3.3.3 So far absorption without enhancement has been considered. The possibility of replacing a pure absorption coefficient by one which is modified to include enhancement is rendered difficult by the absence of an exact calculation method. It will be appreciated from the treatment of fluorescence spectrography that a second element j will have its own intensity excited at successive levels $\Delta(\rho x)$ and the whole radiation excited at any level can excite fluorescent radiation from element i at any other level. The amount excited depends *inter alia* on an efficiency parameter E_{ij} representing the fraction of quanta of the second element j converted to additional fluorescent quanta of element i. The calculation is outlined by Birks and leads to the result analogous to (8.17),

$$\frac{I_i}{(I_i)_p} = \frac{F_i w_i}{(F_i)_p 100} (1 + K_F) \tag{8.18}$$

(Birks uses A and B for i and j).

The correction for fluorescence, K_F, is given by:

$$K_F = 0 \cdot 6 E_{ij} w_j \left(\frac{\mu'_{ij}}{\mu'_j} \right) \left(\frac{V - V_{0j}}{V - V_{0i}} \right)^2$$

Values of E_{ij} (or E_{AB}) are available in the literature (see Birks [9], Appendix I). V is the actual tube voltage and V_{0i} the critical excitation potential for element i, V_{0j} for j. It is thus possible to obtain sufficiently accurate estimates of K_F for particular applications or to provide calibrations based on (8.18). The possibility that two or more elements can contribute to K_F should be taken into account in appropriate cases. It was assumed that $\phi(\rho x)$ curves are roughly independent of atomic number but this may introduce an error up to 5 per cent of the amount analysed. Separate F curves for each atomic number can be calculated. Alternatively the single F curves (as Fig. 8.4(b)) can still be used and the calculated values of the abscissae $((\mu/\rho)' \text{ cosec } \psi)$ corrected by multipliers. This reduces the possible error to 2 per cent or less of the amount analysed.

The take-off angle ψ appears in the intensity formula. If the voltage of electron excitation is increased, the electrons penetrate further and the X-rays are generated in greater depth. They therefore pass through more material before they emerge, the paths depend on cosec ψ as in the equations but also on the absorption of the matrix for the particular radiation. Hence calibration curves for analysis of binary alloys such as those in Fig. 8.1 (e) show, not only the basic curvature due to absorption and/or enhancement effects, but also differences due to applied voltage, and angle or take-off. The effect of take-off angle is small but tends to increase below 30° because in this region surface roughness introduces variations in the true angle of emergence for a given nominal angle. Surface preparation should always be good and consistent.

8.3.3.4   Quantitative Analysis: The principles are similar to those already outlined for general X-ray spectrography.

(a) Standards can be used to provide a calibration.
(b) The pure elements can be used in conjunction with procedures indicated above.
(c) It may be necessary to carry out a preliminary qualitative analysis to establish the elements present—these may be different in different parts of a specimen, or in different constituents. A series of chart scans may therefore be needed giving the wavelengths (and elements) present.

In determining intensities the same statistical principles apply including the relation between peak height and background.

8.3.3.5   The possible effects of the depth of electron penetration on intensities also include dependence on the volume of metal affected and the way in which electrons are scattered and absorbed. Phases of different composition in contact at a grain boundary may lead to errors in analysis if the electron probe strikes one of the phases near the boundary so that irradiation is excited from both sides. A useful treatment of these effects and the factors affecting intensity has been given by Salter [18]—see also Ref. [13] (pp. 379–417).

8.3.4   Some Metallurgical Applications

8.3.4.1   Metallurgical applications have been described in a number of reviews and general articles and reference may be made to these. In addition to publications already referred to [9, 11, 12, 13, 14], there are the accounts by Philibert [19], Cosslett [20], Wittry [21], Banerjee et al. [22], Melford [23, 24, 25] and several others (e.g. [26], [27]). An illustrative selection of these applications is outlined.

8.3.4.2   1. Phase identification in alloys. Numerous metallographic techniques may contribute in one way or another to the identification of phases. The micro-probe has frequently been used for this purpose either alone or in association with other techniques. Identification is easier the greater the difference in composition between the phases, see Fig. 8.5(a). Examples include intermetallic compounds in alloys, the difference in composition and identity of phases in meteorites (nickel–iron alloys). Two similar brittle phases in stainless steel have been distinguished by the fact that one of them (sigma phase) has a consistently higher ratio of chromium to molybdenum compared with the other (chi phase).

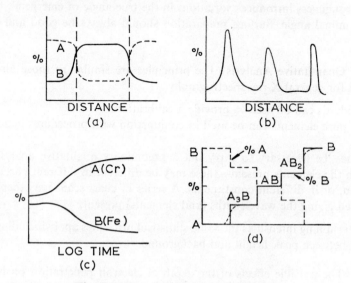

Fig. 8.5 Some applications (schematic).
- (a) variation of composition across second phase or inclusion
- (b) segregation of an element across dendrites
- (c) change in composition of a phase (e.g. $M_{23}C_6$)
- (d) principle of diffusion couple method

2. Identification of inclusions. The identification of particles of oxides, sulphides and other constituents is established. Examples include the general recognition of manganese sulphide in steels. The identification and distribution of this sulphide and lead particles in free cutting steels is one instance where the inclusions play a significant metallurgical role. Inclusions may be found to be complex—consisting of two or more phases in association (e.g. see Melford [23]).

3. Surface Enrichment. A notable application of considerable industrial importance was the discovery that during the heat treatment and processing of mild steel, the more rapid oxidation of iron led to surface enrichment of residual or 'tramp' elements such as nickel, copper, tin or arsenic. The scanning of a piece of steel showed an electron back-scatter pattern with definite evidence of heterogeneity especially between grain boundaries and grain interiors. The X-ray scans for Ni and Cu showed segregation of both

300

elements at austenite grain boundaries [24, 25]. In this way, features visible in the optical microscope but not easily interpreted can be more fully and usefully explained.

In applications of this kind a scan along a line passing through an enriched region would show an increase in the intensity from the enriched element and corresponding decrease in the intensity from the principal element. (These are similar to the scans over diffusion couples, etc., referred to below.)

4. The identification of constituents in oxide scales and of oxide particles formed by internal oxidation.

Examples include the scales formed on iron–chromium alloys, where a point by point analysis along a line through alloy, inner scale, and enter scale, showed marked changes in the relative amounts of Fe and Cr corresponding to different phases in these scales [28].

5. The study of segregation. This may be types of segregation from solidification such as the variation of composition between dendrites and interdendritic regions (Fig. 8.5(b)). Segregation may also be associated with the presence of inclusions (2 above) or the occurrence of different phases deriving from the basic phase relationships for the alloy system involved (cf. 1 above). By rapid quenching of partly solid alloys the compositions on either side of a solidification front can be studied.

6. Diffusion studies. The study of diffusion has led to the provision of information on the distribution of elements in, for example, metal coating processes. The relative amounts of chromium and iron in a chromised layer provides an illustration. The change in composition with time due to diffusion of alloy elements is exemplified by the carbide $M_{23}C_6$ in an '18–8' stainless steel (Fig. 8.5(c)). Interesting fundamental studies use diffusion couples. If two pure metals are placed in contact and then heated to a sufficiently high temperature they inter-diffuse and the region between contains all the intermediate solid solution phases in order as they would appear in the equilibrium diagram at the temperature of the diffusion experiment. The principle is illustrated for a hypothetical system AB, with compounds A_3B, AB and AB_2 in Fig. 8.5(d). Examples are the systems Cu–U where one compound occurs, viz. UCu_5 and Ni–U where several compounds occur. It is found, however, that intermetallic compounds with narrow composition range may not all appear and some care is therefore needed in the use of this technique [29] (e.g. check by diffusion couples between an end member and compound or two compounds).

8.4 X-RAY MICROSCOPY AND MICRORADIOGRAPHY

8.4.1 Some Relationships and Other Techniques

8.4.1.1 In the previous consideration of the electron micro-probe analyser it appeared that the primary interest in this technique was the possibility of accurate quantitative analysis on a micro-scale. It therefore followed as a logical development from a more general X-ray spectrography. It was, however, also explained that the scanning instruments provided a *visual image* of either the electrons back-scattered from the surface of the instrument or the distribution of the radiation intensity for a particular characteristic wavelength. The former type of image might be expected to resemble the optical microscope image in many of its features, and the latter rather less so. It remains to consider what other similar techniques might be available as metallographic tools—using X-rays, electrons or other means. The techniques to be considered here use X-rays. They comprise firstly methods which have been devised for the production of images necessarily taking account of the fact that X-rays cannot be focused with lenses like light or electrons and so the conventional microscope principles are not applicable. It is not surprising that diffraction will also take place under certain conditions and this too will affect the nature of the images obtained—the diffraction contributes usefully to the creation of contrast in some cases (e.g. by reducing the intensity of the direct X-ray beam). The occurrence of Kossel lines is incidental to one of the techniques. Another type of X-ray image can be formed from an enlarged diffraction spot which may be studied for differential contrast effects—probably within a single grain which may also contain sub-grains or polygonised structures.

8.4.1.2 To complete the general viewpoint of imaging techniques other than light optical (Chapter 1, Volume 2) and electron microscope images (Chapter 2, Volume 2) the following comparatively new developments are summarised here.

1. An instrument exclusively for the production of scanned point by point images of secondary electrons. This provides a type of scanning electron microscope using a 'reflected image'. This procedure is similar to the production of an electron image in the micro-probe. An alternative uses lenses to focus the secondary image (rather than scanning) and so resembles the conventional electron microscope in this respect.
2. Alternatively a heated specimen ejects thermionic electrons which can be made to produce images in the same way.
3. If a stream of positive gaseous ions is used to bombard a metal surface, secondary ions as well as electrons are emitted from the surface and can be directly imaged. Alternatively the inclusion of a mass spectrometer enables images due to different ions to be produced separately (rather like the separate characteristic X-ray images in the micro-probe).
4. Field emission microscopy and field ion microscopy represent two further developments which are in a somewhat different category and are summarised in more detail at the end of this chapter.

The above techniques were outlined by Melford in the general review article already noted [23]. Reference may also be made to this article and to the previously listed general works and conference volumes for the X-ray 'microscope' and microradiography techniques now to be considered [11–14].

8.4.2 Experimental Principles

8.4.2.1 The impossibility of focusing X-rays by lenses arises from the fact that the refractive indices only differ from unity by amounts of the order of 10^{-5}. This would result in very long focal lengths and image distances. X-ray 'microscopes' are therefore based not on refraction but on direct geometrical principles including curved mirror focusing.

8.4.2.2 An indication of the essential principles is indicated in Fig. 8.6.

In Fig. 8.6(a) the X-rays from an ordinary crystallographic or similar tube are employed. The specimen in the form of a thin slice or foil is in contact with the film (or plate). Exposure therefore produces an image of zero magnification. This is then developed and fixed so that it can be magnified and

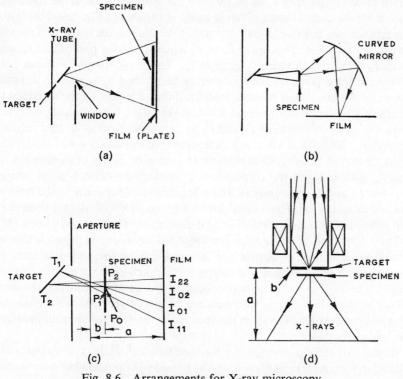

Fig. 8.6   Arrangements for X-ray microscopy.

(a) contact microradiography
(b) focusing by mirror systems
(c) projection from finite source through pinhole aperture
(d) projection from point source

studied by transmission in a light microscope. This technique of *contact microradiography* is of considerable metallurgical use.

Fig. 8.6(b) represents a system which has probably not been used for metallurgical work (and need not be considered in detail here). A curved mirror system—various arrangements have been designed—focuses the rays. The radiation from a single point on a specimen is brought to a point on the image plane by the usual focusing principles of curved cylindrical (spherical or parabolic) systems.

8.4.2.3 Alternative procedures derive from the fact that if an object is placed near to a point source of radiation its image projected on to a screen increases linearly with the distance from object to image, or with distance from point source to object. The principles are those of similar triangle geometry. Two ways of achieving magnification are shown. In the first of these the X-rays come from the target of finite size but a point source is created outside the tube by means of a pinhole aperture in a plate of metal. Radiation from different points of the target passes through the pinhole, through the specimen, onto a film or plate at some distance away. As shown in Fig. 8.6(c) this gives a magnification, but due to the finite size of the pinhole the image of a point is spread over an area. Thus a ray from T_1 would project a point P_0 to I_{01} and from T_2 a ray would give an image at I_{02}. The finite aperture size not only allows the same point on the specimen to be reached by rays from different points on the target, but it also allows for different points on the specimen to be reached by rays from the same point on the target, e.g. a simple point projection of all points between P_0 and P_1 by rays from T_1 gives images between I_{01} and I_{11}. The effects are much exaggerated in the diagram but clearly the aperture should be as small as possible to reduce the image of P_0 between I_{01} and I_{02} (umbra), and the corresponding overlapping of such point images between I_{02} and I_{22} (penumbra). The magnification M is given by the ratio of the two distances a/b. For a consideration of the effects of umbra, penumbra and other limitations to resolution or definition see Cosslett and Nixon [14].

The last method to be illustrated in Fig. 8.6(d) relies on the principle that an intense point source of X-rays can be produced by focusing a stream of electrons with electromagnetic lenses (as in the micro-probe or electron microscope). The focal spot can be small enough to give good imaging and the magnification is again a/b. The specimen can be placed very near to the source. Thermal conduction away from the spot in the target is sufficient to dissipate the heat generated.

Because of their particular use for metallurgical study, the methods indicated in Figs. 8.6(a) and (d) will now be considered in somewhat more detail.

8.4.3 Contact Microradiography

8.4.3.1 In the arrangement shown in Fig. 8.6(a) the specimen is indirectly in contact with the photographic emulsion. The magnification is therefore that which can reasonably be obtained by an optical microscope with transmission. The grain of the photographic emulsion itself limits the practical magnifications. Special fine grained emulsions have resolutions of up to 1000 or more lines per mm and reasonable images can be obtained at magnifications up to 400 times or higher. An ordinary crystallographic tube up to 60 kV rating can be used although higher voltages (150–200 kV) have been employed. A thin section of material has to be placed in contact with the film or (preferably) plate which is cut to a small size of perhaps a square inch or so. The sample and photographic material can then be fixed in a simple camera container with a hole for the X-rays to enter. A metal specimen serves to protect the emulsion from exposure to visual light. Ceramic (refractory) specimens can contain holes or be translucent so a small covering of black paper may be placed in front. The camera can be made of metal parts or even simpler cassettes of metal foil, or cardboard employed. It is, however, better to ensure even and close contact between specimen and emulsion and some such arrangement as that indicated in Fig. 8.7(a) is recommended—numerous variants and improvisations are possible (e.g. see Refs. [30], [31]). The occasional 'outsize' specimen can be examined with one of the improvised arrangements.

8.4.3.2 The preparation of specimens: Metal specimens can be prepared from blocks of metal by sawing off a thin layer. In the case of sheet or strip the original material can be used as a starting point and thinned down further. The process of thinning can sometimes be achieved by chemical or electro-chemical methods (section 7.1). It is, however, generally preferable to use metallographic preparation methods suitably adapted. This avoids the possibility of selective removal of phases, pitting, etc. There is no reason why electro-polishing should not be used for some intermediate stages of thinning if it is ensured that the ultimate result is the same.

One method of thinning by grinding on emery is to mount the thin slice or a piece of the strip or foil to be examined in a shallow recess in a block of steel large enough to hold in the hand. If one side of the specimen has first been prepared to a flat finish at the normal finishing grade (000 emery), this side is mounted into the block and held with a hard wax. The metallographic preparation of the other side then proceeds. If the recess in the block has been machined to the thickness required (or slightly more) the flat surface of the holder then comes into overall contact with the emery paper when the specimen is the desired thickness. Alternatively a simple adjustable holder can be used such as that described by Sharpe [32]. The process of preparation can thus be tedious and time consuming. It should be possible to arrange for all but the last stages to be effected automatically providing the specimen does not become over heated. The preparation will be harder or otherwise according to the hardness of the specimen and problems may arise with materials which tend to flow under preparation. The study of ferrous metals or other harder alloys is generally easier from this point of view.

Sharpe [32] has also provided a useful summary of the kind of defects that can arise during preparation either from incorrect handling or inherent in the material. These defects can be summarised thus:

(a) The specimen has tapered instead of parallel surfaces.
(b) The cross-section varies unevenly due either to
 (i) harder and softer constituents, or
 (ii) overall gradation in hardness arising from uneven heat treatment or some other cause.
(c) The original material may have contained internal stresses so that when it is thinned down the specimen curls up.
(d) Scratched or scored surfaces.
(e) Cracking or even fragmentation of brittle specimens.
(f) Where two metals or phases are in contact as in plated or coated specimens, preparation may lead to detachment.
(g) Constituents can become detached from, or pulled out of the surface, or removed by etching.

This account applies mainly to metal specimens which may vary between 0·002 and 0·005 in (0·05–0·1 mm) in thickness. For ceramic oxides, refractory, sintered ore, and other such materials thin sections prepared by conventional methods for transmission microscopy may serve equally well for microradiography except where different thicknesses are required, e.g. if the optical transmission is low. Typical microradiography samples have thicknesses about 0·005 in (0·1 mm).

8.4.3.3 Contrast: Conventional radiography on the macro-scale depends on whether differences in absorption for different X-ray paths are sufficient to

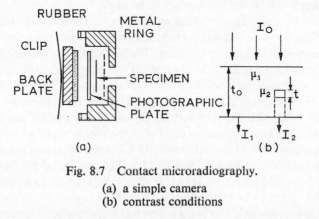

Fig. 8.7 Contact microradiography.
(a) a simple camera
(b) contrast conditions

give a difference in intensity which is detectable on the resulting radiograph by the human eye. The same principle applies to the (optically enlarged) image on a microradiograph, and it can be understood by reference to Fig. 8.7(b). A material of absorption μ_1 contains an 'inclusion' of different material with absorption μ_2. Under the (simplified) conditions shown, irradiation by X-rays of incident intensity I_0, passing through a thin layer of thickness t_0 and *linear* absorption coefficient μ_1, results in an emergent intensity of I_1 given by

$$I_1 = I_0 \exp\left(-\mu_1 t_0\right) \quad \text{(equation (5.26))}$$

This is the absorption law already considered in relation to diffraction (section 5.2.4 and later) and spectrography. When the rays pass through the region containing the second material of thickness t, the path of the X-rays is:

$$(t_0 - t) \text{ in material of absorption } \mu_1$$

$$t \text{ in material of absorption } \mu_2$$

Therefore the emergent intensity is

$$\begin{aligned} I_2 &= I_0 \exp\left\{-\mu_1(t_0-t)\right\} \exp\left(-\mu_2 t\right) \\ &= I_0 \exp\left(-\mu_1 t_0\right) \exp\left(\mu_1-\mu_2\right)t \end{aligned}$$

so
$$\frac{I_2}{I_1} = \exp\left(\mu_1-\mu_2\right)t \tag{8.19}$$

For small values of the exponent

$$(\mu_1-\mu_2)t = \frac{I_2}{I_1} - 1 = \frac{\Delta I}{I_1}$$

(sign $\pm$ according to whether $\mu_1 \gtrless \mu_2$).

The difference in intensity will appear on the photographic image as a difference in blackening. The blackening depends on the intensity and time of exposure according to the definition:

$$B \propto \log\left(I \times \text{time}\right)$$

or
$$B = \gamma \log\left(I \times \text{time}\right) + \text{constant} \tag{8.20}$$

Since the exposure time is the same, two differences in blackening, B_1, B_2, give the relation:

$$B_2 - B_1 = \gamma \log_{10} \frac{I_2}{I_1} = \gamma \log_{10} \exp\left(\mu_1-\mu_2\right)t$$

or
$$\Delta B = 0.4343\gamma(\mu_1-\mu_2)t = 0.4343\gamma(\Delta\mu)t \tag{8.21}$$

γ is a constant, characteristic of the photographic emulsion and is the slope of the major part of the curve of B against log exposure. It does not apply to the initial part of the curve (very low exposures) or for over exposure. Equation (8.21) shows that if some contrast exists it will be more easily seen for higher γ or higher difference in absorption or greater thickness t. The minimum visible to the naked eye is thus a value of ΔB for the photographic

307

material being used for a minimum distance t of a particular type of segregation (e.g. an inclusion in a piece of steel).

Since the absorption coefficients vary with atomic number and wavelength according to equation (8.2) the contrast increases rapidly with difference in atomic number Z. The above discussion and calculation applies to radiation which is monochromatic or nearly so. The calculation is more complicated otherwise and so it is better to use monochromatic radiation, but white radiation gives good contrast for wide differences in Z such as graphite in cast iron.

Another reason for using monochromatic radiation is as follows. One is seldom examining a material involving a contrast between pure elements but differences in relative proportions which may be quite small. The mean absorption is given by equation (5.28). It is therefore an advantage to chose a characteristic X-radiation in the neighbourhood of the absorption edges of the elements present. An example is suggested by Fig. 8.1(d). If Ni$K\alpha$ radiation is used with an iron–cobalt alloy, segregation of either element is much more easily detected because of the large difference in μ for the pure element. Hence the contrast is detected for a much smaller difference in concentration than if the two elements were on the same branch of a single absorption curve given by equation (8.2). This principle is generally useful in metallurgy because many commonly available characteristic radiations are obtained from transition elements in the first long period (up to Cu) and the ferrrous metals and several important non-ferrous alloys can therefore be treated in this way.

8.4.3.4 A development of this principle is to obtain microradiographs with two different radiations which show a reversal of contrast. The reversal occurs because an element is on the high absorption side for one wavelength and low for the other. A specific example is manganese sulphide particles in iron (or steel). The linear coefficients are:

	Cr$K\alpha$	Co$K\alpha$	Ni$K\alpha$
Fe, μ_1	903	470	3120
MnS, μ_2	670	1445	1010
$\Delta\mu = \mu_2 - \mu_1$	-233	$+975$	-2050

Other forms of segregation of manganese or other elements can be revealed providing the concentration differences are large enough.

A further possibility of contrast arises in some materials with a (reasonably) large grain size. The absorption coefficient μ includes a contribution due to scattering and some of the scattered radiation may emerge in a diffracted beam if a grain happens to be correctly oriented for Bragg's Law to be obeyed.

In contact microradiography back-reflected rays subtract from the total transmitted intensity. The outlines of different grains may therefore be revealed even when there is no difference in composition.

8.4.3.5 Resolution is limited by a number of geometrical factors including focal spot size and distances or thicknesses. These are not as crucial as in the pinhole point source methods (Figs. 8.6(c), (d)) but affect the resolution and quality of the image. It will have been seen that because this is a transmission image it contains details projected from different depths through the specimen. This can help to reduce resolution but it can also give a stereographic or 'depth' effect as when grain boundaries can be seen sloping through the specimen. A thicker specimen is otherwise better from the point of view of contrast and a thinner for better resolution.

8.4.4 Projection Microradiography

8.4.4.1 Metallurgical applications of projection microradiography have not been so extensive but the method requires further brief consideration for two main reasons. It was pointed out that the geometry of Fig. 8.6(c) led to problems in resolution. The point source method of Fig. 8.6(d) is preferable. The magnification of $M = a/b$ (see Fig. 8.6(d)) is, however, also affected by the finite size of X-ray source. Even a very sharply focused electron beam will provide a less sharp X-ray source due to scattering in the target foil. Thus an electron beam of 1 μm diameter produces an X-ray source of 2 μm diameter.

A simple construction similar to that used for Fig. 8.6(c) is shown in Fig. 8.8(a). If the source has diameter S then a point in the specimen is represented by an image of size S^* (diameter) given by $S^*/(a-b) = S/b$, or $S^* = S(a-b)/b = S(M-1)$. Two such points P_1, P_2, give images I_1, I_2. If these are separated by distance d the images will just touch if $d/S = (a-b)/a$. If, however, the intensity is not uniform across S but follows a Gaussian or similar distribution, the images I_1 and I_2 will be less intense at the edges and so can still be resolved if they overlap somewhat. The maximum overlap can be taken to be one-half of the image width involving a movement of I_1 and I_2 of $\frac{1}{4}S^*$. This is equivalent to moving P_1 and P_2 nearer to each other each by a distance $(S^*/4)(b/a)$. The distance P_1P_2 is then reduced to

$$d - 2S^*b/4a = S(a-b/a) - \tfrac{1}{2}S(a-b/b)b/a = \tfrac{1}{2}S(a-b/a)$$

and so the maximum resolution is $\frac{1}{2}S(1-(1/M))$ or $\frac{1}{2}S(M-1/M)$. Thus with a 1 μm beam source (2 μm X-ray source) the resolution is of the order of 1 μm.

8.4.4.2 The second reason for considering the point projection method here is the occurrence of the so-called *Kossel Lines*. These are actually due to diffraction by the *divergent beam* of X-rays. They have also been obtained in micro-probe analysers or from an X-ray target where there are X-ray point sources from which the rays can spread out in all directions, and grains of sufficiently good crystallinity. If the number of grains increases the patterns are more confused and may interfere with the interpretation of the microscope image. The lines are also affected by strain or other departures from crystallinity and so may be useful metallurgically for studying such changes. The best patterns are obtained from perfect single grains or crystals.

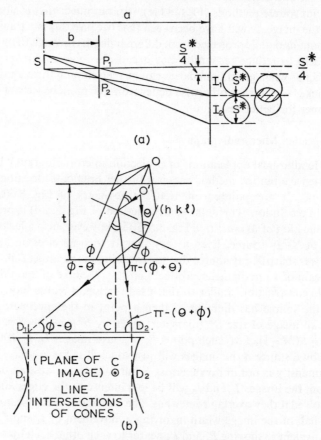

Fig. 8.8 Point projection microscopy.
(a) maximum resolution
(b) origin of Kossel lines (also Kikuchi lines)

The origin of Kossel lines can be understood by reference to Fig. 8.8(b). In this diagram X-rays may radiate from a point O outside the specimen as in point projection microradiography or any divergent beam diffraction arrangement. Alternatively X-rays are generated inside a specimen or target material by electron bombardment (e.g. in a micro-probe analyser) and radiate in all directions from a point such as O'. *In either case* the result is the same—those rays which happen to make the appropriate angle θ with a set of planes (hkl)

will be diffracted. There are two complementary ray paths as indicated, corresponding to $(h\overset{+}{k}\overset{+}{l})$ and $(\overline{hkl})$. The rays are not necessarily in one plane. In the diagram c is so large compared with t that one can ignore the question of finite width and consider that all diffracted rays lie on cones of semi-angles $(\frac{1}{2}\pi - \theta)$ with the plane normas. Hence their intersections with the plane of the image are conic sections and are continuous lines which require a reasonably perfect crystal to be formed.

In Fig. 8.8, c is approximately $a-b$ (more exactly its mean value is $a-b-\frac{1}{2}t$ and $c+\frac{1}{2}t$ is the 'specimen to image distance'). In the image plane, if C is the centre

$$CD_1 = c \cot (\phi - \theta)$$
$$CD_2 = c \cot (\pi - \phi - \theta)$$
$$= -c \cot (\phi + \theta)$$

(The difference in point of emergence in this diagram is, of course, negligible.) Thus

$$D_1 D_2 = \frac{2c \sin 2\theta}{\cos 2\theta - \cos 2\phi} \tag{8.22}$$

and if $\phi = 90°$,

$$D_1 D_2 = 2c \tan \theta \tag{8.22a}$$

The recognition of complementary pairs of curves enables lines $D_1 D_2$ to be located and so C, the intersection of two or more. The distances CD_1 and CD_2 then enable θ and ϕ to be found. Symmetrical patterns arise when a principal axis (e.g. a cube edge) is normal to the specimen and image planes. The planes (hkl) then have a corresponding set making an angle $\pi - \phi$ so that there are curves corresponding to both D_1 and D_2 reflected through C or rotated about this point.

These relationships enable simple patterns to be understood and calculated. They can thus be used to determine orientation, lattice spacings, and sometimes to indicate symmetry. Spacings are calculated from θ in the above equation and an accuracy of 1 part in 1000 is possible.

For further details of divergent beam diffraction reference can be made to a paper by Lonsdale [33]. The corresponding phenomenon in electron diffraction is the formation of Kikuchi lines and the interpretation is similar.

8.4.5 A Technique Using Diffraction Images

If a perfect crystal is used to produce a Laüe diffraction pattern, it would be expected that the intensity would be uniform across a single diffraction spot. If, however, the crystal is imperfect then the diffracted beam will not have a uniform distribution of intensity across it and the image will thus show variations corresponding to the irregularities in the crystal or grain. A technique for the study of crystal texture making use of this effect was described by Berg [34] and was extended by Barrett [35] for metallurgical applications. Further reference should also be made to papers by Honeycombe [36], and Weissman [37].

The geometrical arrangement is indicated in Fig. 8.9. The film is near to the specimen which is placed some distance from the X-ray source (e.g. 30 cm). If necessary the specimen can be positioned first with the aid of a fluorescent screen so that the angle of incidence for particular grains is right for Bragg reflection. A broad or divergent beam can be used. The intense diffracted beam or beams can then be recorded on a film or plate. The image can be studied under an optical microscope with magnifications generally up to 100.

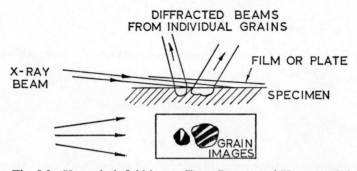

Fig. 8.9 X-ray dark field image (Berg, Barrett and Honeycombe).

The contrast shown in the images depends on the fact that the different parts of the crystal—subgrains or polygonised regions—have slightly different orientations from each other. They do not therefore all satisfy the exact diffraction condition. The photographs obtained are also sensitive to small variations in orientation. The technique has an advantage in its applicability to solid specimens.

It is to be appreciated that this is a 'dark field' technique. Thus if a direct image were obtainable (e.g. if the specimen were thin enough for transmission microradiography) a lighter part of an image would correspond to less transmitted radiation, but the diffracted beam from the same specimen region would show a correspondingly greater intensity so that this part of the image would be relatively darker. This complementary image effect has its counterpart in the light field (direct beam) and dark field (diffracted beam) images produced in electron microscopes (Chapter 2, Vol 2).

8.4.6 Metallurgical Applications of the Various X-ray Microscopy Techniques

8.4.6.1 Some metallurgical applications have already been noted or suggested. These include graphite in cast iron [31], the case of manganese

312

sulphide particles [31, 32] in steel and the application of the dark field diffraction image technique to obtain images of subgrain or polygonisation structures, etc.

8.4.6.2 Many other examples of contact microradiography are cited in the literature and these include:

Dendritic structures as first described for an Au–Al alloy by Heycock and Neville [38] (the first known example).

Phases precipitated in magnesium alloys (Sharpe [32]).

Segregation of chromium in steels and other applications to ferrous materials [31, 39]. These applications also include ores, sinters and refractories. In the case of sinters the shape and form of iron oxide particles—both dendritic and polygonal shapes—are revealed crystallising from a slag matrix.

Some applications to mineral dressing have also been given by Cohen and Schloegl [40].

In addition to the normal study of manganese sulphide particles in steel, one may also cite the importance of this phase in free cutting steels. Contact microradiography is also particularly useful for leaded steels or steels containing both lead and sulphur when the association of the different types of particle is of interest.

Specific examples of projection microradiography have been quoted by Cosslett and Nixon [14]. These include segregation of tin in bronze, and in an aluminium alloy. In the latter case the grain boundaries show up very well through the specimen thickness. Another case is a zinc 5 per cent aluminium alloy where a eutectoid structure is very clearly revealed and thus the different segregation of the alloy elements into the respective phases. Further applications to metals are given by Ruff and Kushner [41]. These include an interesting dendritic formation of iron growing in magnesium, segregation in aluminium–copper alloys, aging and precipitation effects.

8.5   USE OF RADIOISOTOPE TECHNIQUES

8.5.1   Elementary Principles

8.5.1.1   An elementary knowledge of atomic structure is presupposed here but some essential details are summarised:

(a) The atomic nucleus is considered to be built up primarily from protons with unit positive charge and neutrons with zero charge but the same mass. The electrons which surround the nucleus have each a unit negative charge and a mass which is 1/1837 of that of the proton. A neutral atom thus contains a number of electrons equal to the number of protons in the nucleus and this number determines the chemical identity.

(b) Z = atomic number = number of protons (= electric charge of nucleus = number of orbital electrons).

$$N = \text{number of neutrons}$$
$$A = \text{mass number}$$
$$A = N+Z(N = A-Z, \text{ or } Z = A-N)$$

An isotope of the same element requires the same value of Z but different values of N and so of A. The chemically determined atomic weight is the mean of the stable isotopes in the naturally occurring atoms and so depends on the relative proportions of atoms with different A (e.g. Fe: 5·82 per cent of $A=54$, 91·66 per cent of $A=56$, 2·19 per cent of $A=57$ and 0·33 per cent of $A=58$. The resultant mean atomic weight = 55·85). The mass number is not necessarily exactly equal to the atomic weight of the nuclide for reasons which need not be considered here.

It is usual to denote an isotope by the number A and its chemical symbol, e.g. ^{24}Mg, ^{29}Si, ^{28}Al, ^{56}Mn (less frequently by Mg^{24}, etc.). Although it is not strictly necessary to add the atomic number Z as well (since it is already indicated by the symbol) this is often done, e.g. $^{24}_{12}$Mg, $^{29}_{14}$Si, $^{28}_{13}$Al, $^{56}_{25}$Mn (less frequently $_{12}$Mg^{24}, etc.).

One advantage of this notation is that nuclear equations can be more fully descriptive of changes in mass and charge.

It is necessary to distinguish between an *isotope* as defined above, an *isobar* and an *isotone* as in the following table due to Faires and Parkes [42].

TABLE 8.1

	Z	A	N	Examples
Isotope	Same	Different	Different	$^{31}_{15}$P and $^{32}_{15}$P; $^{63}_{29}$Cu and $^{65}_{29}$Cu
Isobar	Different	Same	Different	$^{32}_{15}$P and $^{32}_{16}$S; $^{92}_{40}$Zr and $^{92}_{41}$Nb
Isotone	Different	Different	Same	$^{2}_{1}$H and $^{3}_{2}$He; $^{52}_{24}$Cr and $^{51}_{23}$V

8.5.1.2 Naturally occurring nuclei may be stable or unstable. In the latter case they undergo radioactive transformations. These are the elements with atomic number greater than 82 (lead) apart from a solitary naturally occurring radioactive isotope of potassium. Large numbers of artificial isotopes have, however, been produced by irradiation or bombardment in an atomic pile or a cyclotron or other particle accelerator. For research and technical applications pile irradiation is used to produce the required activity.

8.5.1.3 The changes which can occur in *natural radioactive* atoms are considered first and then the reactions produced artificially.

For this purpose the above relationship between Z, A and N forms a useful basis. If the number of protons Z is plotted against N as in Fig. 8.10(a) then the most stable nuclei are on a curve which deviates, from nearly equal N and

Z for atomic numbers up to about 10, to 50 per cent more neutrons than protons. On such a plot the co-ordinates increase by unity in either direction. It has therefore become accepted to represent each individual *nuclide* by a square. Adjacent nuclides then obey the rules

(a) those in a horizontal line are isotopes,
(b) those in a vertical line are isotones, and
(c) those on a diagonal line are isobars represented by the equation $Z+N = A$. A chart based on these principles was produced by G.E.C. [43] and has a German counterpart which is coloured [44].

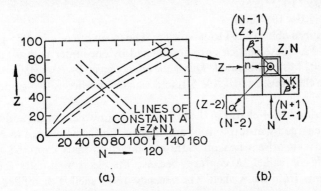

(a) (b)

Fig. 8.10   Isotopes and radioactive processes.

(a) variations of Z and N for most stable isotopes (full line) and approximate range of isotopes
(b) changes produced by natural radioactive decay processes—including K capture of electrons from orbits

The emission of particles due to radioactive disintegration then follows according to Fig. 8.10(b) which represents an enlargement of a region on the $Z-N$ diagram. The decay processes require the admission that the elementary concept of a nucleus as a complex of stable neutrons and protons is inadequate and requires modification. The 'radiations' produced are as follows. Actually three (at least) of these are particles in the accepted sense (non-zero rest mass).

1. Alpha particles (α). These have $A=4$, $Z=2$, $N=2$ and correspond to He nuclei He^{++}.
2. Beta particles (β). These are electrons regarded for purposes of mass balance as having zero mass, i.e. ${}_{1}^{0}\beta^{-}$ for the familiar negative electron. Nuclear processes also involve the positive electron or positron ${}_{1}^{0}\beta^{+}$.
3. Neutrons represented by ${}_{0}^{1}\text{n}$.
4. Gamma rays (γ). These are electromagnetic rays and so correspond to photons of zero rest mass and no charge.

In addition certain aspects of the energy balance in nuclear equations have led to the assumption of a virtually undetectable particle, viz.

315

5. The neutrino (ν) of extremely small mass and no charge. An antineutrino ($\bar{\nu}$) has been postulated.

The appearance of a negative electron can be explained by a neutron changing to a proton and the positive electron by a proton changing to a neutron according to the respective equations:

(a) $_0^1\text{n} \rightarrow {}_1^1\text{H} + {}_1^0\beta^- + \text{neutrino} + 0.78$ MeV
(b) $_1^1\text{H} \rightarrow {}_0^1\text{n} + {}_1^0\beta^+ + \text{antineutrino}$
    (with or without emission of energy)
(1 eV $= 1.603 \times 10^{-12}$ erg)
(1 MeV $= 1.603 \times 10^{-6}$ erg)
(b) is sometimes abbreviated to $\text{p} \rightarrow \text{n} + \beta^+$. The mean life of a free β^+ particle is only 10^{-9} s. These particles are annihilated by encounter with β^- and an energy of 1.02 MeV is liberated. This is the energy given by $E = mc^2$ for the rest mass of the electrons, as in Table 5.1, and usually appears as two gamma-ray photons.

The emission of a +charge in the form of β^+ is equivalent to the addition of a negative charge by absorption of an electron. The same change in nuclear constitution can arise when an extra-nuclear electron is captured by the nucleus. This is called 'K capture' because the electron usually (but not always) comes from the K shell. The vacancy in the shell is then filled by an electron from another orbit and so the characteristic X-ray is emitted.

The spontaneous processes, in so far as they lead to a change in mass or charge, are all represented in Fig. 8.10(b). The product nucleus is then usually in an excited state and emits one or more γ-ray quanta in order to reach its ground state. Energy schemes are referred to below.

8.5.1.4 The production of *artificial radioisotopes* has much extended the variety of nuclear species and isotopes are available over the whole range of atomic number. In addition there are a number of further nuclear reactions which take place in the course of the irradiation or bombardment. The reactions produce the unstable nuclides and so can be intentionally used to produce these isotopes for research purposes. The available reactions are denoted by a system of symbols in which the particle causing the reaction and the particle ejected (or the nuclear reaction) are bracketed together. The particles are:

Bombarding particles		*Ejected particles, etc.*
Alpha particles	α	The bombarding particles (one or
Neutrons	n	more in a reaction)
Protons	p	*or* nuclei produced by fission—f
Deuterons	d or ^{2}H	
Tritium nuclei	t or ^{3}H	*or* fragments broken off, i.e. 'spal-
Gamma-ray photons	γ	lation'

Reactions are exemplified by:

(α, n) i.e. bombarding alpha particle in and neutron out.

(n, γ) i.e. neutron particle absorbed with emission of gamma-ray.

$(n, 2n)$ i.e. neutron particle bombards nucleus, and a second neutron is ejected (or one absorbed and two ejected).

(n, f) i.e. a neutron causes nuclear fission.

It is therefore useful to extend the diagram of Fig. 8.10(b) firstly to show the other possible changes which transform one nuclide to another including some which occur in the atomic pile or cyclotron and not in natural radioactivity. Secondly a further step is to show the nuclear reactions on a similar diagram since in some cases a particle is captured and another ejected so that the net weight change is required. The two diagrams are shown in Figs. 8.11(a) and

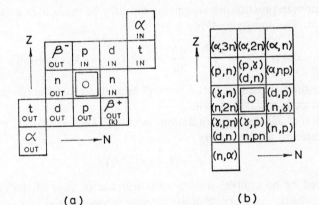

(a)   (b)

Fig. 8.11   Full scheme for processes and reactions (excluding fission and spallation).

(a) possible nuclear processes changes up to $A \pm 4$, $0 =$ original nucleus

(b) nuclear reactions by particle bombardment

(b) and the step from one to the other is then simple. These diagrams include reactions which involve changes in the mass number up to (and including) ± 4, and so exclude fission and spallation. Although these latter processes produce atoms which could be used for metallurgical research they are not readily available but may become more so as the supply of used reactor fuel elements builds up.

8.5.2   Some Quantitative Aspects

8.5.2.1   The most important quantitative relationship is the rate of disintegration since it determines the intensity of the radiation and density of particle emission. Given a number of radioactive atoms, N, it is not certain which of these atoms will undergo radioactive disintegration, but for a given reaction

317

there is a statistical probability that any one will disintegrate in a given time. Alternatively a definite fraction of the total number N will transform in a given time. Evidently the number of atoms, dN, which transform in a small interval of time dt will be proportional to this time and to the total number N (e.g. if N is doubled twice the number will transform in the same time). That is,

$$dN = -\lambda N \, dt \tag{8.23}$$

where λ is called the *decay constant*. Therefore

$$N = N_0 \exp(-\lambda t) \tag{8.24}$$

where N_0 is the number of unchanged atoms at a time $t=0$. Since a given mass of radioactive atoms will generally have been undergoing radioactive decay before any measurements could be made, the law always applies from any convenient starting time and the number N_0 instantaneously present. Furthermore in practice measurements can only be made over a finite period of time Δt. The approximation:

$$\text{decay rate} = \frac{\Delta N}{\Delta t} = \frac{dN}{dt} = -\lambda N = -\lambda N_0 \exp(-\lambda t) \tag{8.25}$$

is only justified if the constant λ is small enough so that the fraction $\Delta N/N_0$ for unit time is relatively small.

Equation (8.25) defines a radiation 'intensity' $I = I_0 \exp(-\lambda t)$ more generally referred to as the *activity*.

$$A = A_0 \exp(-\lambda t) \tag{8.26}$$

There need be no confusion with mass number in view of the context. In practice therefore measurements are made in terms of activity or rate of decay even if a number of disintegration events is determined within a finite time. It is convenient to define the relative decay rates for isotopes in terms of the *half-life*.

This is the time for the number of atoms present, or the activity to reduce to one-half of the initial value. This is a constant depending on λ. Thus

$$N = \tfrac{1}{2}N_0 = N_0 \exp(-\lambda T_{1/2})$$

so the half-life

$$T_{1/2} = \frac{\log_e 2}{\lambda} = \frac{0.693}{\lambda}$$

N and A are reduced by one-half during each successive interval of $T_{1/2}$ and to a fraction $1/2^m$ in m half-lives. (Sometimes T or $\gamma_{1/2}$ is used for $T_{1/2}$.)

Fig. 8.12(a) shows representative exponential decay curves with different values of $T_{1/2}$ for which $N/N_0 = A/A_0 = 1/2$. If the ordinates are plotted on a log scale as in Fig. 8.12(b) straight lines are obtained. The decrease in activity over one interval of time is sufficient to give the slope. The different slopes are

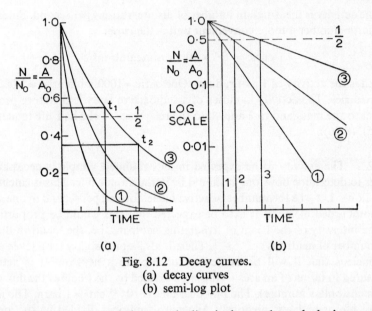

Fig. 8.12   Decay curves.
(a) decay curves
(b) semi-log plot

$-\lambda = -0.693/T_{1/2}$. If $T_{1/2}$ is known the line is drawn through the intersection for this time and the horizontal line at $\frac{1}{2}$.

The determination of activity by taking the number of disintegrations between two times such as t_1 and t_2 (closer together than indicated here) leads to the following relationships.

$$(-\varDelta N) = N_1 - N_2 = N_0[\exp(-\lambda t_1) - \exp(-\lambda t_2)]$$
$$= N_0 \exp(-\lambda t_1)[1 - \exp(-\lambda \varDelta t)]$$

where $\varDelta t = t_2 - t_1$

$$-\frac{\varDelta N}{\varDelta t} = [-\lambda N_0 \exp(-\lambda t_1)]\left[\frac{1 - \exp(-\lambda \varDelta t)}{-\lambda \varDelta t}\right]$$

The term in the first bracket is the activity at $t_1 = A_1$, as defined by (8.25), which is

$$\frac{\varDelta N}{\varDelta t} = A_1\left[1 - \frac{\lambda \varDelta t}{2} + \frac{(\lambda \varDelta t)^2}{3!} \cdots\right] \tag{8.27}$$

The correction term in square brackets depends on $\lambda \varDelta t$ and so on $\varDelta t/T_{1/2}$. Thus if this ratio is 0.01, i.e. 1 per cent the correction is 0.0034, i.e. 0.34 per cent, at twice this ratio the correction is 0.69 per cent. The correction is only necessary if $T_{1/2}$ is small or long counting times are needed as for experiments where the active atoms are much diluted.

(*Note*: Equation (8.27) is alternatively the mean counting rate between t_1 and $t_2 = t_1 + \varDelta t$. That is

$$\bar{A} = \frac{A_1}{\varDelta t} \int_{t_1}^{t_1 + \varDelta t} \exp(-\lambda t)\, \mathrm{d}t$$

319

The activity is measured in number of disintegrations per second. Since this is a large number a more convenient unit is the *curie*.

$$1 \text{ curie} = 3{\cdot}7 \times 10^{10} \text{ disintegrations/s}$$

which is the activity of 1 g of radium. One curie = 1000 millicuries = 1 000 000 microcuries. The activity in curies of a radioactive body is therefore proportional to the mass and to λ and so to the reciprocal of the half-life (equations (8.25), (8.26)).

8.5.2.2 The activity to be expected in an irradiated sample is necessary in order to determine how long it should be in the atomic pile. This is calculated as follows. Let α be the number of active nuclei formed per second in one gram of bombarded element. It is to be expected that this would be proportional to the intensity of the beam of irradiating neutrons, i.e. the 'neutron flux' = ϕ = number of neutrons $\text{cm}^{-2} \text{ s}^{-1}$. There is also a probability for a given atom or nucleus that it will be excited by or capture a neutron. It is actually measured in terms of an area σ (this area divided by the 1 cm^2 of the flux gives a dimensionless number). The practical unit is $10^{-24} \text{ cm}^2 = 1$ barn. The number of atoms in 1 g is obviously Avogadro's number divided by the atomic weight—A in this context. Thus:

$$\alpha = \frac{6{\cdot}06 \times 10^{23}}{A} \sigma\phi$$

As soon, however, as new radioactive atoms are formed they begin to disintegrate and the two processes go on together. If after time t there are N atoms present then

$$\frac{dN}{dt} = \alpha - \lambda N$$

which, with $N=0$ at $t=0$, integrates to

$$N = \frac{\alpha}{\lambda}\{1 - \exp(-\lambda t)\}$$

At the end of this time the material has a *specific activity* S which is, by (8.25), $-\lambda N$. With the above definition of α but substituting the conversion factors for barns and curies, S in curies per gram is given by:

$$S = \frac{0{\cdot}606}{A} \frac{\sigma\phi}{3{\cdot}7 \times 10^{10}} \left[1 - \exp\left(\frac{-0{\cdot}693t}{T_{1/2}}\right) \right] \qquad (8.28)$$

The quantity σ is a function of the irradiated atom, ϕ of the pile conditions, and $T_{1/2}$ of the product. These are tabulated and readily available in the literature. It is therefore possible to establish the initial activity of an isotope produced for research at a suitable level.

320

8.5.2.3 The absorption of radiations or particles passing through matter is also important.

Gamma rays are absorbed and undergo different reactions, viz.:

(a) Photoelectric absorption in which all the energy is transferred to an inner shell electron.
(b) Compton scattering in which some of the energy is transferred and a new gamma-ray quantum of lower energy results.
(c) Pair production—β^+ and β^- are produced—the reverse of the mutual annihilation process.

The general absorption law for gamma rays is, as would be expected, identical with that for other electromagnetic radiations such as X-rays, viz. $I = I_0 \exp[-\mu x]$ which is equation (5.26). μ is the linear absorption coefficient (total of all processes by which the absorption occurs) and x is the distance penetrated.

Alpha particles being charged helium nuclei attract electrons and so ionise any gas through which they pass. Ionisation falls off with distance from source and for particles of a given energy and air (or gas) of a given pressure there is a range R, e.g. 2·8 cm for alpha particles from ^{232}Th and 8·6 cm for ^{212}Po. For solids or substances other than air

the *stopping power* is the ratio $P = R_{air}/R_{substance}$

For example, for aluminium, $P = 1660$.

The range in air relates to the initial particle velocity V_0, to the initial energy E_0 and to the decay constant λ as follows:

$$
\left.
\begin{array}{ll}
\text{(a)} & V_0^3 = 1 \cdot 03 \times 10^{27} R \\
\text{(b)} & E_0 = (3 \cdot 09 R)^{2/3} \\
\text{(c)} & \log \lambda = A \log R + B
\end{array}
\right\} \tag{8.29}
$$

(A, B constants)

For beta particles the corresponding relations to (b) and (c) are

$$
\text{(i)} \qquad R = 0 \cdot 542 E_{max} - 0 \cdot 133 \quad \text{for} \quad E_{max} > 0 \cdot 8 \text{ MeV}
$$
$$
R = 0 \cdot 407 E_{max}^{1 \cdot 38} \quad \text{for} \quad 0 \cdot 15 < E_{max} < 0 \cdot 8 \text{ MeV}
$$

$$
\text{(ii)} \qquad\qquad\qquad \lambda = k E_{max}^5 \tag{8.30}
$$

(k constant for a given radioactive series).

In addition, although one would not expect an equation such as (5.26) to apply an approximate empirical relationship of that kind can be used. It is also useful to define a 'half thickness'.

$$
\text{(iii)} \qquad\qquad\qquad d_{1/2} = 32 E^{1 \cdot 33} \text{ mg cm}^{-2} \tag{8.31}
$$

If a specimen containing beta radiating atoms is thick and especially if the beta particles have low energy the intensity emitted from a surface will be reduced by self-absorption. For counting rates under these conditions the true and observed counting rates n_t and n_o are related by

(iv)
$$\frac{n_t}{n_0} = \frac{1-(\frac{1}{2})^{x/d_{1/2}}}{0.693^{x/d_{1/2}}}$$
(8.32)

When particles from the innermost distance x are all absorbed before they reach the surface this is called the 'infinite thickness' and corresponds to the maximum value of n_o.

8.5.2.4 A useful (semi-quantitative) way of summarising the energy relationships half-life and reactions for an individual isotope is the Decay Scheme. Typical examples are shown in Fig. 8.13. It will be seen that a change in atomic number is shown by a sloping line. Thus beta particle emission alters Z by ± 1 but the mass number is not altered.

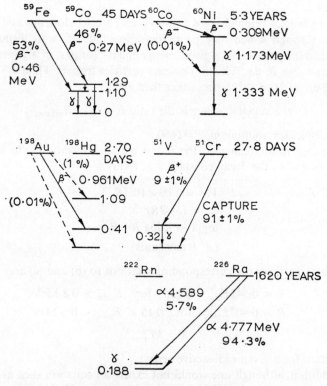

Fig. 8.13 Some decay schemes.

These diagrams also show that a radioactive atom can sometimes disintegrate in more than one way and the relative proportions are shown (in some cases one reaction or process predominates). Double decay systems may, of course, make practical interpretation more complicated.

322

For further reading at a reasonably elementary level reference can be made to the book by Faires and Parkes [42]. Other general references are Leymonie [45], Houseman [46], Glasstone [47], Whitehouse and Putnam [48].

8.5.3 Experimental Aspects

8.5.3.1 Health Hazards: Before engaging in research or experiments with radioactive isotopes it is necessary to take account of the possible hazards to health. Specialist knowledge on the hazards, protection procedures and legal requirements is obtainable from a variety of sources and reference is made to these matters in the volumes noted above. It is therefore less easy to contrive laboratory demonstrations or experiments which can be done by a student. Nevertheless teaching establishments will often have equipment for such work. It is useful to have some explanation of research techniques and an elementary knowledge of the principles.

8.5.3.2 Radiation Detection and Measurement (Counter Methods). Radiation detection has already been considered in connection with X-ray diffractometry and spectrography. For an outline of the principles of counters, counting statistics and sources of error reference can be made again to sections 6.4.6.3/4. The use of counters for detecting alpha or beta particles or gamma radiation is essentially similar.

Geiger counters do not distinguish between ionising radiations but are most suitable for detecting beta particles. Varieties of scintillation counters are available for alpha particles or gamma rays.

Counting techniques are used where it is necessary to determine the location of radioactive isotopes in a system involving diffusion or distribution or movement as in a gas system. These are the typical 'tracer' applications. Accurate determination of activity or radiation intensity is necessary where quantitative proportions are required. In some cases long counting times are needed as when the radioisotope is present in great dilution. In such cases the radiation may be very little above that of the background which must be determined in the same way. The counting times chosen must be such as to reduce the limits of error to a sufficiently small fraction of the difference between active count and background—see section 6.4.6.4 and particularly equation (6.33(b)). It is frequent practice to state the standard deviation σ. For N counts this is $\sqrt{N}$ so that the relative error is $\sigma = \sqrt{N}/N = 1/\sqrt{N}$. This is the random error and can be expressed as a percentage. There is thus a chance of about one in three that $N \pm \sqrt{N}$ will be exceeded. (σ corresponds to $\pm 68\cdot3$ per cent.) 10 000 counts are therefore required to give a ± 1 per cent error. For the cases where background is relatively high, corresponding to (6.33(b)), the random error becomes:

$$\frac{\sqrt{(N+N_B)}}{N-N_B} \times 100 \text{ per cent}$$

In some work the difference $N - N_B$ may be as small as 10 per minute in 200. This would give a random error

$$\frac{\sqrt{(205 + 195)}}{205 - 195} \, 100 = 200 \text{ per cent}$$

Counting must therefore go on for a much longer time. If each count is done for 50 min the numerator is multiplied by $\sqrt{50}$ and the denominator by 50 and the random error is now

$$\frac{200 \times \sqrt{50}}{50} = 28 \cdot 3 \text{ per cent}$$

i.e. the result states $N - N_B = 10 \pm 2 \cdot 8$ counts per minute.

8.5.3.3 Photographic Detection: Autoradiography. Photographic recording is employed in metallurgical applications which come under the general heading or description of *autoradiography*. A solid specimen containing radioactive atoms unevenly distributed is in contact with a photographic film. The radiation emitted will give an image which is darker in the regions where the concentration of the radioactive atoms is greater.

The resemblance to microradiography is close (but not exact) although autoradiography can be used without subsequent photographic enlargement—whereby it resembles ordinary radiography.

The different radiations have different photographic effects. In fact the least penetrating alpha particles affect the photographic emulsion most—an advantage therefore to have thin layers of emulsion. Beta particles penetrate much further but have only 1/300th of the action of alpha particles. Radiation emanates from a point and travels in all directions. Image definition is therefore best when the film is as thin as possible but not too thin or otherwise excessively long exposures will be needed. It is also necessary for the film to be in the closest possible contact with the surface.

For this last reason, special techniques and emulsions have been devised. These include emulsions which can be removed from the backing plate or film and transferred to the surface under examination. Alternatively an emulsion can be melted and painted onto the surface, or the emulsion formed directly on the surface. For further details see the book by Leymonie [45] or the chapter by Ward [30] (in the same volume as Houseman [46]), where original references are given.

8.5.3.4 Radioactive elements can be added to molten metal having first been activated by pile irradiation. Sometimes the active element is combined with another as in the production of radioactive phosphorus additions where ferrophosphorus is irradiated.

Alternatively an alloy itself may be irradiated in the pile. The fact that it is an alloy implies that equation (8.28) must be modified to take account of the concentration. Generally A must be replaced by m/n where m is the molecular weight and n the number of active atoms of the element under investigation. In an alloy with a weight fraction w, the product wA represents the equivalent of m/n.

In such circumstances, more than one or even all the elements in a specimen undergoing irradiation may become activated. In general, one element is likely to be much more affected than the others. In the event of two (or more) being sufficiently active the resultant decay curve will be composite. In this case the activity can be measured as a function of time. If one radioactive element only is responsible for this activity the plot with logarithmic ordinate will be a single straight line as in Fig. 8.12(b). If, however, two elements are responsible a curve will be obtained. As is often the case one of the components may have an appreciably different half-life from the other. The curve approximates to the shorter-lived component for short times and to the longer-lived component at longer times. The straight lines for the respective elements, which are sometimes known, are thus roughly asymptotic, and their slopes naturally correspond to the respective decay constants or half-lives. Direct addition is not possible and would only be used in a direct decay curve of the form of Fig. 8.12(a) but taking account of relative amounts and initial value of A_0 if these can be estimated.

The detection or resolution of two (or more) active elements may be useful on specimens otherwise intended for autoradiography.

8.5.4 Some Metallurgical Applications

The following summary covers representative applications. These include some which are broadly in the category of autoradiography. Others will be seen to concern problems which can be regarded as coming within the terms of chemical metallurgy.

8.5.4.1 *Diffusion Studies.* The basic relationships will first be summarised. For further details reference should be made to standard text or articles written from the metallurgical point of view [49, 50].

Fick's first law states that the mass, or number of atoms diffusing in a given direction, is proportional to the concentration gradient. The flux ϕ through an area A with direction x is given by

$$\phi = \frac{dm}{dt} = -DA\frac{dc}{dx} \tag{8.33}$$

where D is the diffusion coefficient and c the concentration (e.g. weight or atomic per cent or fraction). Fick's second equation follows in a similar manner to the equation for thermal conduction (section 2.4) and allows for variation with time. Thus, if a layer of thickness δx, normal to A, is considered, and a mass δm_1 flows in on one side and δm_2 flows out, there is a net

325

increase of mass $\delta m_1 - \delta m_2$ in a volume $A \, \delta x$. Let this change occur in a time δt. There is a change in concentration of

$$\delta c = \frac{\delta m_1 - \delta m_2}{A \, \delta x}$$

so that by referring to (8.33), the following sequence results:

$$(A \, \delta x)\frac{\delta c}{\delta t} = \frac{\delta m_1 - \delta m_2}{\delta t} = \phi_1 - \phi_2 = -A\left(D\frac{dc}{dx}\right) + A\left\{\left(D\frac{dc}{dx}\right) + \frac{d}{dx}\left(D\frac{dc}{dx}\right)\delta x\right\}$$

from which immediately:

$$\frac{dc}{dt} = D\frac{d^2c}{dx^2} \tag{8.34a}$$

or

$$\frac{dc}{dt} = \frac{d}{dx}\left(D\frac{dc}{dx}\right) \tag{8.34b}$$

if D varies with concentration.

The most general expressions involve partial differentials, e.g.

$$\phi = -DA\left(\mathbf{i}\frac{\partial c}{\partial x}\mathbf{j} + \frac{\partial c}{\partial y} + \mathbf{k}\frac{\partial c}{\partial z}\right) = -DA \text{ grad } c$$

and

$$\frac{\partial c}{\partial t} = D\nabla^2 c$$

Because chemical equilibrium is more correctly defined in terms of activities rather than concentrations, it is more strictly correct to represent the equations in terms of chemical potential gradients rather than concentration gradients. Since concentrations are still generally measured it is usual to express the relationships in terms of concentrations and activity coefficients and the coefficient determined is usually for dilute solution and is denoted by D^*.

Diffusion experiments using isotopes depend on the assumption that the difference in the diffusion coefficient between the isotope and inactive atoms of the same element is negligible. It is also generally necessary that the half-life should be long enough relative to annealing times.

Typical experiments involve the *diffusion couple*. The two metals for which the inter-diffusion is to be studied are butt-welded together or one is deposited on the other. Several arrangements are possible. See Fig. 8.14.

(a) A thin layer of active material at one end of a 'semi-infinite' bar specimen.
(b) A thin layer between two such bars.
(c) Two similar bars joined—one active and the other normal.
(d) A thicker deposit or layer of active material on a semi-infinite bar.
(e) Both materials have thickness small relative to cross-section but the active layer is still relatively much thinner.

When the specimen is heated at a given temperature and for a time t, then the active atoms will have diffused into the inactive metal and the distribution of its concentration with distance will be provided by the appropriate solution of equation (8.34). The solutions corresponding to the above cases are briefly described as follows:

For (a)

$$c = \frac{Q}{\sqrt{(\pi D^* t)}} \exp \left[\frac{-x^2}{4 D^* t}\right] \qquad (8.35a)$$

where Q = initial number of radioactive atoms, x = distance, t = time.

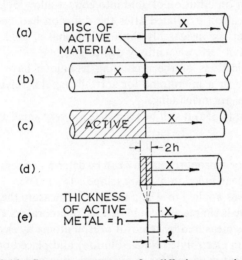

(a) DISC OF ACTIVE MATERIAL

(b)

(c) ACTIVE

(d)

(e) THICKNESS OF ACTIVE METAL = h

Fig. 8.14   Some arrangements for diffusion experiments.

    (a) single thin active layer on 'semi-infinite' bar
    (b) active layer sandwiched between two bars
    (c) two bars joined
    (d) thick active layer
    (e) thin layer—short length of bar

For (b), half the material diffuses in each direction and c = half the above expression in each case.

For (c), the method is only useful for self-diffusion, i.e. isotopes of the same element in each bar,

$$c = \frac{C_0}{2} \left[1 - \mathrm{erf}\left(\frac{x}{2\sqrt{(D^* t)}}\right)\right] \qquad (8.35b)$$

'erf' denotes the error function given by

$$\mathrm{erf}\, z = \frac{2}{\sqrt{\pi}} \int_0^z \exp\left(-y^2\right) \mathrm{d}y \qquad (8.36)$$

For (d) and (e) the solutions are more complicated.

327

In a typical experiment the specimen as for case (a), for example, is prepared—by electrodeposition evaporation or otherwise. It is then heated in a furnace with inert atmosphere or vacuum and annealed at a suitable temperature. After removal of surface layers cuttings are taken on a lathe and these are carefully weighed—the weight being proportional to the thickness removed. By successive removal of sections in this way a series of samples are collected, and these are dissolved in acid and their radioactivity measured. For further details reference should be made to the book by Leymonie [45] or to the original papers.

An interesting variant which might have wider application is represented by an experiment on diffusion of gold into copper alloy [51]. In this case the specimen was neutron irradiated after the diffusion had been effected. The half-life of copper is considerably shorter than that of gold so that within a few days the activity left was entirely due to gold.

Several modifications of the sectioning technique have been proposed, including the use of a microtome for soft alloys, electrolytic solution, and special techniques for machining.

From equation (8.35a) it will be seen that for very small x,

$$c = Q(\pi D^* t)^{-1/2}$$

Thus if the activity A proportional to c can be determined for the surface alone it should be a linear function of $1/\sqrt{t}$ (slope $= 1/\sqrt{(\pi D^*)}$).

This method was applied to γ iron in order to measure the self-diffusion of Fe. The technique is simple and could be used in laboratory experiments. The surface activity is measured by removal of iron atoms by electrolysis. A filter paper is soaked in electrolyte (NaCl solution) and placed on the end of the specimen, a current is passed for a measured time and a constant amount of metal transferred to the paper. (In the experiments on γ iron a current of 20 mA was passed for 3–5 min.) The activity of the transferred atoms is then measured. The technique can be combined with sectioning.

A further development using the thin single layer deposit as the starting point is based on measurements of the activity at this end of the specimen [52]. This is the activity of the whole specimen. Allowing for any differences in activity during experiment or handling time, it is only necessary to consider the change due to the diffusion of active atoms away from this surface with diffusion time. The radiation from the atoms must travel further to the end and so is more strongly absorbed—but the absorption coefficient (μ) must be large enough to make sufficient difference. It is assumed that the exponential absorption law is followed as for electromagnetic radiation, but it may be necessary to check whether this law applies if beta particle emission is involved. The coefficient μ can be determined experimentally by interpolating layers of known, different, thicknesses between a source and counter.

The concentration of activated atoms at a distance x is given by c in (8.35a). The actual number in a layer of unit area δx thick is $c \, \delta x$. The radiation from these atoms travels the distance x to the end and is absorbed according to the factor $\exp(-\mu x)$. Hence the contribution to the activity for the specimen from this layer is:

$$dA = \frac{KQ}{\sqrt{(\pi D^* t)}} \exp\left[\frac{-x^2}{4D^* t}\right] \exp(-\mu x)\, dx$$

with $K=$ constant including the area of cross-section. The activity of the whole 'semi-infinite' specimen measured at this end is the integral from 0 to ∞. The combined exponential term is of the form $\exp[-\{(x/a)^2 + bx\}]$ which can be changed to the form $\exp - [y^2 - \frac{1}{4}a^2 b^2]$ by putting $y = x/a + \frac{1}{2}ab$, a substitution which enables the definite integral to be evaluated. With $A_0 = KQ$ for $t=0$ the result is

$$\frac{A}{A_0} = \exp(\mu^2 D^* t)[1 - \text{erf}\{\mu\sqrt{(D^* t)}\}] \tag{8.37}$$

$$= F(\mu^2 D^* t), \text{ say}$$

('erf' is defined above in (8.36)).

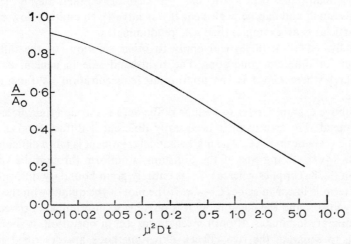

Fig. 8.15   Decrease in surface activity with the arrangement of Fig. 8.14(a). This is a master curve for general application (method devised by Steigman, Shockley and Nix [52]).

F is thus a function which can be determined once and is valid for all metals *providing* it is correct to assume the exponential absorption law. See Fig. 8.15. The technique will not be valid if μ is too small since the absorption will not be sufficiently altered with distance. If, however, μ is too large, the distance from which an appreciable contribution to A can be made without excessive absorption is small so that the experiment then only measures the activity at and near the surface. The method has therefore been developed by Gruzin [53]. From the end of the specimen a thickness x_n is removed and the activity

A_n measured on the new surface. The above integral is now from x_n to ∞, and the absorption factor is $\exp{[-\mu(x-x_n)]}$. Equation (8.37) now becomes

$$\frac{A_n}{A_0} = \exp{(\mu^2 D^* t)}[1 - \text{erf } y_n]$$

where $\qquad y_n = \frac{x_n}{2\sqrt{(\pi D^* t)}} + \mu\sqrt{(D^* t)} \qquad$ (the above substitution)

By differentiating (see equation (8.36)) it can be proved that:

$$\mu A_n - \frac{\partial A_n}{\partial x_n} = \frac{KQ}{\sqrt{(\pi D^* t)}} \exp\left[\frac{-x_n^2}{4D^* t}\right] \qquad (8.38)$$

The terms on the left-hand side are thus determined by experiment and a log plot against x_n^2 enables D^* to be determined. Thus

$$\log\left[\mu A_n - \frac{\partial A_n}{\partial x_n}\right] = C - \frac{x_n^2}{4D^* t} \qquad (8.38a)$$

where C is constant.

If μ is small (^{60}Co with $\mu = 0.3$ cm^{-1} is such a case) then the first term is relatively small and can be neglected. If μ is large (^{63}Ni emitting low energy beta particles is an example) then μA_n predominates.

Actually, equation (8.38) can apply to other solutions of the diffusion equation for different conditions. The right-hand side in general is given by $KC(x_n)$ where $C(x_n)$ is the appropriate concentration as function of distance.

The above examples refer to volume diffusion, i.e. through the whole body of the metal. The treatment is necessarily different if diffusion takes place along the grain boundaries. The mathematical treatment is more difficult. It is however worth noting that if the diffusion is uniform through the volume equation (8.38a) applies whereas if it is entirely grain boundary diffusion the second term is linear in x, i.e. $C - d.x$. If the plot of the quantity on the left is made against x and then against x^2 it is possible to discriminate between the two if one predominates. If both mechanisms are in operation, it should be possible to separate the two effects. Other methods are described in the literature.

The technique of autoradiography is valuable in giving a visual image of the distribution of the diffusing atoms. It is therefore also able to indicate when there is an appreciable difference between grain-boundary and volume diffusion in specimens of suitable grain size. The image will be darker along the grain boundaries if the former mode predominates.

Autoradiography is probably best applied by cutting a wedge-shaped surface on the specimen so that the blackening is progressive corresponding to increasing distance x. A symmetrical blackening curve is obtained with the arrangement of Fig. 8.14(b). It may be necessary to take account of the fact that the blackening arises not only from atoms which are near the surface of contact but also others further away—but not far enough for their radiation to be reduced sufficiently by absorption. A method of correction has been devised [54] so that the method can be used to determine the diffusion coefficients. The autoradiographic method has also been of value in studying grain-boundary structure. As an example the method has shown diffusion along the incoherent boundary of a twin but not along the coherent interface. The dependence of activation energy on the angle separating two slightly misoriented crystals is an indication of a more detailed kind of experiment involving several measurements of D for different temperatures of diffusion.

Other examples of diffusion include diffusion over surfaces and in liquids. The penetration of liquid metals along solid grain boundaries is illustrated by an important case—the diffusion of a molten copper-rich phase along austenite grain boundaries in steels heated in the region of 1100 °C. This is associated with the phenomenon of hot shortness. The association with copper is well known and the copper phase can be seen with the optical microscope but the radioactive study provided confirmation and showed conclusively the effects of other elements (Ref. [45], p. 114).

8.5.4.2 *Segregation Studies.* The autoradiographic method is particularly valuable in this respect and again gives direct visual evidence of the distribution of segregated elements. Small amounts of active elements added to the melt appear in cast structures—segregated according to the mode of crystallisation. Information is obtained about the distribution and shape of columnar, dendritic and equiaxed crystals and the distribution of elements within or between them. Practical examples include the segregation of phosphorus or sulphur [55], and of chromium [56] in steel. Aluminium alloys with iron, copper and zinc [57] respectively have also been studied.

The effect of forging, rolling and other forms of hot working upon the segregation of alloy elements is noteworthy, especially the breakdown of primary segregation patterns. Such for example is a study of the effect of forging on phosphorus segregation [58]. Other important subjects include the segregation of carbon to grain boundaries in stainless steel, and general segregation of metallic elements prior to precipitation. The identity of and segregation of elements into particles regarded as non-metallic inclusions have also formed the subject of some useful metallurgical investigations.

8.5.4.3 Some applications to metallurgical chemistry. Applications range from the study of corrosion or oxidation processes to the distribution of elements between slag and metal in a furnace. The former type of study may well involve further applications of diffusion techniques to the oxides or sulphides.

Other work concerning metal surfaces and their reactions has been in the study of friction and wear. The essential aspect of the latter is the transfer of material from the surface and its detection and assessment by radiation

331

measurement. A reverse process is the deposition of materials on surfaces— as for example condensation from a metal vapour, or electrodeposition.

In the case of a bulk reaction involving large amounts of metal and slag, one may add suitable isotopes to the metal and measure the increase in activity with time in the slag or vice versa. The element or component is added to the phase in which it is generally present in larger concentration, e.g. activated slag oxides are added to the slag or metallic elements to the metal. The experimental techniques are comparatively simple but the necessary precautions for confining the radioactive material must be observed. Practicable experiments depend, as always, on the availability of suitable isotopes.

Another example involving slag and metal employs a technique of more general application which is known as Isotope Dilution Analysis. The bulk of slag in a furnace has been determined by adding a small amount of active material, allowing time for diffusion and then measuring the activity of the bulk. Thus, if a mass M of specific activity S is added to an unknown mass M_x the new mass $M + M_x$ will have a (reduced) activity S_x proportional to the actual concentration of M in the mixture, i.e.

$$\frac{S_x}{S} = \frac{M}{M + M_x}$$

(the activity S applies when $M_x = 0$). Therefore

$$M_x = M\left(\frac{S}{S_x} - 1\right) \tag{8.39}$$

This equation is of quite general application.

8.5.4.4 *Isotope Radiography*. Isotopes provide sources of radiation for the radiographic examination of castings, welds, etc., for defects. The techniques are comparable with the use of X-rays for radiography—regarded as a technique of 'non-destructive testing'—physical in its basis but on the fringe of physical metallurgy. Typical isotopes are ^{60}Co which has been found suitable for steel above 2·5 in (62 mm) thick, and ^{192}Ir or ^{137}Cs for smaller thicknesses. For fuller details reference should be made to the literature on non-destructive testing [59–63].

8.6 AN ADDITIONAL NOTE ON FIELD EMISSION MICROSCOPY AND FIELD ION MICROSCOPY

8.6.1 Two Special Techniques

A brief further mention should be made here of the two related techniques which are not however properly described under the general heading of this chapter. They are difficult to fit into any scheme such as that indicated in Table 1.2. They are included because they fall into place between the subject matter of earlier chapters dealing with atomic aspects of structure and later chapters (Vol. 2) where the microscopic techniques are considered. They also require rather specialised equipment and specimen preparation techniques. They are not likely to find widespread applications in applied or industrial research laboratories but have a place in basic research. For general reading reference may be made to books by Gomer [62] and Brandon [63] who give further references to original papers. See also Cottrell [64].

8.6.2 Field Emission Microscopy

This technique requires a specimen in the form of a wire which is prepared with a fine needle-point by electropolishing. It is then fixed into the centre of a glass vessel which is highly evacuated. There is a conducting phosphorescent layer on the surface so that a high potential difference can be applied. With voltages of the order of 5–10 kV, electrons are emitted and accelerated through the potential field. They travel outwards in an almost exactly radial direction from the tip, and on impact with the screen produce an image which shows variations in intensity corresponding to variations in emission at the tip.

The tip will very often be a single crystal. The image will therefore show variations due to the crystallographic dependence of the work function which along with other variables determines the energy required to liberate electrons from the field within the metal lattice. In cases where the tip has a crystallographic orientation of low indices the pattern shows a recognisable symmetry. An example would be the three-fold symmetry of a cubic crystal with a $\langle 111 \rangle$ orientation. The method has been used to study the surface migration or vapour deposition of foreign atoms which obscure the image of the substrate.

8.6.3 Field Ion Microscopy

The experimental conditions are similar, except that a somewhat lower vacuum is required, but smaller tip radii and higher potentials are necessary. Whilst every precaution is taken to ensure that sources of contamination are reduced as far as possible, a small controlled amount of helium or neon gas is let into the system. The gas atoms are ionised near the tip in the high (positive) potential field and after collision with metal ions on the surface bounce off again. Under suitable conditions the gas ions pass through the vacuum onto the screen and produce an image which reveals the distribution of metal ions and surface steps on the tip with which the gas ions have been in collision. The images therefore show features which depend on the way in which the tip shape cuts through lattice planes. The most regular patterns emerge when the tip is hemispherical, the orientation simple, and a single crystal is involved. Otherwise disturbances caused by sub-grain boundaries, dislocations and vacancies may appear.

333

References

1. KOH, P. K. and CAUGHERTY, B., *J. Appl. Phys.*, 1952, **23**, 427.
2. LIEBHAFSKY, H. A., PFEIFFER, H. G., WINSLOW, E. H. and ZEMANY, P. D., *X-Ray Absorption and Emission in Analytical Chemistry*. Wiley, London, New York, 1960.
3. VON HEVESY, G., *Chemical Analysis by X-Rays and its Applications*, McGraw-Hill, New York, 1932.
4. BIRKS, L. S., *X-Ray Spectrochemical Analysis*, vol. XI of *Monographs on Chemical Analysis*. Interscience, New York, 1959.
5. BEATTIE, H. J. and BRISSEY, R. M., *Anal. Chem.*, 1954, **26**, 980.
6. SHERMAN, J., Symposium on Fluorescent X-Ray Spectrographic Analysis, *A.S.T.M. Spec. Tech. Publn.*, No. 157, 27, 1954.
7. LUCAS-TOOTH, J. and PYNE, C., *Advances in X-Ray Analysis* (1963 Conference at Denver Research Institute. W. M. Mueller, G. Mallet and M. Fay, eds.), 7, 523. Plenum Press, New York, 1964.
8. JOHNSON, W., Sheffield Symposium (Papers presented at 4th Conference) *On X-Ray Analytical Methods*, 73. Philips, Eindhoven, 1964.
9. BIRKS, L. S., *Electron Probe Micro-Analysis*, vol. XVII of *Monographs on Chemical Analysis*. Interscience, New York, 1963.
10. ZEMANY, P. D. *et al.*, Several articles contributed to *Symposium on X-Ray and Electron Probe Analysis*, *A.S.T.M. Spec. Tech. Publn.*, No. 349, 1964.
11. COSSLETT, V. E., ENGSTRÖM, A. and PATTEE, H. H. (eds.), *X-Ray Microscopy and Micro-Radiography*. Academic Press, New York, 1957.
12. ENGSTRÖM, A., COSSLETT, V. E. and PATTEE, H. H. (eds.), *X-Ray Microscopy and X-Ray Microanalysis*. Elsevier, Amsterdam, 1960.
13. PATTEE, H. H., COSSLETT, V. E. and ENGSTRÖM, A., *X-Ray Optics and X-Ray Microanalysis*. Academic Press, New York/London, 1963.
14. COSSLETT, V. E. and NIXON, W. C., *X-Ray Miscroscopy*. Cambridge University Press, London, 1960.
15. HILLIER, J., U.S. Patent 2,418,029, 1947.
16. CASTAING, R. and GUINIER, A., *Electron Miscroscope*, Proc. Delft. Conf., 60, 1949. R. Castaing's thesis, University of Paris, 1951.
17. BOROVSKII, I. B. and IL'IN, N. P., *Exp. Tech. Phys.*, 1957, **5**, 36.
18. SALTER, W. J. M., *A Manual of Quantitative Electron Probe Microanalysis*. Structural Publications, London, 1970.
19. PHILIBERT, J., *J. Inst. Met.*, 1961/2, **90**, 241. (See also *J. Iron Steel Inst.*, 1956, **183**, 42.)
20. COSSLETT, V. E., *Metall. Rev.*, 1960, **5**, 225.
21. WITTRY, D. B., *Metals Engng Quart.* (A.S.M.), 1962 (Aug.), 47.
22. BANERJEE, B. R., BINGLE, W. D. and BLAKE, N. S., *J. Metals* (A.I.M.E.), 1963 (Oct.), 769.
23. MELFORD, D. A., *Iron and Steel Institute Special Report*, No. 80, 1963, 206.
24. MELFORD, D. A. and DUNCOMB, P., *Metallurgie*, 1960 (May), **61**, 205.
25. MELFORD, D. A., *J. Inst. Metals*, 1961/2, **90**, 217 (and other papers in the same volume).
26. CARROLL, K. G., ibid., 1962/3, **91**, 66.
27. HEINRICH, K. F. J., *A.S.T.M. Spec. Tech. Publn.*, 349, 1964, 163 (see Ref. (10)).
28. WOOD, G. C. and MELFORD, D. A., *J. Iron Steel Inst.*, 1961, **198**, 142.
29. ADDA, Y. *et al.*, *Compt. Rend.*, 1956, **242**, 3081; 1956, **243**, 115; 1958, **246**, 113; 1958, **247**, 80; 1957, **245**, 2507; 1960, **250**, 115. (Also *Mem. Sci. Rev. Met.*, 1961, **58**, 716.)
30. WARD, R. G., Microradiography and Autoradiography, Chapter XVII of Chalmers and Quarrell (eds.), *The Physical Examination of Metals*, 2nd edition, p. 825. Arnold, London, 1960.
31. ANDREWS, K. W. and JOHNSON, W., p. 581 of Ref. (11). Also *Iron and Steel*, 1958 (Sept.), **31** (10), 437.
32. SHARPE, R. S., p. 590 of Ref. [11].
33. LONSDALE, K., *Phil. Trans. Roy. Soc.*, 1947, **240**, 219.

34. BERG, W. F., *Z. Krist.*, 1934, **89**, 286.
35. BARRETT, C. S., *Trans. A.I.M.E.*, 1945, **161**, 15.
36. HONEYCOMBE, R. W. K., *J. Inst. Met.*, 1951, **80**, 39, 45.
37. WEISSMAN, S., *J. Appl. Phys.*, 1956, **27**, 389, 1335.
38. HEYCOCK, C. T. and NEVILLE, F. H., *J. Chem. Soc.*, 1898, **73**, 714.
39. BETTERIDGE, W. and SHARPE, R. S., *J. Iron Steel Inst.*, 1948, **158**, 185.
40. COHEN, E. and SCHLOEGL, I., p. 133 of Ref. [12].
41. RUFF, A. M. and KUSHNER, L. M., p. 53 of Ref. [12].
42. FAIRES, R. A. and PARKES, B. H., *Radioisotope Laboratory Techniques*. Newnes, London, 1958.
43. STOCKLEY, J., *Nuclides and Isotopes* and Chart of the Nuclides, 5th edition. General Electric, Schenectady, N.Y., 1956.
44. *Chart of Nuclides*, 2nd edition, Institute of Radiochemistry. Gersbach and Sohn Verlag, Munich, 1963. (In German and English.)
45. LEYMONIE, C., *Radioactive Tracers in Physical Metallurgy*. Chapman and Hall, London, 1963. (English translation by V. Griffiths.)
46. HOUSEMAN, D. H., Radioactive Isotopes and Their Applications in Metallurgy. Chapter XVI of Chalmers and Quarrell (eds.), *The Physical Examination of Metals*, 2nd edition. Arnold, London, 1960.
47. GLASSTONE, S., *Source Book on Atomic Energy*, 2nd edition. Van Nostrand, New York, 1958.
48. WHITEHOUSE, W. J. and PUTMAN, J. L., *Radioactive Isotopes*. Clarendon Press, Oxford, 1953.
49. JOST, W., *Diffusion in Solids, Liquids, Gases*. Academic Press, New York and London, 1960.
50. LE CLAIRE, A. D., *Progr. Metal Phys.*, 1949, **1**, 306; 1953, **4**, 265.
51. MARTIN, A. B., JOHNSON, R. D. and ASARO, F., *J. Appl. Phys.*, 1954, **25**, 364.
52. STEIGMAN, J., SHOCKLEY, W. and NIX, F. C., *Phys. Rev.*, 1939, **56**, 13.
53. GRUZIN, P. L., *Dokl. Akad. Nauk. S.S.R.*, 1952, **86**, 289.
54. KURTZ, A. D., AVERBACH, B. L. and COHEN, M., *Acta Met.*, 1955, **3**, 442.
55. KOHN, A., *Rev. Met.*, 1951, **48**, 219.
56. DE BEAULIEU, C. and PHILIBERT, J., *Compt. Rend.*, 1958, **246**, 3615.
57. MONTARIAL, F., *Publ. Sci. Ministière de l'Air*, No. 334, Paris, 1958. (See also Ref. [45], Chapter 3.)
58. KOHN, A. and DOUMERC, J., *Rev. Met.*, 1955, **52**, 249.
59. HINSLEY, T. F., *Non-Destructive Testing*. Macdonald and Evans, London, 1959.
60. MCGONNAGLE, W. J., *Non-Destructive Testing*. McGraw-Hill, New York, Toronto, London, 1961.
61. MCMASTER, R. C. (ed.), *Non-Destructive Testing Handbook* (2 vols.). Ronald Press, New York, 1959. (E.g. Section 20 on Film Radiography.)
62. GOMER, R., *Field Emission and Field Ionisation*, Harvard Monograph. Oxford University Press, London, 1961.
63. BRANDON, D. G., *Modern Techniques in Metallography*. Butterworths, London, 1966.
64. COTTRELL, A. H., *J. Inst. Metals*, 1962, **90**, 449.

INDEX TO VOLUME 1